Blockchain-Based Systems for the Modern Energy Grid

Blockchain-Based Systems for the Modern Energy Grid

Edited by

Sanjeevikumar Padmanaban

Rajesh Kumar Dhanaraj

Jens Bo Holm-Nielsen

Sathya Krishnamoorthi

Balamurugan Balusamy

ELSEVIER

ACADEMIC PRESS

An imprint of Elsevier

Academic Press is an imprint of Elsevier
125 London Wall, London EC2Y 5AS, United Kingdom
525 B Street, Suite 1650, San Diego, CA 92101, United States
50 Hampshire Street, 5th Floor, Cambridge, MA 02139, United States
The Boulevard, Langford Lane, Kidlington, Oxford OX5 1GB, United Kingdom

ISBN: 978-0-323-91850-3

Publisher: Charlotte Cockle
Acquisitions Editor: Graham Nisbet
Editorial Project Manager: Aleksandra Packowska
Production Project Manager: Anitha Sivaraj
Cover Designer: Miles Hitchen

Typeset by MPS Limited, Chennai, India

Contents

List of contributors

R. Anusuya Information Technology, PSG College of Technology, Coimbatore, Tamil Nadu, India

L. Ashok Kumar Information Technology, PSG College of Technology, Coimbatore, Tamil Nadu, India

Balamurugan Balusamy Student Engagement, Shiv Nadar University, Delhi-National Capital Region, Delhi, India

Santhosh C. Department of Computer Science and Engineering, Kongu Engineering College, Erode, Tamil Nadu, India

César Cazal (Andrés) Institute for Automation of Complex Power Systems, RWTH Aachen University, Aachen, Germany

S. Chellam Velammal College of Engineering and Technology, Madurai, Tamil Nadu, India

R. Devipriya Department of Artificial Intelligence and Data Science KPR Institute of Engineering and Technology, Coimbatore, Tamil Nadu, India

Rajesh Kumar Dhanaraj Computer Science, and Engineering, School of Computing Science & Engineering, Galgotias University, Greater Noida, Uttar Pradesh, India

A. Shamila Ebenezer Department of CSE, Karunya Institute of Technology and Sciences, Coimbatore, Tamil Nadu, India

R. Hariharan Department of Information Technology, Vel Tech University, Chennai, Tamilnadu, India

M. Jajini Velammal College of Engineering and Technology, Madurai, Tamil Nadu, India

K. Jayanthi Department of Computer Science, KGiSL Institute of Information Management, Coimbatore, Tamil Nadu, India

J. Kalaivani Department of Computer Science and Engineering, SRM Institute of Science and Technology, Chennai, Tamilnadu, India

N. Kamaraj Thiagarajar College of Engineering, Madurai, Tamil Nadu, India

A. Kameshwaran Department of Information Technology, Dr. M.G.R. Educational and Research Institute Chennai, Tamilnadu, India

S. Deepa Kanmani Department of IT, Sri Krishna College of Engineering and Technology, Coimbatore, Tamil Nadu, India

N. Kanya Department of Information Technology, Dr. M.G.R. Educational and Research Institute Chennai, Tamilnadu, India

D. Karthika Renuka Information Technology, PSG College of Technology, Coimbatore, Tamil Nadu, India

C. Kotteswari Department of Computer Science and Engineering, Velalar College of Engineering and Technology, Erode, Tamil Nadu, India

S.A. Krishna Department of Mechatronics, Kongu Engineering College, Erode, Tamil Nadu, India

Sathya Krishnamoorthi Department of Computer Science and Engineering, Kongu Engineering College, Erode, Tamil Nadu, India

Ravinder Kumar Ajay Kumar Garg Engineering College, Ghaziabad, Uttar Pradesh, India

Vijayakumar M. Department of Computer Science and Engineering, Sri Shanmugha College of Engineering and Technology, Dharmapuri, Tamil Nadu, India

S. Malathy Department of Computer Science and Engineering, Kongu Engineering College, Erode, Tamil Nadu, India

M. Mohanasundari Department of Computer Science and Engineering, Velalar College of Engineering and Technology, Erode, Tamil Nadu, India

Kanimozhi N. Department of Information Technology, Kongu Engineering College, Erode, Tamil Nadu, India

P.L. Narayanan Department of Information Technology, Dr. M.G.R Educational and Research Institute Chennai, Tamilnadu, India

M. Nikhil Raghava Rao Vignan Institute of Technology and Science, Deshmukhi, Telangana, India

Whig Pawan Vivekananda Institute of Professional Studies, New Delhi, New Delhi, India

Ferdinanda Ponci Institute for Automation of Complex Power Systems, RWTH Aachen University, Aachen, Germany

Sivaraj R. Department of Computer Science and Engineering, Nandha Engineering College, Erode, Tamil Nadu, India

Vani Rajasekar Department of Computer Science and Engineering, Kongu Engineering College, Erode, Tamil Nadu, India

R.S. Rashika Department of Information Technology, Dr. M.G.R Educational and Research Institute Chennai, Tamilnadu, India

Michel Rwibasira Jain University, Bengaluru, Karnataka, India

Saranya S.S. Department of Computer Science and Engineering, Kongu Engineering College, Erode, Tamil Nadu, India

Dahlia Sam Department of Information Technology, Dr. M.G.R. Educational and Research Institute, Chennai, Tamil Nadu, India

M. Santhiya Department of Electrical and Instrumentation Engineering, Kongu Engineering College, Erode, Tamil Nadu, India

K. Sathya Department of Computer Science and Engineering, Kongu Engineering College, Erode, Tamil Nadu, India

Kaushik Sekaran Department of Computer Science and Engineering, ICFAI Deemed University, IcfaiTech, IFHE CAMPUS, Hyderabad, Telangana, India

Desh Deepak Sharma MJP Rohilkhand University, Bareilly, India

Adlin Sheeba Department of CSE, St. Joseph's Institute of Technology, Chennai, Tamil Nadu, India

R. Suchithra Jain University, Bengaluru, Karnataka, India

Rishabh Tiwari Department of CSE, Dr. M.G.R. Educational and Research Institute, Chennai, Tamil Nadu, India

C.N. Vanitha Department of Computer Science and Engineering, Kongu Engineering College, Erode, Tamil Nadu, India

K. Vanitha Department of Computer Science and Engineering, Faculty of Engineering and Technology, JAIN (Deemed-to-be-University), Bangalore, Karnataka, India

Preface

As editors, we feel privileged to have been asked to edit the first edition of *Blockchain-based Systems for the Modern Energy Grid*. The electricity grid, also known as the energy grid (e-grid), is the electricity distributed from producers to consumers. The traditional centralized system had energy produced at one site and distributed the same to all consumers. Consumers then started producing energy on their own for their needs, such that excess energy was fed into the grid system. This deformation of a centralized system gave rise to autonomous grids, where consumers can produce energy for their needs, and excess of energy is driven into the grid for other consumers. Such entities are now referred to as "prosumers." Energy grids can be distributed to more expansive areas; however, they are now susceptible to cyber-attacks. Furthermore, failure in any one part of the grid will cause the entire grid to shut down. Any disruption in the grid will cause an uncontrolled flow of electricity.

To overcome the halting problems in decentralized grids, energy grids are now getting transitioned into the Internet of energy (IoE). Entities of various sizes are involved in generating energy and consuming energy. The producer can consume the generated energy, and the surplus can be traded to the neighboring consumers. Blockchain records all the transactions and verifies them in a decentralized manner and eliminates the need for a central authority. Blockchain will help the IoE in managing the identity of devices, securing the transactions of energy consumption data, and payment for energy. Certificates for eco-green energy can be issued securely and are resilient to forgery with the use of blockchain.

With IoE, wireless communication protocols need to be considered for transmission of data among peers. The use of blockchain in energy sector may help increase the market for energy trading. Consumers can now find producers of energy locally and trade with them securely and efficiently. The economy will have a positive impact owing to the energy market being efficient with minimal loss that occurs in long-distance transmission. The balance between supply and demand for energy is achievable with the use of blockchain. Integrating blockchain with IoE will make the network resistant to any failure in subcomponents of the network.

At the end of each chapter, authors have clearly identified important research gaps and needs for future research. This book includes chapters that provide the latest scientific knowledge on requirements for smart energy grids, transforming smart grid into IoE, communications taking place in IoE, security framework for IoE, utilizing blockchain in renewable energy, and their issues and challenges. Authors have also included case studies involving blockchain in electrical vehicle charging, energy trading markets, and energy-exchange platforms that will benefit the readers in understanding the application of blockchain in energy sector. We believe that the authors have done an outstanding job in presenting the latest information in their respective fields and hope this edition will bring about a prospective shift in energy market globally.

Sanjeevikumar Padmanaban
Rajesh Kumar Dhanaraj
Jens Bo Holm-Nielsen
Sathya Krishnamoorthi
Balamurugan Balusamy

Introduction—blockchain and smart grid

1

A. Kameshwaran[1], Dahlia Sam[1], R.S. Rashika[1], N. Kanya[1], P.L. Narayanan[1] and R. Hariharan[2]
[1]Department of Information Technology, Dr. M.G.R Educational and Research Institute Chennai, Tamilnadu, India, [2]Department of Information Technology, Vel Tech University, Chennai, Tamilnadu, India

1.1 Blockchain

Blockchain contains information that is defined in a chain of blocks, where a time-stamped digital document becomes the record linked by a cryptographic hashing to the previous block only in a technique that is not possible to backdate with an intent to tamper without also breaking the chain. A double-entry problem that is so common in paper-based record-keeping systems is not possible here and to do away with which problem in the central server is thereby taken care of. In a secure transfer, the blockchain is used in deals for property, contracts, money, etc., where third parties like the government or bank will not act as intermediaries. It is difficult, or rather impossible, to change the data once it is recorded within the blockchain.

Software protocols are used in blockchains (for email, it's SMTP). However, the internet will run with the help of blockchains which can be said to be meta technologies: technologies that affect how other technologies run. Without blockchains as the firmware between software and the internet, several pieces of tools can be compromised, like software applications, database applications, and/or some connected systems.

A network can be replicated among new members having access to the document when regular intervals can be added continuously to generate new blocks in a blockchain network. In a lightweight device, a blockchain can allow one to search quickly for data to be included in a block's contents without altering any block. A chain can be maliciously impossible to modify because every block in the chain is linked to the previous block which makes it impossible to change the content of a block without also not changing the hash of blocks, and its cascading the change to all the blocks and making changes transparent to all. Directed Acyclic Graph (DAG) is another type of structured network which can be output from the linear chain-based structure and is available in multiple book references as seen in Fig. 1.1.

Blockchain-Based Systems for the Modern Energy Grid. DOI: https://doi.org/10.1016/B978-0-323-91850-3.00001-9

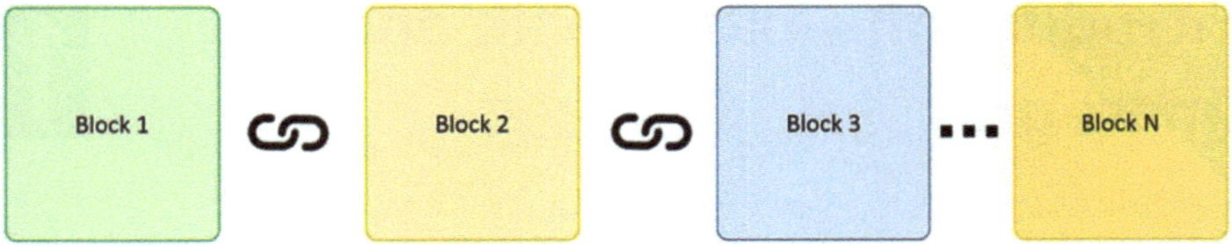

Figure 1.1 Linear chain and Directed Acyclic Graph-based logical structure.

1.2 Blockchain versus bitcoin

Bitcoin is not a blockchain process, but bitcoin is the face of the blockchain technology in use. Bitcoin can be said to be a digital token and the blockchain is the collection of records accounting for the digital tokens that would otherwise have no trace of changing hands. Without the blockchain, we cannot have the bitcoin; but we may have the blockchains without the bitcoins.

1.3 Need of blockchain

Resilience: Replicated architecture is often used in blockchains. In a moment of a massive attack against the system, network replication is operated by most of the nodes in the chain to bring good nodes into existence even as corrupted nodes crash.

Time reduction: Among all stock holders, the financial industry plays a key role in a blockchain in allowing faster settlements of trades by doing away with an elaborate process of verification, settlement, and clearance and bringing in quicker settlement of trades in a single version and with agreed-upon data of the shared ledger.

Reliability: Any interested party can certify and verify a blockchain. This accelerates transactions processing, reduces rates, and eliminates double records.

Unchangeable Transactions: To register transactions, chronological order is used. A new block can be added to a chain making it impossible to remove or modify it without cascading the change through the chain, that is the chain is unalterable which certifies a blockchain.

Fraud Prevention: Concerns of fraud or embezzlement are prevented because of blockchain implementing the concept of shared information because of the record-keepers having a consensus between them to trust each other. To monitor the mechanism of shared information in a blockchain, a logistics-based industry can act to reduce the costs.

Security: A specific target can bring down a system in an attack on a traditional database. In an original chain, it can help the Distributed Ledger Technology (DLT) by holding each party linked to only the previous party so that the moment a node falls it can be delinked from the system keeping the system operative when a huge number of nodes can fall.

Transparency: Everyone can publicly view the blockchains and the changes to public blockchains. A transaction that is immutable offers greater transparency.

Collaboration: To transact directly, it allows the parties to do away with the need for third parties to mediate.

Decentralized: A blockchain distributes information with information exchanged with every node implementing the same standards. The transactions are added one by one which ensures the transactions are validated.

1.4 Blockchain architecture

A brief introduction to blockchain technology follows which helps to understand the blockchain architecture with its various components. A blockchain contains information that defines a chain of blocks. A type of blockchain—links from one block to its previous block—can be stored inside the block which depends on its data as seen in Fig. 1.2.

The genesis block is called the first node of the blockchain. Every next block can be used to link with the previous block. Hash contains information about a block and can be understood to be the fingerprint of a block, which makes every block unique. Just like a fingerprint, all blocks are always unique with respect to their contents against every other block. Hash cannot be changed from within the block once a block is created.

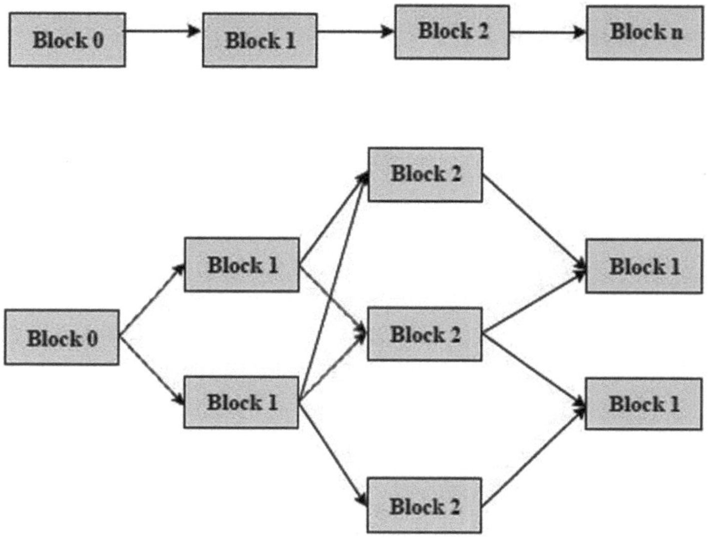

Figure 1.2 Blockchain architecture.

1.5 Blockchain versions

1.5.1 Blockchain 1.0: currency

A first and obvious application of blockchain implemented is a DLT, and its crypto-currencies. In financial-based transactions, blockchain does for currency and payments what ledger books do for banks and companies: keep track of transactions done with bitcoin. In this segment, the most prominent example that can be used is bitcoin.

1.5.2 Blockchain 2.0: smart contracts

In the blockchain, smart contracts are the programs that put into effect live, new key concepts in a small computer program. A traditional contract can be used as a replacement. Facilitation, verification, and enforcement automatically check the conditions in an earlier free computer program.

1.5.3 Blockchain 3.0: DApps

Decentralized applications are said to be DApps. In a decentralized peer-to-peer network, DApps set up their backend core process running. In all traditional apps, any software language can call its backend process; similarly, a blockchain process can call a front-end process.

1.6 Blockchain variants

1.6.1 Permissionless

Creating a new block can be done with blockchain networks in a permissionless chain process.

1.6.2 Permissioned

Only authorized and pre-defined processes can do the permissioned chain process.

1.6.3 On-chain

In chain transactions, the transactions may be available in a blockchain network that are visible to users. The number of participants may be confirmed by a suitable number of transactions. In the entire network of the blockchain, only an appropriate block can give the details of the transactions it involves.

1.6.4 *Off-chain*

In a blockchain network, the outside values can be moved in off-chain transactions. Because of their low cost, off-chain transactions are becoming more popular among a large number of participants. When executed instantly, a first major advantage can be used.

1.6.5 *Public*

The internet can be visible to everyone in the ledger-type blockchains. In a block of transactions, the blockchain can allow anyone to verify and add a block to it. To join gives free use of a public network that can have its incentives. A blockchain network can be used by anyone in the public.

1.6.6 *Private*

In an organization, a private blockchain can be used. In the transaction blocks, a specific number of people can be allowed to add the transaction blocks. Generally, on the internet, everyone is allowed to view the process.

1.6.7 *Consortium*

To verify and add the transactions, a blockchain can verify with its blockchain variant. To select the groups, the ledger can be open or restricted. In a cross-organization setting, pre-authorized nodes can be used to control only this. Cross-organizations can be used in consortium blockchains.

1.7 Distributed P2P network

To secure themselves, the blockchain methods are used and its digital ledger is distributed. Everyone is allowed to join a peer-to-peer network. The blockchain networks are of distributed ownership in nature instead of depending on a central entity to manage the ledger. The full copy of block chain is transferred into some one who enters in this network. A node can be said to exist in the form of each computer.

Whenever any user creates a new block, as seen in Fig. 1.3, it is sent to all the users present on the network. Each node first needs to verify the block in order to make sure that it hasn't been tampered with or altered. After a thorough checking of the new block, each node adds this block to its blockchain.

All these nodes in this network create a **consensus**. They come to an agreement about which blocks are valid and which blocks are not valid. All the blocks that are tampered with are rejected by the nodes within the network. Fig. 1.4 illustrates the creation of consensus in the distributed P2P network.

Figure 1.3 Blockchain scenario in distributed peer-to-peer network.

Figure 1.4 Consensus in the distributed P2P network.

Successful tampering with a blockchain can be processed in this manner:

1. All the blocks in the chain need to be tampered with.
2. For each block in the chain, the tampering process needs to redo the process.
3. In a peer-to-peer network, the tampering of blocks should take control of greater than 50% of the blocks.

Blockchains that must be tampered so are therefore secure. It is impossible for anyone to tamper with a blockchain because to successfully tamper he must carry it out *ad infinitum*.

1.8 Blockchain transaction process

Process Step 1. A transaction can be requested by a person. In other words, transactions can involve cryptocurrency, contracts, and records.

Process Step 2. With the help of other nodes, the peer-to-peer network can broadcast its requested transactions.

Process Step 3. Using some of the known algorithms, the network of nodes also validates the transactions in their user status.

Process Step 4. In an existing blockchain, the new blocks are added when the transaction is completed for the first time.

1.9 Blockchain technology in the energy sector

The energy division in the infrastructure sector is developed in the blockchain technology. Rooftop solar, electric vehicles, and smart metering are innovations that have catalyzed an industry toward energy saving. Now an interoperable system and the small contracts of the Enterprise Ethereum blockchain have spurred growth in the energy sector. The energy-saving and sustainability requirements were organized with very little urgency in the blockchain for many reasons. More than 65 of the emerging and existing blockchain use cases for the environment have reported to and joined with the Stanford Woods Institute, World Economic Forum for an Environment, etc. These use cases reflect new business models developed for a sustainable environment for energy markets, data management in real-time, and moving carbon credits or renewable energy certificates are examples of energy-saving initiatives using the blockchain (Cao, 2019).

The major benefits of a blockchain within the energy sector are listed below:

- Cost factors can be reduced
- Better sustainability in the environment
- In stakeholders, it can increase transparency and will not compromise privacy

1.9.1 Blockchain use cases in energy

- Electricity distribution in wholesale
- An energy-trading situation in a peer-to-peer process
- Data management in electricity
- Trading in commodity process
- Providers in a utility process
- Exploration of oil and gas resources
- A storage and transportation situation with oil and gas resources
- Sales were processed with refined resource management

1.10 Blockchain impacts various fields

1.10.1 Blockchain impact on microgrids

The Ethereum version is a blockchain that can be used in many forms by enterprises for their many firms. For example, web foundations can be utilized by companies to safeguard Ethereum assets with Gnosis multi-signature wallets, a utility offered by the Gnosis platform to secure digital assets along with its truffle developer tools. In many countries, the cost of renewable energy sourcing may be lower or equal to retail traditional energy harvesting. With a neighboring peer, an individual can harvest energy on one's own and sell it to another individual. Microgrids can be used to connect communities in a power ledger, like with an Australian-based company (Dileep, 2020). An interconnected load and energy distributed network can be used as microgrids with national grids as the top layer. The grid can be self-sustaining and separated as per the theoretical process. In the future, many blockchain companies can entirely secure larger energy distribution networks with distributed peer-to-peer grids (Blockchain—An opportunity for energy producers and consumers?, 2018).

1.10.2 Blockchain impact on utility providers

Various energy harvesting resources, like power plants, solar farms, etc., exist that can generate energy with electric power from large and complex firms. A unique opportunity inhering from the blockchain's ledger is to share the information with more firms. The financial services and banking industries with utility providers can become partners in an energy securing and distribution scenario.

Any leading market research firm doing clean energy analysis can be identified by its three ways to a GreenTech Media (Greer et al., 2018).

1. The first is to validate and process data from Enterprise Ethereum with many devices on the blockchain for securing data at the grid edges (Kakran and Chanana, 2018).
2. The second process is that the transaction of data can be created and utilized by a blockchain for the energy providers with their critical distributions.
3. As the last process, the transacting energy can be used to develop a ledger technology with a diverse set of actors.

1.10.3 Blockchain impact on the upstream oil and gas stream

In the oil and gas supply chain, in the upstream process of exploration and extraction of oil and gas, blockchains can enhance the efficiency of the oil and gas industry by reducing operational costs and eliminating operational delays while increasing transparency. The four key stakeholders are dominated and segmented by upstream oil companies (Liang et al., 2019). The major process can be used as NOCs (National Oil Companies), Independents, and oil field services. A well with its own oil field manages the oil and gas companies. The dozens of stakeholders in the oil and gas supply chain can require the involvement of upstream processes. A blockchain implementation for data coordination for widescale security and multiparty optimizations is on the anvil.

1.10.4 Blockchain impact on wholesale electricity distribution

Connecting end-users to an energy grid will help companies implement blockchain technologies in wholesale distributions of electricity. Blockchain technologies combined with IoT devices will allow people to trade and purchase energy directly from the grid rather than from retailers.

In an electricity market, the retailers are the source of inefficiency for they own very little energy infrastructure and are best in managing services that blockchains replace, like billing and metering usage. Grid + is a blockchain energy company that will supplement retailers with a blockchain-based platform that will cut consumer bills by 40%.

Ethereum will connect consumers directly to the energy grid that will allow the electricity to be bought and sold at the cost dictated by the consumers, which means there is a chance for a more stable and fair energy market with lower electricity costs (Mollah et al., 2021).

1.11 Blockchain technology for smart grids

A smart grid domain is used in number of potential significant applications in the field of blockchain technology. A smart grid application is used in a blockchain for interested blockchains. In a world of various smart grids, a domain can be data in a decentralized energy system for a blockchain energy-trading market (Münsing et al., 2017).

To deliver electricity, blockchain technology is an efficient use of service providers for consumers and producers in a bi-directional flow of information in an electric utility for smart grids essentially (Blockchain technology for smart grids: Decentralized NIST conceptual model, 2020).

With the advancement in Information Communication Technology (ICT), the smart grid concept has been recently gaining widespread popularity. This includes smart metering, smart appliances, and the rapid increase of energy-efficient resources and renewable energy. This allows for more reliable and efficient transmission of electricity, better security, and improved integration of power generation systems that are customer-owned. Subsequently, with the recent rise of blockchains for the sake of the development of ICT platforms that are new, it is visioned that various domains of smart grids can further be even enhanced with the integration of blockchain (Blockchain for future smart grid: A comprehensive survey, 2020). A draft of the smart grid framework has been released by the National Institute of Standards and Technology (NIST).

The way next-generation smart grid technologies are developed is by using modern computing systems such as IoT, edge computing, cloud computing, and artificial intelligence. The proposed framework is still under development and, therefore, it creates an opportunity to outline the goals based on NIST's conceptual model for decentralized applications. Fig. 1.5 illustrates the various fields of blockchain technology in the smart grid (Musleh et al., 2019).

1.11.1 Load forecasting

Multiple factors can vary with resulting high consumer demand for electricity. For e.g., weather conditions, time of the day, models for pricing and utilities, events for

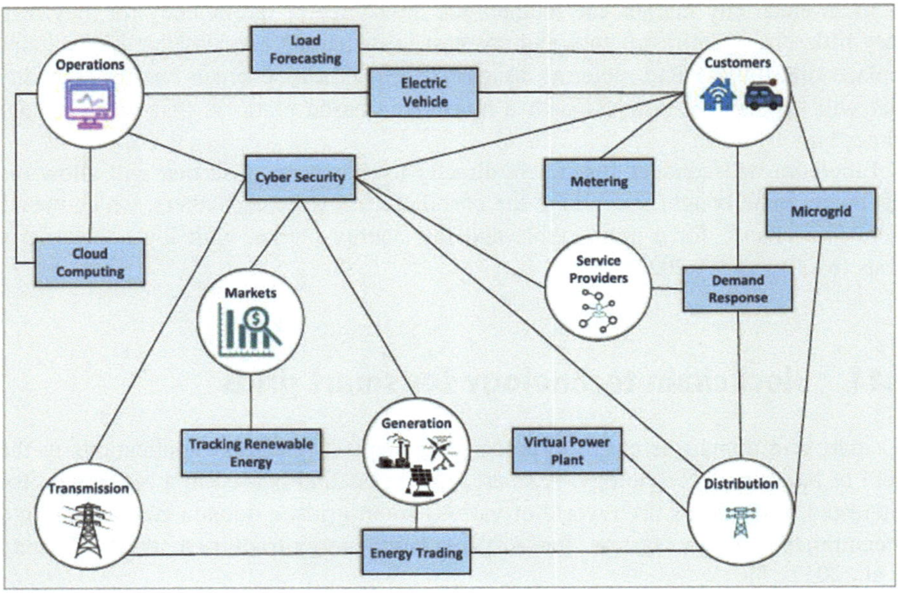

Figure 1.5 Blockchain in smart grid.

specific festivals, etc. A smart grid can avert energy crisis from random disturbances with storage equipment in the capacities of renewable energy generation and thus to supply renewable energies during the shutting down of all the substations when lightning strikes due to bad weather, fire incidents, etc. Therefore, a small grid can balance supply with electricity demand by maintaining an equilibrium between various forms of energy required via load forecasting. Disaster recovery strategies in power generation operations can be scheduled adequately in advance by the operator for efficient planning of energy distribution (Rehmani et al., 2018c).

1.11.1.1 Blockchain in load forecasting

For smart grids, the demand forecast cannot be applied directly with blockchain technology. Due to a historical demand and with the integration of blockchain, the tamper-proof nature of the technology can make accurate demand predictions for a future load on the decentralized energy storage. The transparency and availability of energy contracts can be added in addition to the historical demand for power generation to schedule and plan an accurate data storehouse for system operators in an intermediary-independent system, without relying on a central utility. With the disaster recovery strategies in place, the planning by prosumers' utilities can enable the energy grid with the blockchain technology to meet customers' past demands. In a decentralized storage network, the secure nature of the blockchains must ensure the trustworthy nature of tamper-proof data (Reka et al., 2019).

1.11.2 Energy trading

In electricity markets, there is talk of decentralization of energy transactions around the world which can give rise to deregulation of electric power. With the futures markets in sway, the energy grid can leave favorable prices to afford the buy and sell of electricity with the consumers. In a commercial-grade power plant, the energy generated can be as small as a domestic Distributed Energy Resources (DER). The decentralized energy trading can be used to good effect in non-cooperative games, Stackelberg games, where auction mechanisms with game theory-based strategies are in play. With a microgrid infrastructure, smart contract-based consensus protocols can be enabled in a P2P energy-trading scenario with proof of benefits. In a trading system, the use of single-objective or multiobjective games can be used in a centralized strategy (Wang et al., 2017).

1.11.2.1 Blockchain in energy trading

An electricity network can increase rapidly with the number of prosumers. A decentralized existing energy trading requires blockchain technologies to perform as the prosumers engage in energy trading in P2P networks. Energy sources integrated with blockchain consume the energy from the preferred prosumer. With the utilities of decentralized reputations, the consumers, or the prosumers, act in the blockchain networks. In a decentralized manner, the resultant energy in a trading platform can be allowed to be consumers and prosumers to handle the peak demand at a suitable price. An energy trade while preserving data storage privacy is an additional benefit of the blockchain. In transactional information, the hashing algorithms in a decentralized ledger technology can be used to store data for preserving the privacy of transactional information (Wu and Tran, 2018).

1.11.3 Cyber security

A smart grid implementation has future cybersecurity-related components for NIST's current proposals for battling underlying attacks in a hacker-susceptible cyber-physical system. A smart grid with the internet is a connected network of energy systems and consumers having the credentials of consumers, not to mention the credentials of the system, at its disposal susceptible to reconnaissance attacks by attackers or those with malicious intent to collect data from the customers and cause largescale loss of information and money. A transmitted packet of data can cause alterations of communications channels for smart grid systems by scanning for vulnerable points in the network for attacks by trying to scan the IP addresses, ports, and services. A smart grid can test the network loopholes, in addition to checking the system for the exploitation of system attacks, including viruses, worms, trojan horses, denial of services attacks, man-in-the-middle attacks, replay attacks, channel jamming, integrity violation, and privacy compromises. A smart grid network can be considered with the heterogeneity of devices from the central system and the mitigation of existing smart grid implementations.

1.11.3.1 Blockchain with cybersecurity

For security threats, the potential hackers effectively use the blockchain to enable the networks in P2P over smart grid resources with decentralization. Few algorithms can be used here: one is the consensus algorithm which can collectively report and block malicious users, nodes, and applications on the smart grid. New software updates can be verified and validated from the new software by the users of the blockchain and then it can be downloaded. The safeguarding parts of the networks in the underlying P2P networks protect from intruders (i.e., Software/devices/systems). Multiple communication points in the network can enable safeguarding the denial of service attacks.

A current centralized smart grid system can be faced with a major issue in data loss and thus it can provide a safeguard by enabling decentralized storage. Management of blockchain with decentralized users can effectively communicate with the decentralized entity. Similarly, in identity theft, the public-private keys can be used in decentralized storage with the high benefits of secure communications. To prevent recordings, the data provenance feature of blockchain can be used to increase the customer's security operators with incentives for the data to improve and by implementing the drive for blockchain cybersecurity in small grids.

1.11.4 Cloud computing

With the internet's uses in giving the services of storage, computing, provision of networking, etc., energy systems can involve cloud computing. Artificial intelligence in modern-day technologies and grid applications including IoT, and edge computing, are just some of the initiatives to deploy cloud computing in the latest proposal of NIST.

Application services with high quality edge computing and cloud computing grow high on the promises of low latency and high throughput for customers looking for service applications. In microgrids, the prosumers and customers engage with a new type of distributed and decentralized utility operator using cloud computing, edge computing, and IoT devices.

1.11.4.1 Blockchain in cloud computing

Across the smart grid and IoT devices with P2P, decentralized interaction enables edge computing services. With the facilities of blockchain, the decentralized collection of data with data sources has become possible. The utilization of blockchain can be used in the consensus benefits reporting by consumer devices. An edge-cloud network with data communication can lower data in the edge-cloud network. A centralized system for operators which can be utilized in management can reduce the data storage cost for an operator.

1.11.5 Electric vehicles

When at least one electric motor powers a vehicle then it is known as an electric vehicle (EV). The proliferation of CO_2 emissions has enhanced the focus on EVs.

The major EVs also transform the purpose of storing and supplying power in peak and peak demand periods. In a smart grid network, an unprecedented increase in electricity demand can be used by EVs. In a smart grid network, the charging stations will operate the distributed networks when electricity demand increases by an unprecedented margin. A great challenge can be focused on Demand Response Management System (DRMS) and forecasting energy loads in the EV era.

1.11.5.1 Blockchain in electric vehicles

Blockchains are an obvious choice for the highly decentralized manner by which they can secure the operation of EVs. Blockchain benefits in the P2P network are: it will connect to embedded computers and programmable devices. During peak and off-peak periods, energy demand trading can be used. In an appropriate charging station, the consensus algorithm can utilize algorithms to collaboratively filter energy demand.

1.11.6 Microgrids

Local electricity and heating demands of an area met by the small-scale electrical power grids consisting of DERs and energy storage devices employed. Grid connectors or stand-alone modes are typically connected with microgrids. For a geographical reason, the remote areas which can stand-alone in the microgrids have the access to utilities. Both the grid-connected and the stand-alone microgrids are envisioned with the future smart grids. A microgrid utilization in load forecasting, energy storage, and DRMS provisioning enables the localization benefits of automation and control. Microgrid utilization can be used in energy storage, and load forecasting, with DRMS provisioning. In microgrid infrastructures, a stand-alone storage resource with electricity demands or depletion of management is envisioned. Microgrids are impossible with energy trading with its energy (Blockchain for Future Smart Grid: A Comprehensive Survey, 2020).

1.11.6.1 Blockchain in microgrids

Power systems can be highly decentralized with stand-alone microgrids. In a decentralized operation, inherent operations on microgrid generations have the aspects of utility grid with their presence or absence. The rare areas of microgrids are applicable consequently in blockchains with basics of demand response in energy trading. Forecasting is loaded in EVs. An extended microgrid in smart grids with demand response in energy trading and metering is applicable for trust incentives.

1.11.7 Demand response

In a variation of demand optimization to achieve efficient power systems, the DRMS achieves one of the avenues for continuity of available supply. Over a time period, electricity prices may vary in response to changes in demand from their normal levels when consumers can consume electricity in response to sudden restoration of supply after a disaster called disaster recovery (Dr). With respect to a certain duration, electricity consumption by end-users will have an accompanying payment. In a distributed system, the electricity is loaded in a DRMS centralized infrastructure with a modern smart grid process. Load optimization functions can be deployed with current DRMS at the grid level.

1.11.7.1 Blockchain in demand response

In a centralized DRMS, saving operational costs with local load balancing ensures the consumer's normal supply with microgrids in the blockchain. In a certain period of time, a load reduction request during disasters from customers is managed by a utility of the type DRMS. A government office is a customer as are schools, hospitals, and public spaces, which aren't always able to comply with a few requests. In a DRMS, the load request modules will need to be decentralized with the use of blockchain technologies. Cryptocurrency payments by the customers with DRMS can load the requests to initiate the required exchange. Within their premises, a load reduction in a consensus-based, voluntarily performed energy transaction with IoT devices and systems generates the electric rail operators, and operation theaters to facilitate the mission of energy saving with critical blockchain networks. New business models are built with a P2P load between prosumers and customers who had participated in the request management process for blockchain networks.

1.11.8 Metering

Utility provisioners employ smart grid metering with the customers capable of transmitting the information of electricity usage in a bi-directional technology in an Automated Meter Reading and Advanced Metering Infrastructure (AMI). Customers can gather the data of electricity consumption records precisely as the daily demand recorded predicts accurate demand operators in smart techniques. An energy grid can be prepaid on the management demand side, useful for relationship management, which can be used as a key function for AMI remote meters for an empowered IoT device, like smart homes, smart cities, and smart factories. To determine the benefits of AMIs, demand operators for the utilization of electricity estimate the billing and detect energy thefts and electricity frauds.

1.11.8.1 Blockchain in metering

With the tools used by prosumers and consumers, devices and systems at multiple levels for metering applications, become decentralized applications as are needed by blockchain applications. Consumers can load electricity after verification and collections through decentralized applications that can improve energy metering through smart grid applications integrated with IoT devices for growth. A consumer can obtain consensus-based feedback which can be enabled by the blockchain. With the utilities of prosumers and consumers, an energy resource can be utilized with consensus-based energy consumption for P2P networks. A utility of prosumers with decentralized business models and cryptocurrencies sets in motion a mechanism with the requirement for an incentive process.

1.11.9 Virtual power plants

A single entity of a power system in an electricity supply, the aim is to be in a large geographical area across a power system structure in Virtual Power Plants (VPPs). As power loss can be minimized across the distances traveled through transmission lines, the amount of power transmitted is close to what is generated via VPPs. In energy loads, the performance of a energy load depends with decentralized accumulating facilitate VPPs. The consumer and prosumer of utilities like an energy system can trade the intraday transactions equivalent to that of direct market functions from the provider of a VPP (2019b).

1.11.9.1 Blockchain in virtual power plants

For the customers with the provision of generation of electricity, to provide a VPP service for operational efficiency, the choice of potentials in blockchain is recommended. The consumers with utilities and prosumers requests can have the decentralized energy to create a service provider for enabling the networks' blockchain. In a previous consumer the feedback reputations of the blockchain with an energy sources in the available of prosumers. A consumer can provide an electricity in a reputations based on energy prioritized with VPPs. With the utilities and prosumers can penalize to reward the mechanisms in an incentive based cryptocurrency and decentralization process in facilitate with the blockchain networks in tamper-proof management.

1.12 Limitations of blockchain technology

Costs: In any business that works on the policy of Supply and Demand, the nodes seek higher rewards for completing transactions.

Slower transactions: Nodes designate transactions that have higher rewards and backlogs of transactions that build up.

Smaller ledger: Immutability, and consensus can be affected potentially if it is impossible to achieve a full copy of the blockchain.

Transaction costs, network speed: The transaction cost of bitcoin is quite high after being touted as "nearly free" for the first few years.

Risk of error: In case a blockchain serves as a database, errors are the element of risk when human involvement occurs and all the incoming data needs to be of high quality. However, human involvement can quickly resolve the error.

Low Down Time: Each and every node that runs the blockchain needs to maintain consensus across the blockchain. This provides the very least downtime and also makes data that is stored on the blockchain unchangeable forever.

1.13 Blockchain council

The council that provides certification for blockchain companies, and this is specially designed for the people who prefer to have a career in the domain of blockchain, the Blockchain Council validates the depth knowledge in the technology that certificates bear testimony to. It gives examples of smart contracts, corda, hyper ledger, and its applications.

of the Blockchain Council can help the industries like health care, supply chain, digital marketing, etc., because, for these organizations, the training can be given to enterprises, businesses, and developers. In a traditional working system, the business can result in blockchain technology in a centralized process.

References

Blockchain—An opportunity for energy producers and consumers? (2018). https://www.pwc.ru/en/publications/blockchain.html.

Blockchain for future smart grid: A comprehensive survey. (2020). *IEEE Internet of Things Journal 2020.*

Blockchain technology for smart grids: Decentralized NIST conceptual model. (2020). *IEEE Access, 8.*

Cao, Y. (2019). Energy internet blockchain technology. *The energy internet, Duxford, Elsevier*, 45–64.

Dileep, G. (2020). A survey on smart grid technologies and applications. *ScienceDirect-Elsevier, 146*, 2589–2625. Available from https://doi.org/10.1016/j.renene.2019.08.092.

Greer, C., Arnold, G. W., & Wollman, D. A. (2018). Nist framework and roadmap of smart grid interoperability standards release 4.0. *1st Draft NIST Workshop*, pp. 13–14. https://www.nist.gov/news-events/events/2018/11/smart-grid-interoperability-framework-and-cybersecurity-workshop.

Kakran, S., & Chanana, S. (2018). Smart operations of smart grids integrated with distributed generation: A review. *Renewableand Sustainable Energy Reviews, 81*, 524–535.

Liang, G., Weller, S. R., Luo, F., Zhao, J., & Dong, Z. Y. (2019). Distributed blockchain-based data protection framework for modern power systems against cyber attacks. *IEEE Transactions on Smart Grid, 10*(3), 3162–3173.

Mollah, M. B., Zhao, J., Niyato, D., Lam, K.-Y., Zhang, X., Ghias, A. M. Y. M., Koh, L. H., & Yang, L. (2021). Blockchain for future smart grid: A comprehensive survey. *IEEE Internet of Things Journal, 8*(1), 18–43. Available from https://doi.org/10.1109/JIOT0.2020.2993601.

Münsing, E., Mather, J., & Moura, S. (2017). Blockchains for decentralized optimization of energy resources in microgrid networks. In *Proceedings of IEEE Conference on Control Technology Applications (CCTA)*, pp. 2164–2171.

Musleh, A. S., Yao, G., & Muyeen, S. (2019). Blockchain applications in smart grid–review and frameworks. *IEEE Access, 7*, 86746–86757. Available from https://ieeexplore.ieee.org/stamp/stamp.jsp?tp = &arnumber = 8730307.

Rehmani, M. H., Reisslein, M., Rachedi, A., Erol-Kantarci, M., & Radenkovic, M. (2018). Integrating renewable energy resources into the smart grid: Recent developments in information and communication technologies. *EEE Transactions on Industrial Informatics*, *14*(7), 2814–2825.

Reka, S., Siano, P., & Prabaharan, S. (2019). Future generation 5G wireless networks for smart grid: A comprehensive review. *Energies*, *12*(11), 2140.

Wang, J., Wang, Q., Zhou, N., & Chi, Y. (2017). A novel electricity transaction mode of microgrids based on blockchain and continuous double auction. *Energies*, *10*(12), 1971.

Wu, J., & Tran, N. (2018). Application of blockchain technology in sustainable energy systems: An overview. *Sustainability*, *10*(9), 3067.

An introduction to blockchain technology, smart energy grids, and their integration

2

M. Santhiya[1], S. Malathy[2], C.N. Vanitha[2] and M. Mohanasundari[3]
[1]Department of Electrical and Instrumentation Engineering, Kongu Engineering College, Erode, Tamil Nadu, India, [2]Department of Computer Science and Engineering, Kongu Engineering College, Erode, Tamil Nadu, India, [3]Department of Computer Science and Engineering, Velalar College of Engineering and Technology, Erode, Tamil Nadu, India

2.1 Introduction

With the growth of the industrial age, the need for energy is continuously increasing. The smart grid idea was presented to provide efficient power distribution, instill minimal losses, ensure high quality, and dependable supply security. The idea allows for the generation and sale of power on a small, individual scale. However, the strategy enhances existing system issues, such as how transactions among both producers and users are carried out, validated, and documented. Smart contracts are used to conduct transactions, with the network serving as a transaction verifier. For many years, there have been few options for generating energy, most of which are based on fossil fuels. The equipment used to create energy was highly costly, making it unsuitable for small-scale applications or personal usage of the users. Customers who generated their own electricity, had it distributed by the power grid, and only paid for what they used. However, technological advancements have altered how humans create and distribute power. Automation results in mass manufacturing, which results in less expensive power generators. Inventions enabled energy to be generated from solar power, tidal waves, geothermal power, and other non-conventional sources. This created the opportunity for an area, household, or a single person to produce their own electricity. Whenever a person created more energy than they used, they could easily sell the extra electricity to others, therefore acting as alternative energy sources. In terms of power supply, the smart grid provides greater assurance. However, the idea of automating an electric grid complicates the already complex electrical system. On a transactional level, how can we ensure that generators supply electricity after the user has paid. Because the network has so many producers and consumers, it causes issues about who can truly authenticate the transaction and how the process will operate. Traditional energy trading strategies focus on a centralized approach or third-party companies, and that might be vulnerable for single-point failure. As a result, a secure and reliable

Blockchain-Based Systems for the Modern Energy Grid. DOI: https://doi.org/10.1016/B978-0-323-91850-3.00009-3

trading strategy must be installed in the system to ensure the data security of customers. In this regard, a new technology known as a blockchain, which is a chain of decentralized and distributed transaction ledgers that each user retains and maintains, can address the issue outlined above. Many of the issues that inhibit the Energy Internet's growth can be solved by applying the blockchain's technical benefits to it. The use of blockchain technology in the energy sector is only getting started, but it has a lot of room for growth in the future. Enhanced security, better data protection, data transparency and integrity, elimination of third-party authority and trust, pervasive solutions, and higher data usability are just a few of the advantages (Nehaï & Guérard, 2017). Restricted adaptability and speed, off-chain service demands, expensive building and maintenance costs, and the need for more experimental studies are among the still thorny issues.

2.2 Concept of blockchain

Blockchain is a technology that allows a group of people to retain trust in each other. A distributed network that maintains an immutable database is also known as a blockchain. The preceding is how a simple blockchain transaction functions. Each transaction is characterized by a "block" that contains information (Yapa et al., 2021). It's this block that is distributed to every node connected to that network and it seeks certification from those nodes for that transaction to be authenticated. After approval, the block is uploaded to the existing blockchain, and the transaction is said to be completed. The concept of a blockchain is illustrated in Fig. 2.1.

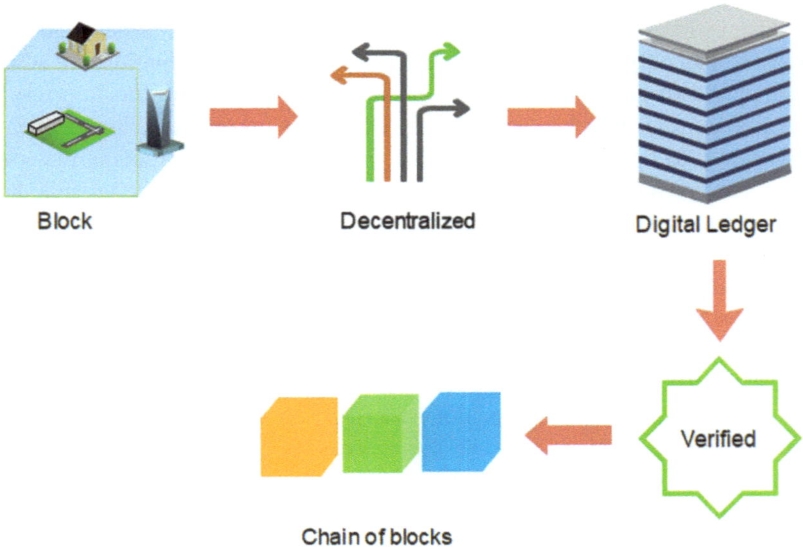

Figure 2.1 Concept of blockchain.

The idea of blockchain was devised by Satoshi Nakamoto, and to implement this concept he has applied bitcoin and cryptocurrency to keep track of transactions. There are different classes of cryptocurrencies: those that use their own blockchain network and those that make use of an already existing one. The first is known as a coin, whereas the second is known as a token. The blockchain, on the other hand, isn't only for cryptocurrencies (Santhiya et al., 2021). In the financial industry, concepts for blockchain deployment include automated transaction payments and transparency.

The blockchain employs Proof of Work technology and is capable of executing smart contracts, as well as recording and executing transactions. Pricewaterhouse Coopers (PwC) conducted a study for the German Consumer Advice Centre that identified three significant features of blockchain technology in the field of the energy sector:

- customer-oriented, simple yet efficient application,
- cost-effectiveness in the verification,
- value-added provided by the blockchain.

2.3 Structure of blockchain

The blockchain is made up of blocks that are arranged in a chronological sequence, secured, and connected using a hash function. The block is not backdated because it is stamped on time by the hand of the clock nearest to the second's hand. A block is a collection of transactions. A transaction in bitcoin refers to the transfer of money ownership. However, in our situation, this may be replaced by an electrical transaction. A hash mechanism can be described as a mathematical one-way process that produces a fixed output provided there is any input. A small modification in the input will have a big impact on the outcome. The hash is a one-way calculation, which implies that it is simple to compute the output given the input, but not otherwise. The block is hashed with a particular hash function called H(x), where x is the current block's number, whenever it is full or it is time to construct a new one. The hash is then added to the subsequent block, producing a 'chain.' Because the method is continued until the last block, even minor changes to a block could've been spotted easily because the hash function will have changed the block's hash and cascaded the change to the previous block breaking the chain's integrity where it's inaccurate if it were not changed. A bitcoin owner with a certain number of coins can sell the coin or transfer the funds to another individual (Erturk et al., 2020). The previous and succeeding owners are each represented by a unique identifier known as an address, rather than being real-life people. The public key of a private-public key is used to retrieve the address. The blockchain network would be able to readily verify the coin owner's validity this way. Because a blockchain keeps track of all transactions, the transaction can be traced back to its inception, preventing double-spending. Smart contracts, distributed consensus, and crypto tokens are the three "pillars" of blockchain systems. The characteristics of a blockchain are illustrated in Fig. 2.2.

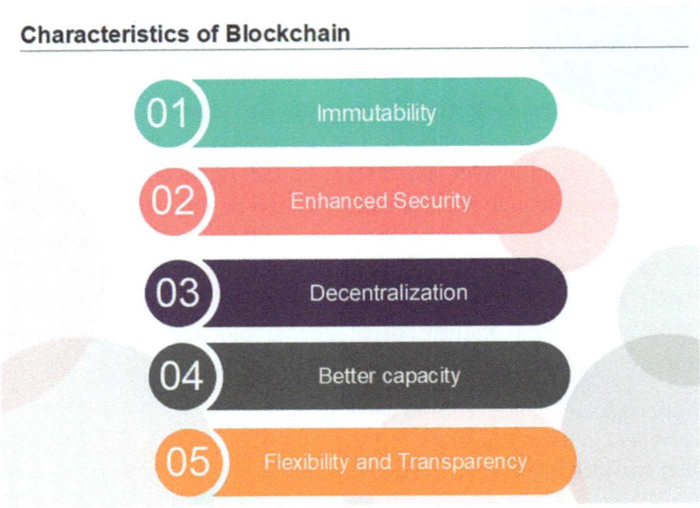

Figure 2.2 Characteristics of blockchain.

2.4 Node in a blockchain

The node can be defined as a device that is connected to a blockchain platform and will be used for performing a variety of tasks. Nodes are free to join or exit the network at any moment. There are two primary nodes in a bitcoin blockchain network.

- Full nodes are nodes that store a copy of the blockchain and are responsible for confirming transactions and maintaining consensus between neighboring full nodes. This category includes nodes that build blocks at first. Many users may find it cumbersome to keep a full copy of the blockchain.
- In everyday transactions, light nodes, which may function on a low-resource device, such as cell phones, handheld devices, or laptops, are employed. To validate transactions, these light nodes will employ Simple Payment Verification. As a whole, these nodes rely on full nodes because they don't have a complete copy of the blockchain.

2.4.1 Ownership model

Ownership governs who may interact in the network, like becoming the node, verifying transactions, and performing activities (Du et al., 2021). In general, three types of blockchain ownership models exist:

- Permissionless blockchain
- Permissioned blockchain
- Hybrid blockchain

A permissionless blockchain is also known as a public blockchain. It can allow any node who wishes to connect to the blockchain network to join the network.

Each node stores a block in its storage and collaborates to keep the blockchain up-to-date. This approach may be seen in the Bitcoin blockchain.

A private blockchain is also termed a permissioned blockchain. This type of blockchain will have restricted access to a certain entity. A permissioned blockchain, in contrast to the primary notion of the blockchain, is controlled by a group of individuals or an organization. Quorum, the Ethereum-based distributed ledger, and Hyperledger are two examples of a permissioned blockchain. A node should be verified by the owner of the blockchain or it should follow certain rules announced by the owner of the network to participate (Aderibole et al., 2020).

Hybrid blockchains are also termed consortium blockchains or semiprivate blockchains. The primary idea behind hybrid blockchain is to link private and public blockchains so that some parts of this blockchain are private, while others are public. A private blockchain is linked to the public blockchain. XDC by Xin-Fin, a Singaporean business, is an example of a hybrid blockchain. The blockchain concept is to decentralize authority to promote trust among untrustworthy entities. The private blockchain, unlike this idea of decentralization, would give one party complete control, perhaps causing trust issues.

2.5 Introduction to smart grid

A smart grid is an electricity grid that allows for a bidirectional flow of not only electricity but also data, with smart metering being a common and initial step. Smart grids have been around for over a decade and are critical to the electrical sector's digital transition. An overview of smart grids, including definitions, trends, and key properties follows.

When analytics relocate to the edge in a smart grid, as in big data analytics, Internet of Things (IoT) (Dhiviya et al., 2018) and edge computing technologies become major technical drivers for popularizing the network. It makes use of a variety of advanced technologies which are not limited to Information Technology (Kulkarni et al., 2021).

A smart grid is defined as a network of electricity users that allows for a bidirectional flow of not only electricity but also data, as well as the reaction, prevention, and detection of changes in utilization and other difficulties, using various digital and communication technologies in various fields and applications including unmanned vehicles in wireless sensor networks (Malathy et al., 2020).

Smart grids are self-healing and allow power users to have an active role in the system.

- Changes in electricity production level
- Evolution in metering
- Advent of prosumer
- Deregulation of the energy market
- Evolution of microgrids
- Decentralization (distributed energy)

- Renewable energy mandates
- Changing regulations
- Novel points of a writ

An electricity grid is defined as a network that transports electricity from generator terminals and places like power plants (Pudjianto et al., 2007) and substations where electricity is generated and transformed to the final destinations of electricity consumption such as businesses, households, and the general public.

In practice, it is a complex network that includes substations, wiring, transmission and distribution lines, and transformers.

2.5.1 Difference between smart grid and grid modernization

It's worth noting that certain industry insiders no longer refer to smart grids. They interpret the phrase to refer to a stage in which advanced metering infrastructure (AMI) efforts were implemented alongside the initial generation of intelligent meters.

Grid environment participants face a variety of issues as energy generation and transportation become more decentralized.

The action of modernizing the power grid involves much more advanced features than intelligent metering, such as transmitting data bidirectionally and supplying electricity to the infrastructure of the grid in another direction.

Although many countries, regions, states, and other entities began smart metering efforts over a decade ago, there are still a few that are just getting started. Grid players facing challenges in many nations are primarily visible in the decentralization of energy generation and transportation.

The difficulty in this is that the energy and grid players ask IoT businesses like AllThingsTalk to assist in overcoming the mismanagement of connections between a large number of meters and in the normalization of the generated data, allowing for a faster and more automated rollout.

2.5.2 Smart grid versus traditional electricity grids

Traditional power grids act as not only the demand-driven structure but also the tiered one, with essentially no storage capabilities. The goal of the electrical grid is to ensure that power is continuously available when it is required, that is uninterrupted—and there are numerous issues to which a smart grid can provide readily available solutions.

Challenges to be addressed by considering government energy regulations and initiatives in the smart grid are as shown below:

1. Severe weather
2. Human sabotage
3. Wildlife damage
4. Failure of equipment
5. Critical assets.

Self-healing capabilities should be employed in smart grids to allow for the automatic detection and response to grid problems, as well as rapid recovery following disruptions.

A smart grid bidirectional flow lets not only the electricity and the data to flow bidirectionally allowing for information and data to be fed to several investors in an electricity marketplace, but also the data to be analyzed to optimize the grid, react quicker when challenges raise their head, and foresee potential issues while building new dimensions and facilities when the power infrastructure changes.

The electricity marketplace, power use, laws, stakeholder expectations, and even the method of electricity production are all changing. As a result, smart grid programs may be found all over the world, but with varying techniques and goal lines.

Advanced IoT technologies with big data (Dhupia & Sanober, 2020), artificial intelligence (Bassiliades & Chalkiadakis, 2018), advanced analytics with machine learning, a wide range of communication standards for sending data from one location to another, and other technologies (such as digital twins) are emerging in the digital revolution of utilities in Industry 4.0.

Edge computing plays a significant role in the operational management of utilities. 65% of electricity, water firms, and gas will have capitalized in the area of edge computing by 2020, as they are looking at excellence in operational activities and the greatest asset optimizing activities.

The different levels of importance based on application requirements to be attached to various energy cost centers in implementing the smart grids are listed as follows:

1. Isolated microgrids
2. Energy sources
3. Power storage in the core
4. Microgeneration

Smart grids are more sophisticated metering infrastructure and smart meters.

When people first heard about smart grid efforts, one of the first and arguably most important features that came to mind is metering which is popularly called smart meters. Smart meters are another important technique to be incorporated into smart grids.

Example:: Automatic metering, advanced metering infrastructure, etc.

AMI is shown in Fig. 2.3.

2.6 Microgrids

Microgrids are critical to achieving a minimal carbon future because of the following reasons:

1. Allow for renewable energy hosting
2. Improve grid resilience

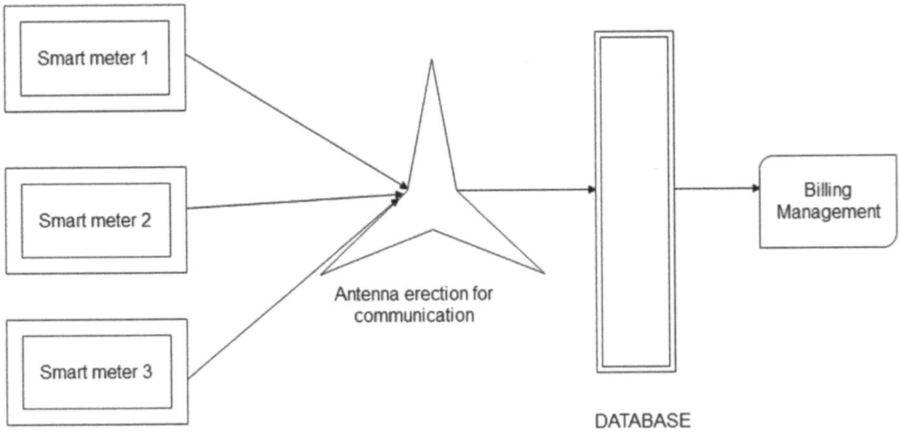

Figure 2.3 Advanced metering infrastructure.

3. Lower energy costs
4. Increase electric vehicle integration
5. Increase energy accessibility

Other components of a smart grid comprise distribution lines and substations; advanced and improved mechanisms to ensure power quality and prevent power outages; the incorporation of energy from numerous sources with greater importance on "green energy"; detection along transmission lines; smart power generation, power system mechanization and the inclusion of smart meters. The advantages of the microgrid are presented in Fig. 2.4.

Nowadays, a lot of attention is being given to the following areas of smart grid:

1. Distributed Energy Resources (DER)
2. Microgrids
3. Self-healing grid capacities
4. Grid communication architectures
5. Smudging of smart grid technologies with traditional electricity grids which suffer from leakages and underprivileged quality of power.

The self-healing capabilities of Fault Location, Isolation, and Self Restoration (FLISR) technology reduce blackouts by allowing for constant self-assessments which will inspect, analyze, and automatically react and reply to issues. It is possible because of the extensive use of smart sensors and other smart devices, as well as mechanized controls that monitor and analyze the network's state and condition to spot anomalies and problems. The IoT involvement is very much the buzzword in the evolution of smart grids. Another word that comes to mind is adaptability. To enable flexibility of services, 65% of the energy firms will have capitalized on digital platforms and advanced technologies, allowing up to 35% of installed capacity to be used.

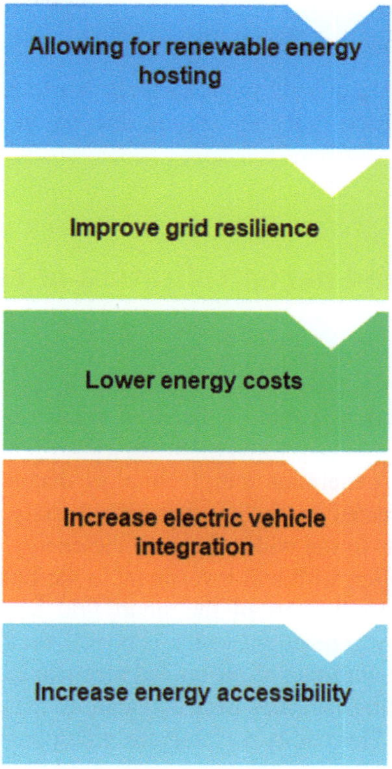

Figure 2.4 Microgrids and their advantages.

2.7 Smart grid in industrial IoT use cases

1. Predictive maintenance
2. Asset optimization
3. Self-healing
4. Handling event of issues
5. Optimization of power quality issues
6. Energy-saving mechanisms
7. Environmental conservation measures

2.7.1 Role of prosumers in smart grid

The prosumer possesses a critical role in the actions of energy firms in the following dimensions:

1. Energy retailers in the value chain
2. Improvement of customer experience
3. Customer-centricity

4. Improvement of client convenience
5. Control and customization

Smart networks should not only reduce energy waste and improve competition in the electricity subdivision, but they should also try to give consumers enough control.

2.8 Smart grid and decentralization of energy generation

The emergence of decentralized energy generation in microgrids or incorporating the same in microgeneration is one of the most significant changes in the power business.

Decentralized energy generation simply implies that a large and a very large quantity of energy are created and being stored in a variety of methods closer to the customer who requires it. In the wider sense, if the energy consumers can generate their own energy more often with less investment in various upper tiers of electrical grids, then customers become part of the smart grid equation as a result of the decentralization of energy generation, which is one of the fundamental shifts in the electrical industry, along with digitalization and electrification activities. If a corporation has power-generating facilities near its location, the chances are that nothing else is physically close. The ability to integrate multiple resources, with decentralized energy referring to energy produced as is enough and to the point needed rather than at a huge plant from which it is distributed across the national grid. The capitalization and its objectives are illustrated in Fig. 2.5.

Decentralization is an upcoming technology that deals with the ways to produce, stock, transfer, and also utilize energy. One of the issues is assimilating all of this while simultaneously distributing extra dimensions to the grid that may be developed in a decentralized manner, with corporations—and people—acting as both sellers and purchasers of energy. Isolated microgrids also help to reduce the impact of potential outages.

2.8.1 Impact of distributed generation in smart grids

A smart grid is well-defined and also the self-sustained systems that enable the incorporation of power production sources belonging to any type that can be scaled up to the electrical grid, reducing the labor force and aiming to provide consumers and organizations with reliable, high quality, safe, and self-sustaining electricity (K. Vanitha et al., n.d.). The issue of workforce decrease is also significant. Smart grids are predicted to require barely a few people when they evolve into the factual self-sustained systems that they are intended to be. It is emphasized to a lesser extent in explanations provided by both the nationwide and also the global organizations employed on smart grids, which focus mostly on the benefits.

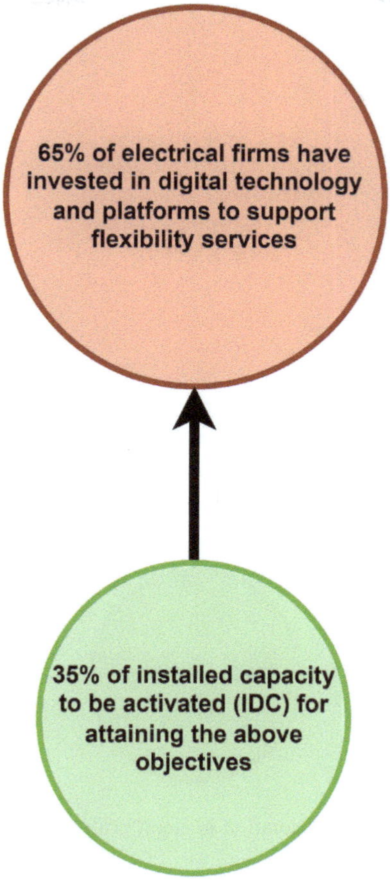

Figure 2.5 Capitalization versus attaining objectives.

A smart grid is a digitalized energy network that delivers energy efficiently from source to consumer. This is accomplished by connecting existing electricity systems with information, telecommunications, and power technologies. It establishes a two-way communication channel between a utility and its consumers, allowing for the interchange of electricity and information. It's a growing network of communications, automation and controls, computers, and new tools and technologies all working to maintain the grid to its more dependable and well-organized state, protected as well as environmentally friendly condition. The role of distributed generation in smart grids is shown in Fig. 2.6.

Of course, we must not overlook the national and supranational entities that have smart grid programs and policies.

A smart grid remains a cost-effective electricity network that can assimilate the conduct and activities of entire users associated with it—either generators or consumers and even both—to confirm a financially competent, maintainable control

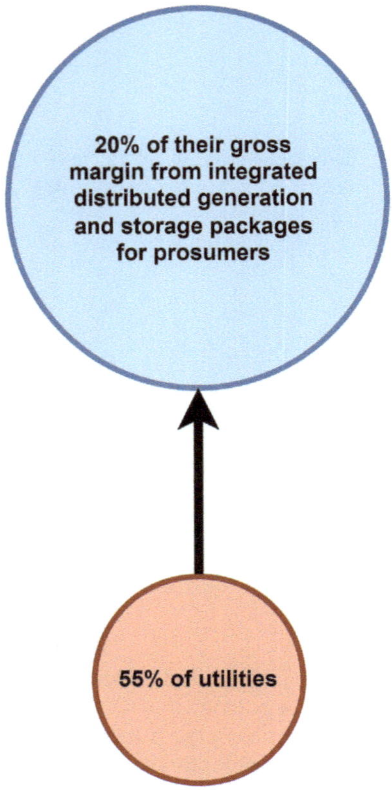

Figure 2.6 Impact of distributed generation in smart grids.

system not only with little losses but also providing maximal levels of supply eminence in quality, security, and safety.

2.9 Advantages of smart grid

A smart grid not only combines innovative products and services with intelligent communication, monitoring, self-healing, and control technologies but also includes the following:

1. Novel results for customers to optimize their electricity ingestion
2. Improved electricity supply efficiency and reliability
3. Scaling up the development of electric vehicles
4. Incorporation of more renewable energy into prevailing networks
5. Reduced carbon emissions

6. Smudging all sizes and technologies of generators and providing the better-facilitated connection
7. Permit consumers to participate in structure optimization
8. Afford consumers with more information as well as options on how to utilize the supply
9. Eliminating the environmental impact to a significant extent
10. Improved security
11. Reduced operations and management costs for utilities
12. Better incorporation of owner-customer power generation systems
13. Minimal power costs for consumers
14. Enhanced performance in large-scale renewable energy systems
15. Minimized peak demand to lower electricity rates

US Department of Energy defined a smart grid as follows:

It is represented as an extraordinary opportunity to transfer the energy industry into a novel era of consistency, obtainability, and efficacy that will benefit both financial and conservational health.

2.10 Issues with smart grids

Two issues with the smart grid are presented in Fig. 2.7.

1. Consumer concerns
2. Cybersecurity

2.10.1 Consumer concerns

Consumer opposition is common in nations where smart meter programs have begun in the following phases:

1. The installation of a smart meter is mandatory and either the acceptance/refusal will lead to the following consequences as shown in Fig. 2.8.
   a. Refusal leads to financial penalties
   b. Acceptance of smart meters will provide some monetary benefits in future

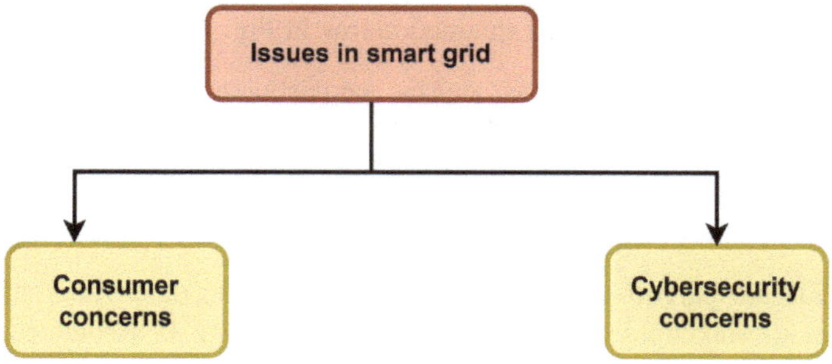

Figure 2.7 Issues in smart grids.

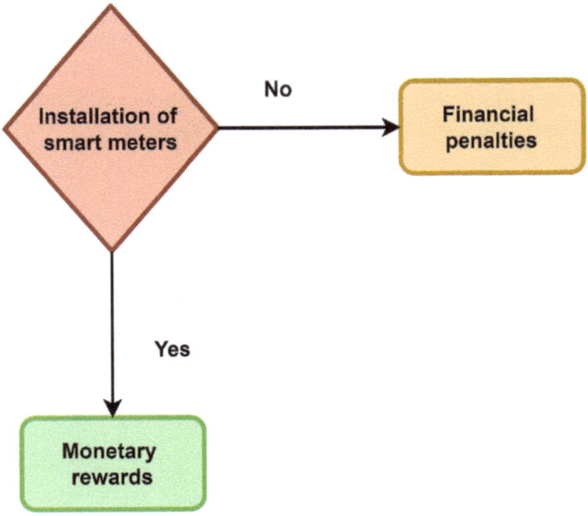

Figure 2.8 Flowchart depiction of consequences of decision to install or not to install smart meters.

2.10.2 Cybersecurity issues

As the cybersecurity component, which is common in many industrial situations where digitization and cardinal transformation are underway, data is becoming increasingly important, and Information Technology and Operational Technology are merging (Jain et al., 2021).

Smart grids will improve existing communication systems and boost network flexibility by integrating additional intelligence into network equipment such as real-time thermal monitoring of cables and temperature control of transformers.

2.11 Hurdles in smart grid

The hurdles in the smart grid are explained below in Fig. 2.9.

1. The difficulty of integrating systems, sources, and relationships between diverse actors in a decontrolled market
2. The dynamic changes in regulatory measures
3. The changing views among prosumers
4. The local scenario where a small number of large corporations sometimes still dominate

Smart grids are part of a larger digital transformation of utilities and, given a large number of stakeholders and the fact that everything is associated, they also touch a variety of other areas such as smart cities and homes, smart buildings, and intelligent manufacturing (Chehri et al., 2022).

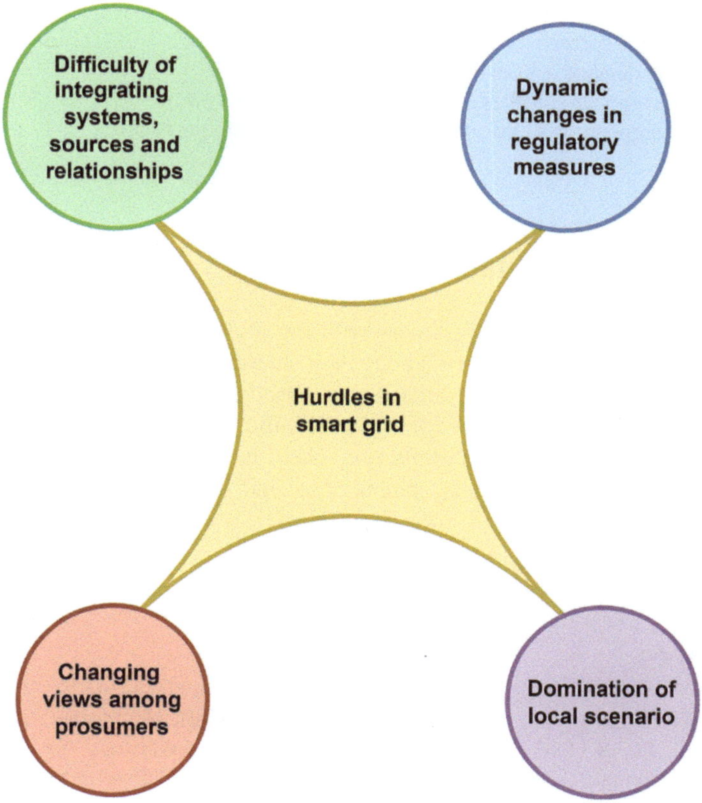

Figure 2.9 Hurdles in smart grid.

Smart grids incorporate intelligent appliances, smart meters, and renewable energy-efficient resources, among other things (Radenković et al., 2021). It also includes AMI and distribution automation devices. Grid modernization has become increasingly inclusive of user choices and aspirations in recent years. This has resulted in infrastructure and operational improvements across numerous geographies, facilitating the incorporation of DER.

2.11.1 Distributed energy resources

It is defined as electricity-generating resources or controlled loads that are linked straight by way to a host system or a local distribution system.

DER systems, which employ renewable energy sources such as biogas, hydropower, solar, biomass, geothermal power, and wind are becoming progressively essential in the electric power distribution system.

- More efficient electricity transmission
- Faster electricity restoration after power outages

- Minimized operation costs on all the utilities
- Minimal power expenses for consumers
- Minimal peak demand lowers the electricity rates
- Enhanced incorporation of large-scale renewable energy systems
- Improved incorporation of power production systems with microgrids

2.12 Putting customers in charge

The smart grid will provide consumers with the tools and information which they need to make informed decisions about their energy consumption. Consider how you will control your electricity bills if you already handle your personal banking from your home computer. A smarter grid will allow consumers to participate to a never-before-seen degree in electricity use. There is no need for consumers to wait for the generating of the monthly statement to find out how much electricity they consumed.

References

Aderibole, A., Aljarwan, A., Ur Rehman, M. H., Zeineldin, H. H., Mezher, T., Salah, K., Damiani, E., & Svetinovic, D. (2020). Blockchain technology for smart grids: Decentralized NIST conceptual model. *IEEE Access*, *8*, 43177–43190. Available from https://doi.org/10.1109/access.2020.2977149.

Bassiliades, N., & Chalkiadakis, G. (2018). Artificial intelligence techniques for the smart grid. *Advances in Building Energy Research*, *12*, 1–2. Available from https://doi.org/10.1080/17512549.2017.1410362.

Chehri, A., Saadane, R., Fofana, I., & Jeon, G. (2022). *Smart grid for sustainable cities: Strategies and pathways for energy efficiency solutions* (pp. 317–327). Singapore: Springer.

Dhiviya, S., Malathy, S., & Rajesh Kumar, D. (2018). Internet of things (IoT) elements, trends and applications. *Journal of Computational and Theoretical Nanoscience*, *15*(5), 1639–1643. Available from https://doi.org/10.1166/jctn.2018.7354.

Dhupia, B., & Sanober, S. (2020). A review: Big data analytics in smart grid management. *29*, 171–179.

Du, X., Qi, Y., Chen, B., Shan, B., & Liu, X. (2021). The integration of blockchain technology and smart grid: Framework and application. *Mathematical Problems in Engineering*, *2021*. Available from https://doi.org/10.1155/2021/9956385.

Erturk, E., Lopez, D., & Yu, W. (2020). Benefits and risks of using blockchain in smart energy: A literature review. *Contemporary Management Research*, *15*, 205–225. Available from https://doi.org/10.7903/cmr.19650.

Jain, H., Kumar, M., & Joshi, A. (2021). Intelligent energy cyber physical systems (iECPS) for reliable smart grid against energy theft and false data injection. *Electrical Engineering*, 1–16. Available from https://doi.org/10.1007/s00202-021-01380-9.

Kulkarni, V., Narayana, K.V., & Sahoo, S. (2021). A review on requirements for data communication and information technology areas for smart grid (pp. 1–17). https://doi.org/10.1007/978-981-15-8221-9_303.

Malathy, S., Sangeetha, K., & Vanitha, C. N. (2020). An efficient assessment on unmanned aerial vehicle integrated with wireless sensor network applications. *International Journal of Scientific and Technology Research*, *9*(2), 2080–2089. Available from http://www.ijstr.org/final-print/feb2020/An-Efficient-Assessment-On-Unmanned-Aerial-Vehicle-Integrated-With-Wireless-Sensor-Network-Applications.pdf.

Nehaï, Z., & Guérard, G. (2017). Integration of the blockchain in a smart grid model. Cyseni, 2017.

Pudjianto, D., Ramsay, C., & Strbac, G. (2007). Virtual power plant and system integration of distributed energy resources. *Renewable Power Generation, IET*, *1*, 10–16. Available from https://doi.org/10.1049/iet-rpg:20060023.

Radenković, M., Bogdanović, Z., Popović, S., Despotović-Zrakić, M., & Labus, A. (2021). Utilization of consumer appliances in smart grid services for coordination with renewable energy sources (pp. 147–167). https://doi.org/10.1007/978-3-030-74644-5_7.

Santhiya, M., Saranya, S., Vijayachitra, S., Lavanya, C.B., & Rajarajeswari, M. (2021). Application of voter insertion algorithm for fault management using triple modular redundancy (TMR) technique. In Proceedings of the 3rd International Conference on Intelligent Communication Technologies and Virtual Mobile Networks, ICICV 2021 (pp. 578–583). Institute of Electrical and Electronics Engineers Inc. https://doi.org/10.1109/ICICV50876.2021.9388450.

Vanitha, K., C.N. Vanitha, M. Mohamed Musthafa, & S. Malathy (n.d.). Efficient semantic interrogation scheme over cryptographic data in cloud. In 2020 International Conference on Inventive Computation Technologies (ICICT) (pp. 271–275). https://doi.org/10.1109/ICICT48043.2020.9112383.

Yapa, C., de Alwis, C., Liyanage, M., & Ekanayake, J. (2021). Survey on blockchain for future smart grids: Technical aspects, applications, integration challenges and future research. *Energy Reports*, *7*. Available from https://doi.org/10.1016/j.egyr.2021.09.112.

Transformation of smart grid to internet of energy

3

Kanimozhi N.[1], *Saranya S.S.*[2] *and R. Devipriya*[3]
[1]Department of Information Technology, Kongu Engineering College, Erode, Tamil Nadu, India, [2]Department of Computer Science and Engineering, Kongu Engineering College, Erode, Tamil Nadu, India, [3]Department of Artificial Intelligence and Data Science KPR Institute of Engineering and Technology, Coimbatore, Tamil Nadu, India

Transformation of Smart Grid to IoE

3.1 Introduction to smart grid

The world is changing. Technology is improving day-by-day. As days pass, knowledge acquisition becomes a stronger hunger for scientists. At the same time, the businesses are facing many issues in meeting buyer demands because customer needs and products are separated by islands of misinformation. The solution is to provide both-way communications. That's why the smart grids came into existence: an electrical grid that utilizes communication and knowledge in order to gather the information that energy provider seeks and buyer demystifies. This data is acted upon by the collected knowledge and information. Nowadays, smart grids are used to decrease the challenges of the traditional grid by improved communications moving bi-directionally and by monitoring different sensors, where the power utilization between consumers and different sensors is maintained at equilibrium. The smart grid which has essential elements can be defined as a self-sufficient system (Ahmed et al., 2019). In common with traditional networks, the smart grid is a blend of the two networks: the traditional network and the bidirectional communication network. The smart grid is not unique in its value proposition to the customer. It depends on the customer's views. The smart grid releases the automatic and scattered energy depending on information passed between two flows. Maximum, distributed computing is used to balance the demand of energy flow at every device. Given communication is very important between each device to provide consistent power voltages and the generated power is delivered to the customer with fewer losses, it follows that it was very challenging for the smart grid to integrate the different systems (Bari, Jiang, Saad, & Jaekel, 2014; Ahmed et al., 2019). The smart grid contains entities components for generating and consuming the power. The entities are connected by a set of connections. The grid generates electricity and provides energy to the consumer for his use (Table 3.1; Fig. 3.1).

Blockchain-Based Systems for the Modern Energy Grid. DOI: https://doi.org/10.1016/B978-0-323-91850-3.00003-2

Table 3.1 Comparison between exiting smart grid and traditional smart grid.

Traditional grid	Smart grid
Communication is one-way	The communication is bidirectional
The energy distributions must be monitored physically	The energy distribution is monitored by digital technology
If there is any repair in the infrastructure, it is restored manually by the technicians	Sensors are used to detect the changes in the infrastructures, the smart grid immediately reports the changes to the technician
Sensors are not used frequently	Using of sensors is frequent
The customer choices are fewer	Here the infrastructure is shared, allowing the customer to have more choices to receive the power
The traditional infrastructure leads to failure, hence the customer doesn't receive the power-on time	In a smart grid, the power can be routed to any problem area. Hence the failures can be managed
The power can be distributed only from the power plant to the customers	The power can be distributed in two ways

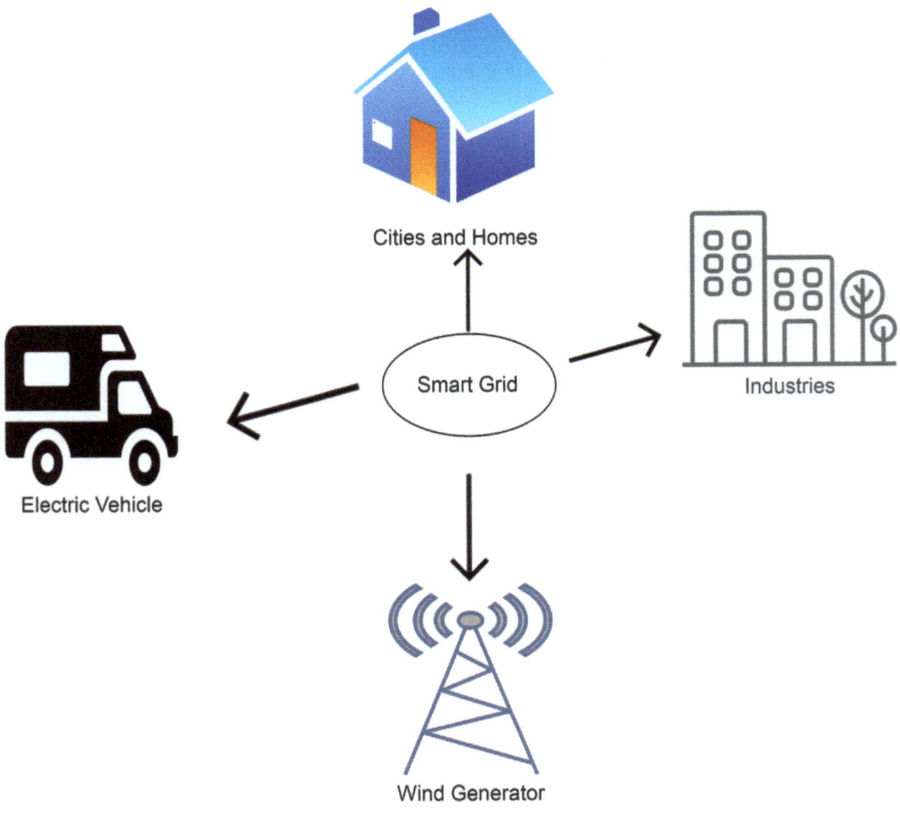

Figure 3.1 The Smart grid architecture.

3.2 Structure of traditional grid

The traditional electric grid's power system is slow in nature. It doesn't provide security to the customer as far as electricity thefts and transmission errors are concerned. It fails to provide communication between suppliers and consumers. The traditional grid fails to provide real-time data usage between consumers and suppliers in some supply systems (Miglani, Kumar, Chamola, & Zeadally, 2020; Narayanaswamy, Karthika, & Balasubramanian, 2022; Ourahou, Ayrir, EL Hassouni, & Haddi, 2020).

The traditional electrical system has power lines, distribution lines, and stations to send power from suppliers to the consumers. The electrical energy generated at high voltage in the power plant is delivered to the customer at low voltage Fig. 3.2.

3.3 Features of smart grid

The smart grid is growing smarter day-by-day. It's very essential to know the importance of a smart grid so as to appreciate its use. The smart grid provides a wide range of uses for both customers and utilities. It can store more information than normal meter readings which allows the use of data analysis.

3.3.1 The important features of smart grid

1. The smart meters are used along with a protocol to record the spending of energy. The updated smart meter helps the customers to monitor the activities of each end user to help them know which action increases the energy consumption in the day-to-day usage

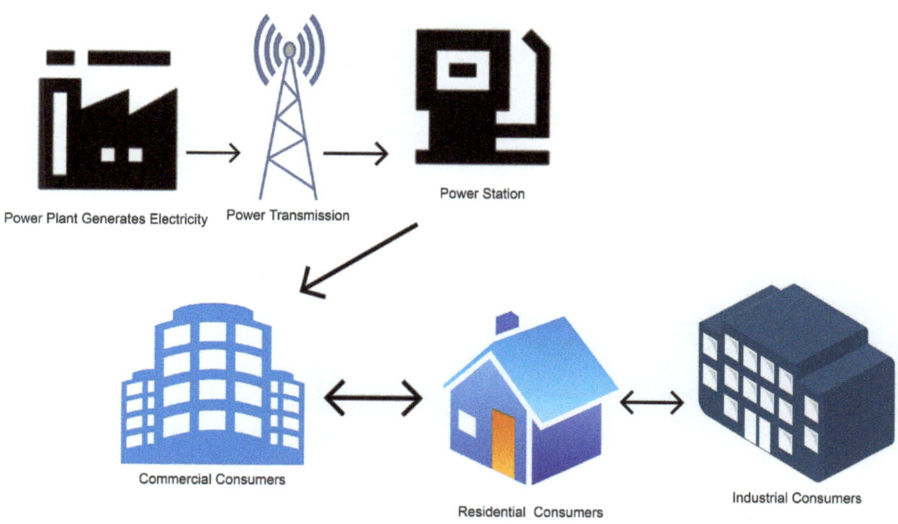

Figure 3.2 Overview of traditional smart grid.

records. The customer gets notifications of energy usage through phone messages or e-mails when their usage peaks at a high cost.

2. The smart grid is used to detect the health of power using time sensors.
3. The electrical energy can be controlled. With the smart grid, the power is reduced to a great extent. The generated power can reduce carbon emissions.
4. This provides better services in connections fulfilled and in operations facilitated of different sizes and different technologies.
5. The customer can get an improved level of consistency, protection of delivery, and integration.
6. The integration of renewable energy is one of the main features of the smart grid. This undergoes some challenges.
   a. The environmental impact of electricity supply is reduced.
   b. Voltage variation is another type of challenge. An example is solar and wind energy sources which fluctuate due to unsteady solar speed and not normal wind radiations.
7. The smart grid's bidirectional interaction features make using the smart grid technology more easy. The grid can be examined only by this technology. The main consideration is to ensure protection between the interactions.
8. The smart grid provides the quality power needed by the present world.
9. The smart grid technology grants the self-healing facility of energy generator from the power plant reaches to the end user which should ensure security protection; the grid should provide automatic detection of anomalous circumstances, like high current, high voltage, etc.
10. The smart grid distributes energy generated using the trio of digitalization, communication technologies, and IT though it provides some challenges for the distributors, as follows:
    a. To confirm the cost-effective power generation and distribution. Hence, the cost boundary for the customer cannot exceed the boundary.
    b. The grid protection gives safety against some natural calamities.
    c. To provide warranties against security where power fails to reach from the generation point to the consumer's endpoint.
11. The smart grid reduces consumer utility power bills.
12. Smart grid energy is clean energy. The smart grid technology aids in decreasing carbon emissions by a certain percentage (Fig. 3.3).

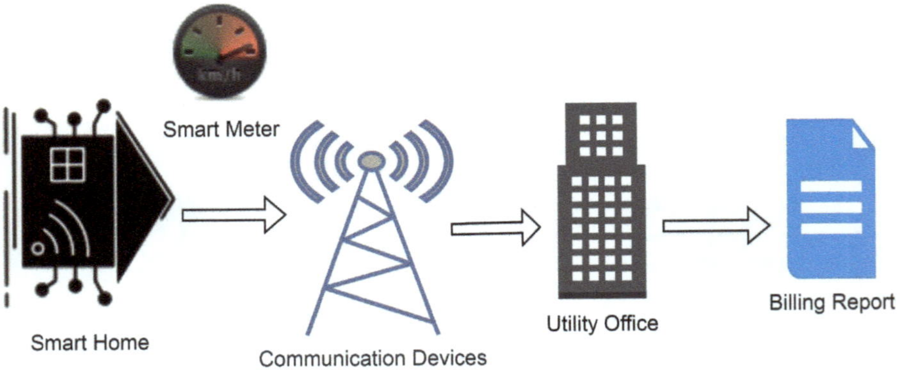

Figure 3.3 Block diagram of smart meter.

3.4 Smart grid components

The smart grid components are smart electrical devices used to produce electricity. An enormous amalgam of technology is developed to use these smart electrical devices. Some of the main components are:

3.4.1 Smart substations

Traditional electricity does not use any device controller. However, the industry around the world is changing to the digital platform. By moving to the digital platform, the different components of the station can be improved, supplemented with additional features that fit them with functions with a high rating and with improved reliability (Miglani et al., 2020; Narayanaswamy et al., 2022; Putri & Maizana, 2020). The smart stations are placed anywhere to cut out a subdivision from the main station. These observe the performance of each station to identify the essential data from the mass of data to find out the status of breaks in the circuit, security of transformers, etc. The substations are used to transfer energy where too many locations must be accommodated with a consistent release of energy. The smart substations are also used to divide the electricity path into many routes. The smart substation uses heavy devices and an enormous workforce to operate them with transformer, network relay, capacitor, etc.

3.4.2 Super connecting cables

The long distance power transmission needs extra strength cables. By using super connectivity, the transmission loss is very low because there is no electrical resistance. In traditional energy transmission, the transmission loss will add up to five percent of the total loss. Another major benefit of super connectivity is the size of the cables which are very compact.

3.4.3 Storage components

It's essential to store the generated energy due to the inconsistency of the renewable energy generated, the gap between high utilization and high accessibility. The energy storage technology includes advanced batteries, flow, compressed air, capacitors, supper-connecting magnetic storage, etc.

3.4.4 Monitoring control

The transmission system includes self-healing, self-adaptive systems. It helps in monitoring the health of transformers in real-time while monitoring the tower to check the high tension wires, providing great flexibility for all the day-by-day operations with enhanced consistency of power transmission.

3.5 Smart gird evolution

In 2007, the definition of a smart grid was given by the Energy Independence and Security Act of 2007 (EISA) of the US government. The characteristics of a smart grid are given as:

1. The electric grid's consistency, security, and competence are improved by using some controlling technology and digital information.
2. It is a combination of smart electrical devices and consumer-related devices.
3. The exploitation and incorporation of renewable resources, distribution of resources, and generation of resources.
4. The lively optimization of grid operations and resources with full security.
5. The exploitation of real-time technologies, automatic technologies, and interactive technologies to optimize the physical operation of devices.
6. The recognition and reduction of difficulties in implementing smart technologies and services.

The policy of the US government is to provide modern electrical transmission and distribution to meet the reliable and secure platform to accomplish each of the following tasks together called the smart grid. The following points list the technical advantages of a smart grid (Ourahou et al., 2020; Shahzad et al., 2020a). In smart grid applications, the digital platform is used to improve security, reliability, and competence. The security of the system depends on the dynamic optimization of physical resources.

3.6 Internet of energy

3.6.1 *Introduction to internet of energy*

Due to advancements in energy and power generation, the energy has been distributed over a peer-to-peer network. The energy producers and consumers must utilize and upgrade the infrastructure to automate the power supply from source to destination in a secure way. Internet of energy (IoE) permits power manufacturing to transport electricity with greater efficacy and cleanliness with less wastage. The energy resources like wind turbines, Photovoltaic cells, and renewable energy resources are incorporated into the smart grid. The sensors that are mounted on the consumer's environment are supposed to be monitored in real-time having been deployed in the smart grid to sense the energy consumption information transmitted via the internet. The potential benefits of the IoE embrace efficiency, minimal wastage, and significant price investments. The IoE facilitates smart internet things to integrate a variety of energy resources. The producers and manufacturers of energy must have an updated energy condition to ease the energy flow. Exclusive of IoE implementation, energy may possibly be lost during transmission via crossways lines which leads to inefficient transformation.

The process of integrating the smart grid with the internet of things (IoT) has led to gaining control over communication, power supply, energy flow, and energy

Table 3.2 Internet of energy (IoE) versus smart grid. The difference between smart grid and IoE.

S. no.	Metric	Internet of energy	Smart grid
1.	Technology	It focuses on the distribution of energy through smart meters and supervision of energy transaction details	It facilitates intellectual information and technology
2.	Limitations	It covers a wide range of systems	It covers only local systems
3.	Energy flow directions	Bidirectional	Unidirectional
4.	Control	Internet (World Wide Web)	Communication technique

utilization to automate power systems. In the course of the smart grid, the user can predict future demands and forecast the business's progress in the energy market. Implementation of the IoE reduces inefficiencies in transmitting energy making it more productive. The digitalized version of energy utilization allows users to constantly monitor the energy supply irrespective of where they belong. The system must be decentralized, secured, and transparent to all who are connected to the peer-to-peer network. The decentralized platform permits the manufacturers and producers to divide proportionately their electricity utilization information in an extra reliable and efficient manner.

The ideology of the IoE is in dispensing and tracking the power transmission to a massive scale efficiently. Instead of focusing on the unidirectional manner of power transmission from machine to machine in a top-down approach, the power internet promotes Power Distribution Networks that are linked with electricity routers to preserve bidirectional power routing in a peer-to-peer networking approach. Since, its complicated heterogeneous infrastructure is comprised of numerous energy and communique technologies. However, the smart grid transportation is activated using the contribution of power distribution. The structure of the system can be manipulated by any nodes of the network. The process of manipulation in smart grid systems is comparable to the energy internet in which the devices are connected to any platform (Table 3.2).

Internet of energy versus smart grid.

3.7 The architecture of the internet of energy communication network

The IoT plays a vital role in IoE communication network. It spends significant efforts to integrate anything at any time with the worldwide communication network. Whereas IoT infrastructure is being utilized by various applications like weather prediction, telecommunication services, supervision of smart grid, and Defense Force services, through enhanced communication technologies like 5G, a massive amount of information is transmitted from the physical layer to the higher-level layers of Internet of Energy Network architecture. The grids are connected to the sensors and actuators of IoE to sense and acquire data from physical objects (Shahzad et al., 2020a; Shanthi, Suvitha, & Suganthe, 2020; Uslar et al., 2019; Velliangiri, Manoharn, Ramachandran, & Rajasekar, 2021). The integration of the smart grid with the IoT offers a wider perspective of energy transactions through leading technologies like big data and cloud computing. The peer-to-peer infrastructure promotes machine-to-machine communication in a decentralized nature where the devices will communicate with each other and monitor the data that passes through it. Energy Internet gives an exceptional performance like decentralized electricity exchange, where electricity router interplays at distribution nodes. Alternatively, energy routers make sure of fault detection and protection because of their speedy and correct measurements.

3.8 Transformation of smart grid to internet of energy

An increase in demand for power generation due to population numbers means electronic power-based devices and appliances are overloaded in the traditional electricity infrastructure. The conventional energy distribution techniques suffer to fulfill the expected demand and supply to the consumer demand on time. In order to satisfy the expected power supply and demand, the smart grid has to be more reliable, fault-tolerant, efficient, and transparent. Due to the increase in power devices that are scattered everywhere, the distribution of energy must be decentralized and distributed in nature. The existing energy transmission infrastructure is not able to accomplish complete efficiency in distributed renewable energy resources.

3.9 Objectives of internet of energy

Due to conventional grid infrastructure not being attuned to managing dynamic power generation and supply, the increase in demand for power expected and energy storage and trading system suffer inflated functional and serviceable costs (Shafiullah, Oo, Ali, & Wolfs, 2013; Shahzad et al., 2020a). The solutions dictate bidirectional energy distribution and trading systems. The smart infrastructure supports and is equipped with unique and bidirectional chargers at charging power stations where consumers can buy and sell their own self-generated power whenever it's required. An increase in demand for renewable sources has brought a drastic change in the paradigm of the energy market from the centralized system into the decentralized system for sharing resources based on demand without a single point of failure. The centralized authority of the traditional power management system leads to security and privacy-related issues. When power management smart grids are integrated with smart devices like the IoT, it elevates the energy market to another production level. This intelligent device promotes smart communication, transfers real-world information, and finally helps in taking a decision over data (Fig. 3.4).

The following list of features appertain to the IoE objectives:

1. Reduction in electricity / power bills
2. Minimal operational / functional cost
3. Efficient resources management
4. Providing and provisioning multi-source supply to wide areas
5. Easy to understand
6. Integration of widely scattered areas in renewable energy production
7. Reducing wastage of energy during transmission
8. Monitoring and managing power supply based on demand with no delay in supply
9. Self-organization and management
10. Smart communication infrastructure

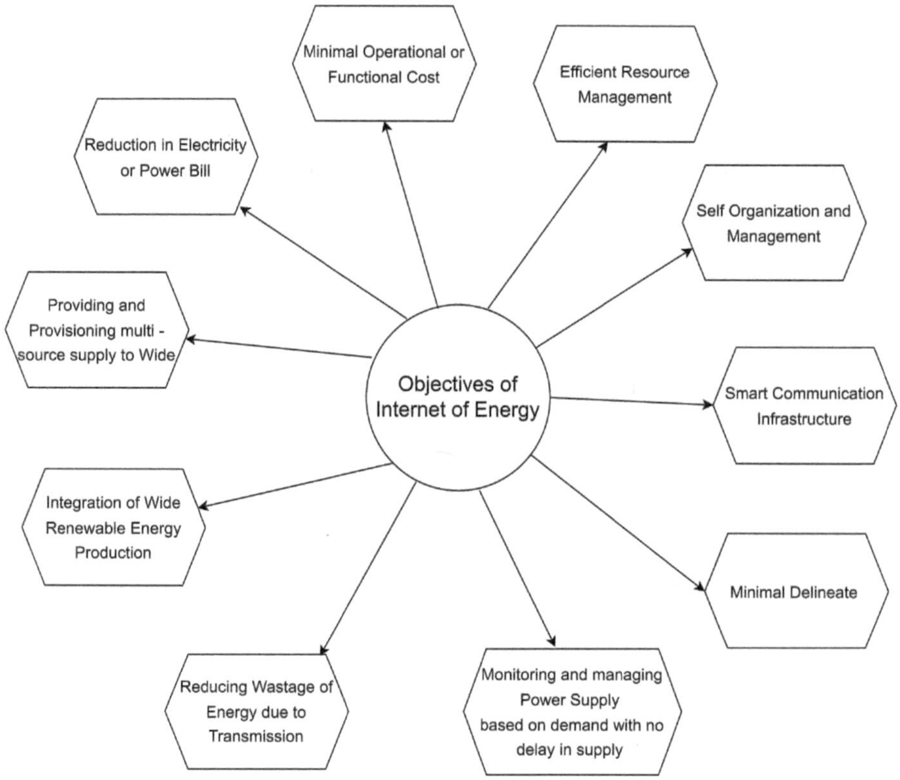

Figure 3.4 Purpose of internet of energy. The diagrammatic representation of objectives of IoE.

3.10 Issues in internet of energy

Due to the intelligent communication infrastructure used on the IoE technology, consumers and prosumers who control and monitor energy transmissions have to deal with broader communication protocols and different standards. Communication devices that are connected to each other must be interoperable. IoE enhances connectivity in heterogeneous ways to create a single infrastructure. Due to the increase of the IoE and its utilization, malicious attacks also increase in possibility (Shafiullah et al., 2013; Shanthi et al., 2020). There are issues raised in the integration of several communication devices which have different forms of energy.

1. Power transaction management
2. Transaction/settlement mechanism
3. Energy trading based on peer-to-peer architecture
4. Demand pricing
5. Hacking and attacks

3.10.1 *Power transaction management*

(Dranka & Ferreira, 2020; Justo, Mwasilu, Lee, & Jung, 2013; Saranya, Santhosh, & VijayaKumar, 2021). The data generated from different infrastructure electrical devices/components on the IoE is transmitted to a cloud environment for monitoring real-time data. To maintain and manage the enormous quantities of data that have been collected from smart meters embedded in home appliances, the data is transmitted to the network for data analytics. In a traditional power management system, the collected energy resource utilization has to be transmitted and maintained by a third-party authority. There may be a possibility of leakage in the personal information of consumers like location, vehicle number, and lifetime of a battery if the data were maintained by a centralized authority. Additionally, it may lead to a single point of failure. The main advantage of IoE is its flexibility to select a familiar aggregator who possibly collects data from various sources and maintains trust among all. It would have been almost impossible in traditional systems as it requires integrity, confidentiality, scalability, and reliability of data. Almost all communicating devices must be interoperable and have common and standard protocols (Bari et al., 2014; Dranka & Ferreira, 2020; Ahmed et al., 2019).

3.10.2 *Transaction/settlement mechanism*

The electricity usage of various nodes like smart grid, microgrid, smart vehicles, and smart home appliances involves the paying of the settle-down amount for the utilization of electricity consumption. The services of the Internet of Energy-based power management system promote prediction of expected load, desired and future status of prices, variation of energy units, and scheduling optimal energy usage. The prosumers and consumers who are involved in the energy trading market has to negotiate for buying and selling renewable energy resources using existing payment methods like a credit card, debit card, etc. The transaction of energy utilization may possibly reveal private as well as account information of consumers and prosumers who are all involved in the transaction with other parties without knowledge of users and their concerns. The utilization pattern of each consumer reveals private information such as the amount of electricity, identity of the consumer, and location (Saranya et al., 2021; Shafiullah et al., 2013). The privacy concern in traditional centralized architecture may lead to the involvement of third-party authorities. The nodes which are connected in a network have a trust concern among the involved parties. The decentralized architecture solution to privacy concerns supports payment through a common platform without the intervention of the third-party.

3.10.3 Energy trading based on peer-to-peer architecture

The home appliances and smart devices are connected and embedded with smart meters to measure and monitor the usage of electricity which has been transmitted through lines over a long distance. Traditional energy management involves the manual implementation of agents, trading energy, bidding, broker, and bank for money transactions with low reliability and high complexity. Due to the increase in population, IoE suffers from a heavy load and requires higher operational transaction costs (Saranya et al., 2021; Shafiullah et al., 2013; Shahzad et al., 2020a).

3.10.4 Demand pricing and hacking

The implementation of dynamic pricing is one of the best solutions to manage and control the demand and supply of the current energy market. The prosumers and consumers can negotiate with each other to bid for energy unit prices. If demand is higher than production, prosumers can increase the unit price of energy or electricity, where the decision has to be taken dynamically based on demand given the scenario of low consumption of energy with higher production decrease costs whereas higher consumption and lower production may increase energy cost. The smart meters embedded in home appliances and energy devices help in analyzing the pattern of users' consumption (Ahmed et al., 2019; Bari et al., 2014; Dranka & Ferreira, 2020; Justo et al., 2013; Kuzlu et al., 2014; Malathi et al., 2022; Miglani et al., 2020; Narayanaswamy et al., 2022; Ourahou et al., 2020; Putri & Maizana, 2020; Roopa et al., 2020; Saranya et al., 2021; Shafiullah et al., 2013; Shahzad et al., 2020a; Shanthi et al., 2020; Uslar et al., 2019; Velayutham et al., 2021; n.d. Velliangiri et al., 2021).

References

Ahmed, S., Gondal, T. M., Adil, M., Malik, S. A., & Qureshi, R. (2019). A survey on communication technologies in smart grid. In Proceedings of the IEEE PES GTD grand international conference and exposition Asia, GTD Asia 2019 (pp. 7–12). Institute of Electrical and Electronics Engineers Inc. Available from https://doi.org/10.1109/GTDAsia.2019.8715993.

Bari, A., Jiang, J., Saad, W., & Jaekel, A. (2014). Challenges in the smart grid applications: An overview. *International Journal of Distributed Sensor Networks, 2014*. Available from https://doi.org/10.1155/2014/974682.

Dranka, G. G., & Ferreira, P. (2020). Towards a smart grid power system in Brazil: Challenges and opportunities. *Energy Policy, 136*. Available from https://doi.org/10.1016/j.enpol.2019.111033.

Justo, J. J., Mwasilu, F., Lee, J., & Jung, J. W. (2013). AC-microgrids vs DC-microgrids with distributed energy resources: A review. *Renewable and Sustainable Energy Reviews, 24*, 387–405. Available from https://doi.org/10.1016/j.rser.2013.03.067.

Kuzlu, M., Pipattanasomporn, M., & Rahman, S. (2014). Communication network requirements for major smart grid applications in HAN, NAN and WAN. *Computer Networks, 67*, 74–88. Available from https://doi.org/10.1016/j.comnet.2014.03.029.

Malathi, D., Ponnusamy, V., Saravanan, S., Deepa, D., & Ahanger, T. A. (2022). A design framework for smart ration shop using blockchain and iot technologies. *Intelligent Automation and Soft Computing, 32*(1), 605–619. Available from https://doi.org/10.32604/iasc.2022.022083.

Miglani, A., Kumar, N., Chamola, V., & Zeadally, S. (2020). Blockchain for internet of energy management: Review, solutions, and challenges. *Computer Communications, 151*, 395–418. Available from https://doi.org/10.1016/j.comcom.2020.01.014.

Narayanaswamy, T., Karthika, P., & Balasubramanian, K. (2022). *Blockchain enterprise: Use cases on multiple industries. EAI/Springer innovations in communication and computing* (pp. 125–137). Springer Science and Business Media Deutschland GmbH Issues 978-3-030–76215–5. Available from https://doi.org/10.1007/978-3-030-76216-2_8.

Ourahou, M., Ayrir, W., EL Hassouni, B., & Haddi, A. (2020). Review on smart grid control and reliability in presence of renewable energies: Challenges and prospects. *Mathematics and Computers in Simulation, 167*, 19–31. Available from https://doi.org/10.1016/j.matcom.2018.11.009.

Putri, S. M., & Maizana, D. (2020). Optimal smart grid management system in campus building. *Jurnal Nasional Teknik Elektro, 9*(3). Available from https://doi.org/10.25077/jnte.v9n3.757.2020.

Roopa, C., Suganthe, R. C., & Shanthi, N. (2020). Blockchain based certificate verification using ethereum and smart contract. *Journal of Critical Reviews, 7*(9), 330–336. Available from https://doi.org/10.31838/jcr.07.09.71.

Saranya, S. S., Santhosh, C., & VijayaKumar, M. (2021). Blockchain endorsement technology—A review of future smart paradigms. *International Journal of Aquatic Science, 12*(3), 975–981. Available from http://www.journal-aquaticscience.com/article_135423_f7ba958d09cbaafe5a780021aae46b28.pdf.

Shafiullah, G. M., Oo, A. M. T., Ali, A. B. M. S., & Wolfs, P. (2013). Smart Grid for a Sustainable Future. *Smart Grid and Renewable Energy, 04*(01), 23–34. Available from https://doi.org/10.4236/sgre.2013.41004.

Shahzad, Y., Javed, H., Farman, H., Ahmad, J., Jan, B., & Zubair, M. (2020a). Internet of energy: Opportunities, applications, architectures and challenges in smart industries. *Computers and Electrical Engineering, 86*. Available from https://doi.org/10.1016/j.compeleceng.2020.106739.

Shanthi, N., Suvitha, R., & Suganthe, R. C. (2020). Blockchain based e-voting approach in P2P network. *Journal of Critical Reviews, 7*(9), 337–342. Available from https://doi.org/10.31838/jcr.07.09.72.

Uslar, M., Rohjans, S., Neureiter, C., Andrén, F. P., Velasquez, J., Steinbrink, C., Efthymiou, V., Migliavacca, G., Horsmanheimo, S., Brunner, H., & Strasser, T. I. (2019). Applying the smart grid architecture model for designing and validating system-of-systems in the power and energy domain: A European perspective. *Energies*, *12*(2). Available from https://doi.org/10.3390/en12020258.

Velayutham, Y., Bakar, N. A. A., Hassan, N. H., & Samy, G. N. (2021). IoT security for smart grid environment: Issues and solutions. *Jordanian Journal of Computers and Information Technology*, *7*(1), 13–24. Available from https://doi.org/10.5455/jjcit.71-1595835783.

Velliangiri, V., Manoharn, R., Ramachandran, S., & Rajasekar, V. R. (2021). Blockchain based privacy preserving framework for emerging 6G wireless communications. *IEEE Transactions on Industrial Informatics*. Available from https://doi.org/10.1109/TII.2021.3107556.

Blockchain for energy transactions

4

Sathya Krishnamoorthi[1], *Vani Rajasekar*[1] *and Balamurugan Balusamy*[2]
[1]Department of Computer Science and Engineering, Kongu Engineering College, Erode, Tamil Nadu, India, [2]Student Engagement, Shiv Nadar University, Delhi-National Capital Region, Delhi, India

4.1 Introduction

Blockchain is a decentralized network with peer-to-peer communication. Peers are the active participants of the network. Peers are also called nodes performing transactions inside the network in the presence of other peers. Other non-participating peers in the network act as examiners and verifiers of every transaction happening in the network. Blockchain obtained its name from the creation of a chain with blocks. The energy sector is ever-blooming with a constant demand for energy in every walk of life. To make the energy grid sturdy against forgeries, energy utilities are introducing the blockchain into the energy grid. This chapter will explore the techniques involved in securing the energy transactions happening in the energy grid with the blockchain framework.

4.2 Blockchain benefits

The blockchain is considered secure against the tampering of transactions. Transactions are bundled securely into blocks and added to the existing chain of blocks (Alladi et al., 2019). The addition of new blocks into the existing chain occurs with the inclusion of the hash value of the most recently added block before the new block. The hash value generated and added to the block serves the purpose of tamper resistance. The blocks are created and verified by peers in the network using various consensus protocols. Some of the most commonly used consensus protocols are:

- Proof of Work
- Proof of Stake
- Proof of Burn
- Proof of Capacity
- Proof of Elapsed Time

Implementation of blockchain in any vertical application reaps huge benefits in terms of security. Some of the key benefits are:

Blockchain-Based Systems for the Modern Energy Grid. DOI: https://doi.org/10.1016/B978-0-323-91850-3.00013-5

- Tamper resistance
- Confidentiality of transactions
- No central authority
- Equal participation of peers
- Verifiability of transactions
- Accountability

The blockchain can be considered as a decentralized network, with each node possessing a copy of a distributed ledger. The distributed ledger is a record of all transactions that happened and got verified in the network. It is the responsibility of consensus protocol to ensure all nodes have a recent similar copy of the ledger (Andoni et al., 2019). Practically, some nodes may get disconnected for a shorter or longer duration from the network and may not possess the recent copy of the distributed ledger. The consensus protocol must provide ways for them to get hold of the recent copy from other peers.

4.3 Evolution of energy grid

The traditional energy sector produced energy from non-renewable resources like petroleum and natural gas. The producers were giant companies fueling the entire consumer market. The energy was flowing in a one-way direction from producer to consumer. Consumers varied from large-scale industries to smaller households and all kinds of vehicles (Basden and Cottrell, 2017). With the advancement in the production of energy from non-renewable resources to renewable sources like solar, wind, water, and so on, the consumers evolved by changing roles to producers. Consumers turned producers are not large-scale producers; however, they are able to suffice their needs. With numerous consumers turning into producers, there was an unbalanced flow of energy in society. To overcome the wastage of energy, the producers introduced energy grids to the consumers. The energy grid is a common reservoir of energy where consumers and producers are interconnected. Whoever produces energy puts the energy into the grid, and whoever needs energy can take it from the grid. Thus, with energy-saving and waste-consciousness, led by a two-way flow of energy and information, came optimal use and cost savings.

Though the energy grid may seem to be the perfect solution to optimize energy use, the popping up of new technologies like the internet of things (IoT), supervisory control and data acquisition, cyber-physical systems, and so on, caused the energy to flow uninterrupted, along with information. This flow of information was needed to balance the demand and supply that prevailed in the energy grid. Information systems now integrated with the energy grid are capable of intelligent delivery of energy where the energy needed is estimated from usage patterns of consumers, proactive upkeep of grid by predicting the future outage due to component failure, competitive pricing by charging the consumers based only on the amount of energy consumed, plan for contingencies by predicting the dynamic change in need for energy, strategize against excess or deficiencies in energy supply by trading energy for a better price, streamline energy use by keeping track of energy produced and consumed by any entity and so on.

4.4 Features of the smart energy grid

Smart energy grids deployed with information technologies like the Internet, cloud, IoT, and so on are capable of showing intelligence in managing and distributing the energy wisely. The traditional grid is made smarter by making two-way communication between the utility (the generator of energy) and the consumer. Two-way communication lets both the utility and consumer communicate their demands and generate capacity. Sensing capability deployed into the transmission lines along with other smart devices like smart meters fitted at the outlet of every consumer, smart transformers, and so on, is able to predict the dynamic demand of energy based on usage patterns. Weak transmission units are identified and restored before they fail, and/or alternate routes of energy distribution are made in the event of unexpected network failure. Consumers can also be producers of small amounts of energy, and it can be returned to the grid to meet spiked demands. Records of energy saved and spent have to be monitored and recorded to bill the consumers. Features of Smart grids with information and communication technology (ICT) are listed in Table 4.1.

4.5 Requirements of information and communication technology in the energy grid

The energy grid was formed initially to connect the utilities and consumers on a smaller scale. With the globalization of industry and communication, the energy sector started to spread its wings to meet the global demands. The major issue faced by the energy sector is the sole dependence on energy generation from very scarce non-renewable resources like fossil fuels and natural gas. Though ICT tools might not help in resolving this total dependency, their technology can help counteract the growing demands if energy by integrating renewable energy sources into the grid and efficiently transmitting and utilizing the energy.

Switching to renewable sources of energy is recommended by nations in favor of not only meeting the growing demand but also facing the threat of green gas emissions. Fossil fuels tend to emit carbon damaging the ozone layer, concomitant with the overwhelming problem of global warming. Dependency on scarce resources is also reduced when the spike in demand is met by renewable energy sources. ICT tools are needed to utilize this precious energy in an efficient manner by working smartly.

Table 4.1 Features of information and communication technology.

Efficient distribution of energy
Faster restoration in the event of failure
Reduction in operation and management cost
Optimal demand supply balancing
Facilitates deployment of renewable energy sources into the grid
Improved trading and security

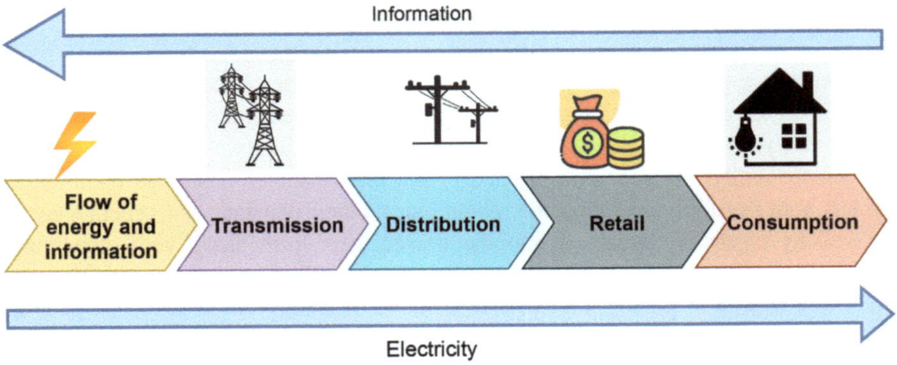

Figure 4.1 Value chain of the conventional electricity grid.

Fig. 4.1 shows the value chain of the conventional electricity grid. In the smart grid, the value chain remains the same, component-wise, but added with the deployment of ICT tools. Below is the use of ICT tools in various components of the smart grid value chain.

Generation:
- Smart meters
- Vehicle-to-grid and grid-to-vehicle

Transport:
- Sensor-based networks
- Embedded systems and software
- Machine-to-machine communication
- Communication Protocols

Storage:
- Vehicle-to-grid
- Grid-to-vehicle
- End-user interfaces

Retail:
- Smart meters
- End-user interfaces

Consumption:
- Smart building
- Smart devices
- Data centers and cloud computing
- Data analytics and intelligence.

4.6 Opportunities in smart grid

The smart grid aims to have efficient energy storage and transmission with the support of ICT tools. The most common areas where ICT is most sought are listed below:

4.6.1 Battery

Batteries are storage containers of energy. In a smart grid, there are times when energy supply surpasses energy demand (Brilliantova and Thurner, 2019). The excess energy need not be wasted and can be stored in batteries. One potential solution is to use batteries from electric vehicles instead of devising and deploying dedicated batteries into the grid. With the batteries of electric vehicles, energy can be stored off the grid during huge supply and can be placed back on the grid during huge demand. The key issues to be resolved include an easy interface for human interaction to keep track of energy stored, consumed, and returned, eco-friendly disposal of batteries and increased demand for raw materials to build up batteries.

4.6.2 Large-scale grid

There is a need for the extension of the grid on a larger scale extending beyond local regions and national boundaries. The global extension of smart grids can interconnect large-scale producers, renewable energy sources, and large- and small-scale consumers. Such connections are capable of meeting the higher demands of energy from renewable sources without depleting the scarce non-renewable resources.

4.6.3 Balancing demand and supply

Only with the sole dependence on ICT tools demands and supplies of energy can be known. Automated systems utilize these data to perform analysis and predict anticipated demands in certain regions at specific times of the day, predict wear and tear of grid components that need maintenance, etc. ICT tools like smart meters, along with communication channels, can help grid operators to better know their consumers and learn their usage patterns. Incentives may be provided to consumers and industries when they have less consumption of energy during the peak hours of the day.

4.6.4 Pay per use policy

When smart grid infrastructure has the capability to monitor and record the energy consumption and production history of consumers, the pricing of the energy can be done precisely. Consumers are billed only for the extra energy they consume beyond their production capacity.

4.7 Potentiality of blockchain in the energy sector

A blockchain is expected to solve many challenges faced in energy applications. It gives a platform for the advancement of IoT applications, peer-to-peer (P2P) trading of energy, and efficient transmission of energy with decentralized grid infrastructure. It also seeks to provide better customer service, and reduced operating costs to the energy utilities, among other conveniences. With the blockchain as an accompanying framework, smart grids can now have small-scale renewable resources, ICT tools, smart devices, consumer participation, balancing demand-supply, and a decentralized distribution system. In a smart grid, prosumers are utilities that play the role of both producer and consumer of energy. The blockchain will help the energy management system with the creation of tamper-proof contracts in energy trading on a P2P basis and helps in securing the transactions of energy. Smart contracts can be put into use to automate the trading process between the supplier and consumer on a demand-basis without the worry of forged transactions. A blockchain can bring transparency to the energy market when consumers can choose their energy suppliers based on the pricing scheme. Energy is delivered through the physical grid infrastructure, and information related to consumption, generation, and trading is taken care of by the ICT.

The blockchain can be used in various scenarios of the energy sector, depending on the need. Use cases may vary from simply recording user consumption history to maintaining a large-scale grid with distributed ledger technology for a secure grid (Kumari et al., 2020). Some of the scenarios where blockchain can be of major benefit are listed below:

- ***Energy billing:***
  Smart contracts and smart meters in the grid infrastructure enable consumers to be billed fairly on a pay-as-you-go basis. Billing can be made securely and transparently with micro-payments or prepaid options.
- ***Sales and marketing:***
  Sales can be made by fully understanding the consumer's usage pattern of energy. A blockchain can be put into use along with artificial intelligence (AI) to predict the usage pattern, understand the future demands of the customer, and tailor their marketing strategies to match the need of the customer. This facility will improve the revenue of energy utilities and make them prepared for future demands.
- ***Energy trading market:***
  A blockchain can change the way the energy market operates conventionally. It brings about a paradigm shift in making energy trades in the case of wholesale markets, commodity trading, and green energy trading with green energy certification and validation.
- ***Automation of production and consumption:***
  A decentralized network of energy grid can lead to the formation of microgrids with self-contained entities. The entities can produce energy on their own and consume on their own and still be monitored using smart meters and other smart devices. Trading can be done within a local region to have a sustainable environment.
- ***Data management and security in smart grid:***
  Smart grids deploy numerous smart devices like smart meters, smart sensors, network monitoring devices, automation systems, control units, and so on. There will be an enormous amount of data in transit at any point in time (Gai et al., 2019). These data can be managed well and secured with the potentiality of blockchain.
- ***Network management in smart grid:***
  A blockchain can protect the network components in the smart grid and offer flexibility in the use of resources. This reduces the expense of maintaining the network resources.
- ***Identity management:***
  Transactions and consumer data can be made secure with tamper-proofing by applying blockchain technology. Consumer identity privacy and confidentiality of the data can be achieved with the use of cryptographic techniques.

- ***Resource sharing:***
  In a decentralized network, users can share their resources like storage capacity, charging infrastructure for electric vehicles, and so on. Sharing needs to be tracked and secured with blockchain to enable accountability and fair pricing of customers.
- ***Competitive advantage:***
  Energy providers can make use of smart contracts for automated billing of customers. This provides a competitive advantage to energy utilities that enable consumers to switch between energy suppliers.
- ***Transparent transactions:***
  Transparency of records is one of the salient features of blockchain. Immutable records of energy transactions and transparent nature of keeping records are necessary for auditing and regulatory compliance (Li et al., 2019).

4.8 Decentralized energy trading market

The usage of blockchain in the energy market shifts the flow of transactions from a centralized framework toward a decentralized framework. Transactions are no more carried out and verified by a single entity like banks. Transactions are made more transparent to all the peers in the network, thus, curbing the forgery in payments and energy consumption. A blockchain has the most useful security features of smart contract and system interoperability that makes it an easy to adopt technology in the energy sector.

Energy trading markets are constituted of utility suppliers, utility providers, supporting vendors, customers, distribution agents, and so on. The use of distributed ledger will make the tracking of supplies and products efficient and secure (Guan et al., 2021). Continuous tracking of supplies guarantees the chain of custody for all the materials and energy being processed in the energy network. With the emerging growth of renewable energy sources, the blockchain also provides a helping hand to the management of renewable resources along with the validation of green certification. Altogether, a blockchain is expected to have a greater impact on our environment by nurturing renewable energy sources with growing energy opportunities.

4.8.1 Blockchain in legacy energy sectors

Legacy energy sectors like petroleum and gas companies have huge networks of entities like refiners, suppliers, tankers to store, distributors, vendors, regulatory bodies, contract job schedulers, and banking (Malathy, Sangeetha, Rajesh Kumar, & Vinayakumar, 2022). All these entities need to work in harmony with blind trust in each other. Petroleum corporates are making massive investments in integrating blockchain technology into their network to have secure transactions and enable them to grow to the corners of the world. A blockchain will also expand its markets, leading to the integration of small-scale energy providers into their network on the agreement, and dynamic adaptation to fluctuating energy demands, among others. Most important among them is the protection of the privacy and trade secrets of their company. A blockchain ensures to protect them by provisioning data access permission only to the selective parties and making the blockchain a private network. A blockchain in the energy sector focuses on providing the following benefits:

- Cost reduction in operation and distribution
- Environment sustainability by nurturing renewable energy sources
- Providing privacy and transparency among the parties

4.9 Use cases of blockchain-enabled energy markets

4.9.1 *Wholesale energy distribution*

Wholesale trading is the common practice followed in any market, where products (goods) are purchased in larger quantities by retailers other than end-users. Retailers typically can be distribution agents, shop owners, large customers, or intermediary agents (Miglani et al., 2020). Bulk quantity is bought at a lower price and sold to end-users at a higher price for a marginal profit. Energy markets also have wholesale trading of energy in any one of the forms of electricity, natural gas, oil, or renewable energy. Wholesalers in energy markets may comprise financial intermediaries, energy traders, and large customers. With the inclusion of other intermediary parties between the producer of energy and the consumer, the price of energy bought by the consumer is increased radically (Jamil et al., 2021).

4.9.2 *Peer-to-peer energy trading*

The peer-to-peer energy market is a network of peers enabling individuals to buy or sell energy among themselves without the need of a central third party. Almost 59% of blockchain applications in the energy sector are built for P2P trading (Organisation for Economic Co-operation and Development, 2012). The absence of central wholesale authorities makes this system more favorable and reaches millions of users. As more users are getting into the production of renewable energy from resources like solar, wind, biomass, and so on, P2P allows them to have greater control and flexibility over their activities. With the help of the blockchain platform, users can decide to/from whom energy should be distributed and by what process all the energy transactions are securely stored in the blockchain database (Plaza et al., 2018).

Some benefits of P2P energy trading include:

- Users with no energy production can still buy energy from other participants
- No transfer of energy from central wholesaler to consumer
- Reduced price in energy
- Full transparent records of energy transactions enable a secure platform

Some countries are in the process of reaping those benefits of P2P energy trading across the nation by forming microgrids. Microgrids are smaller networks of multiple P2P energy networks. When P2P energy markets at different locations are interconnected, we have P2P energy trading in a larger area.

4.9.3 Electricity data management

Blockchain, when used in wholesale energy distribution or P2P trading, records all the transactions. Distributed ledgers in a blockchain network are immutable and publicly accessible. Energy market blockchain ledgers store crucial data like marginal cost, fuel price, energy price on a specific day, consumer energy consumption history, billing of energy, consumer payment details, and so on. Most often, these data are either intentionally or unintentionally modified, leading to huge losses financially. With the blockchain ledgers, these data are immutable yet publicly accessible to people.

4.9.4 Commodity trading

Trading of the commodity takes place in every business. Energy sectors like oil, gas, and electricity trade their commodities either in wholesale units or retail units. Companies of such industries spend millions on deploying secure proprietary software to handle all commodity trading data (Pop et al., 2018). Replacement of proprietary software with blockchain applications can handle all trading data securely, making them immutable and available to all its partners. Most importantly, blockchain applications can be scaled easily as the sector grows with the least introduction of other resources compared to what is achievable by the use of proprietary software.

4.9.5 Blockchain for utility providers

Utility providers are the entities that produce electricity from various sources like power plants (nuclear, thermal, hydro, etc.), solar power farms, windmills, and so on. Utility providers share information among themselves so as to enable them to generate energy efficiently and generate revenue wisely. Blockchain's shared ledger is a great weapon to be used in sharing information among utility providers. Some potential benefits of using blockchain include validation of data from smart devices present at the edge of the grids, the ability to create a transaction of data useful in the distribution of electricity, and the distributed ledger, which can be deployed to store the records of energy transactions sold/shared to diverse entities like consumers, wholesalers, utility partners, and so on.

4.9.6 Blockchain for the oil and gas industry

Oli and gas industry has thousands of companies working together from the point of extraction of natural fuel to the point of consumption of fuel by the end consumer. The use of blockchain technology in such a vast industry could potentially reduce the cost in terms of data management, labor cost, data visibility, delay in settlements, and communication between the partners (Rahmadika et al., 2019). A blockchain is now being used in all three sectors of the oil and gas industry, namely upstream, midstream, and downstream industries.

4.10 Blockchain projects in the energy industry around the world

It is evident from the previous sections that blockchain is paving its way into every industry, and the energy market is no exception to it. Many organizations around the world—be it governmental, non-profit, or private—have integrated the blockchain platform into their trading systems. These blockchains though having similar features, are custom-tailored to meet the requirements of energy partners and have greater flexibility (Premalatha and Sathya, 2020). This section will explore some of the blockchain applications put into use in various parts of the world.

4.10.1 Brooklyn microgrid

Brooklyn Microgrid (BMG) is a blockchain-based grid on P2P energy trading platform. This platform is designed specifically to enable local prosumers to share their solar-derived energy. Prosumers are local entities capable of being both producers and consumers of energy. Prosumers can be local residential users or small commercial energy producers present in the locality of New York, as shown in Fig. 4.2. BMG was established with the goal of enabling the production and consumption of solar energy within the city in a cost-effective way. This project was developed by

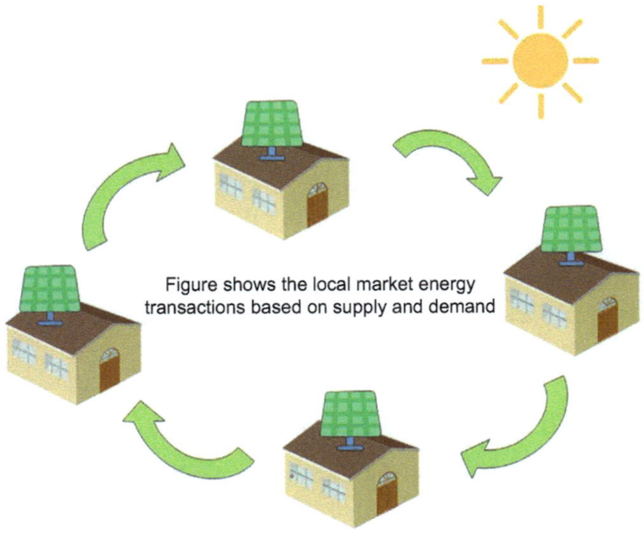

Figure 4.2 Brooklyn Microgrid.

LO3 Energy Inc. It uses TransActive Grid with a smart meter capable of measuring the energy units produced and sold. Buyers and sellers of energy negotiate the price of energy using a mobile app. The smart meters of the buyer and the seller then communicate with each other directly, and payments are made securely between them.

4.10.2 ME SOLShare

ME SOLShare is a microgrid community in Bangladesh. It uses blockchain technology to form a community of producers, consumers, and prosumers, as shown in Fig. 4.3. Each user is provided with a DC meter SOLBox. This meter provides users with access to community trading by leveraging their wireless connection module. SOLBoxes communicate with each other to perform P2P trading. Smart contracts are deployed to automatically perform digital transactions when the smart meters measure the amount of energy produced and sold to peers. The digital wallets of users are linked to the smart meters, and smart contracts perform the transactions. The transactions are verified and settled by the banks within a day. ME SOLShare makes life easier and increases productivity.

4.10.3 Enerchain 1.0

Enerchain 1.0 was the first blockchain-based platform designed for energy trading by wholesalers in Europe. Enerchain 1.0 was developed by Ponton with 44

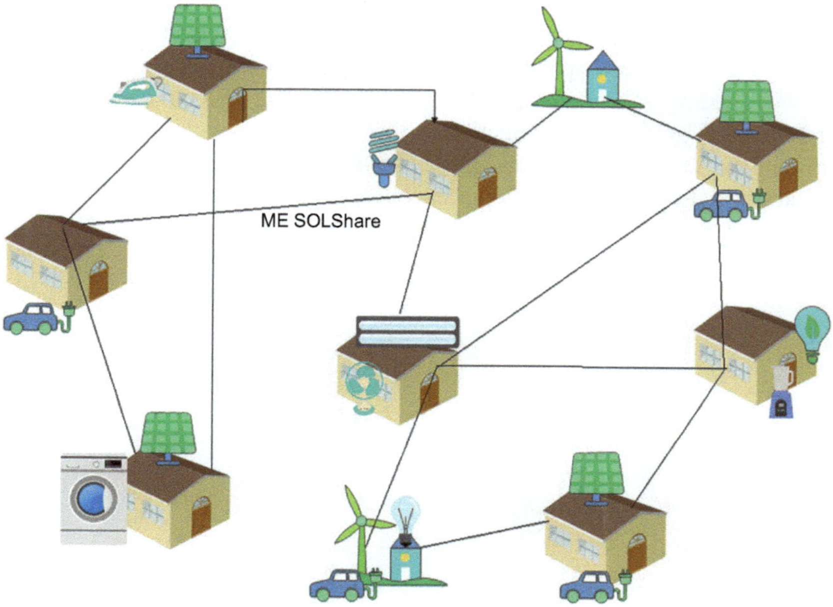

Figure 4.3 ME SOLShare.

companies involved in energy trading. It enables traders to communicate the distribution data, have a low entry fee, and trade energy products like flexibility, load curve, certificates, etc (Rajasekar et al., 2021). It is designed to be a distributed trading infrastructure enabling a decentralized European energy market to trade various energy products at different levels of scale. It is well known for its fastest block creation time of up to 1 second for a block.

4.10.4 Tobalaba

Tobalaba is an Ethereum-based platform developed by the Energy Web Foundation of Rocky Mountain Institute and Grid Singularity. Tobalaba was developed to simplify the certification process of renewable energy resources. To enable the certification process, Origin, a mobile app, was created. It monitors all the smart meters attached to the solar panels of the microgrid and keeps track of all necessary data like type of energy, time of production, the measure of energy produced, location of production, and so on. It is helpful in the procurement of renewable energy resources and flexible participation of different energy assets.

4.10.5 Sun exchange

Sun Exchange is a crowd-funded project in Africa, as depicted in Fig. 4.4. It enables buyers from around the world to buy photovoltaic cells. The producers of energy in Africa can then install solar panels on their homes and start producing energy. They get paid in cryptocurrencies for the energy produced, and it is used for paying the rent of solar panels. African rural communities are greatly benefited from crowdfunding as most of them are not connected to their country's energy grid nor funded by their government. To make it more beneficial to society, the buyers rent the solar cells to public buildings like schools, hospitals, factories, and other commercial buildings to scale up the energy production.

Figure 4.4 Sun Exchange.

4.10.6 ImpactPPA

ImpactPPA is a blockchain-based platform enabling consumers to pre-pay for the energy they plan to consume later using a mobile device. The working in ImpactPPA is depicted in Fig. 4.5. Developers, investors, consumers, government, and utility providers are transparently connected to the blockchain network. Smart meters are used in the blockchain network to know the energy consumption rate, and consumers can pay in advance for the energy needed in the near future.

4.11 The decentralized energy transaction model

Energy trading has been in practice for longer periods but with a central authority that controls and monitors the energy transactions happening in the network. The central authority may pose a risk of a single point of failure. Also, the cost of central operating authority piles up on the billing of consumers. A blockchain-based energy trading market is completely decentralized with no new central authority (or agent), and the energy transactions are recorded in the public ledger (Wang et al., 2021). This decentralized network is resilient to the failures of nodes, and the cost of the energy consumed is negotiated by the producer and consumer rather than by a third party.

The decentralized energy transaction model, as shown in Fig. 4.6, requires the consumers and producers to be connected through a network for the transaction of both energy and payment. Smart grids can be implemented with smart meters to

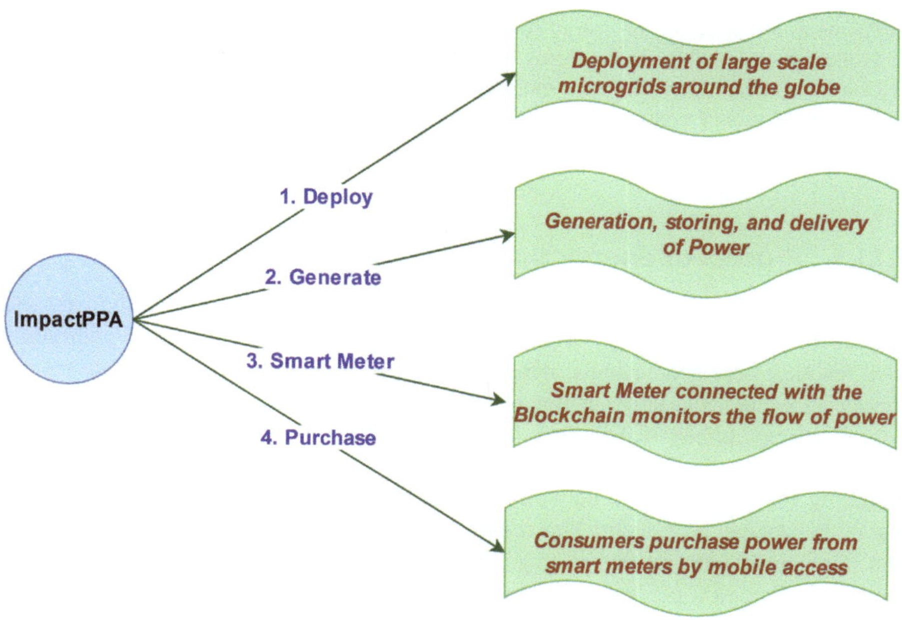

Figure 4.5 ImpactPPA.

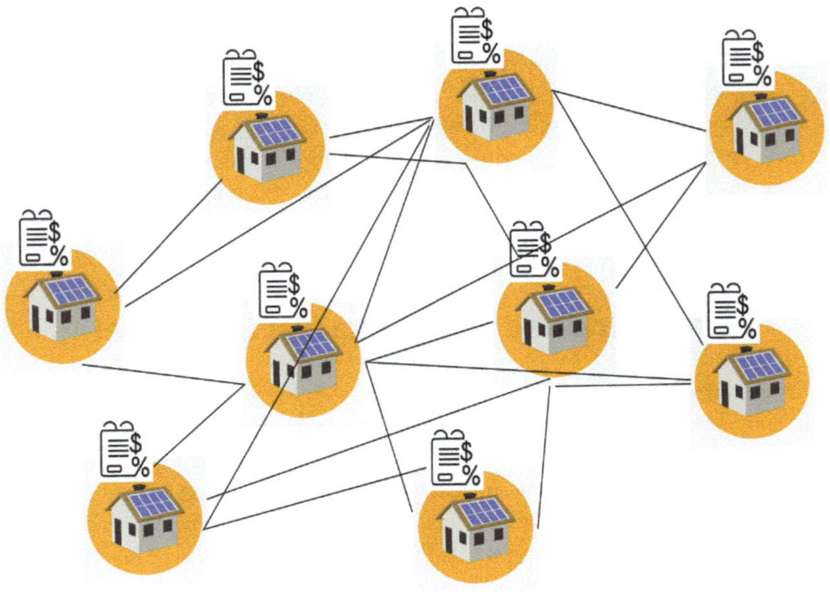

Figure 4.6 Decentralized energy transaction model.

monitor the energy flow and gather data. The smart grid with blockchain enablement secures the energy network, ensuring:

1. Payments are made for the energy consumed
2. Producers don't sell the same energy twice to different consumers
3. Consumers don't spend the same money twice on different producers
4. Energy and money transactions are made transparent and tamper-proof
5. No need for a central authority (agent) to distribute energy and manage the transactions

4.11.1 Process of energy transaction

The producers and consumers are now connected with a smart grid. Producers are sources of renewable energy from resources, given some of the most commonly used resources are solar panels, windmills, and so on. When a producer has excess energy generated, the smart meter at the producer site informs the excess production of energy to the distribution node. The distribution node is responsible for resolving any trading issues in the network. The consumers have their smart grids searching for energy in the network. The consumer can choose the producer from whom energy can be bought. The consumer and producer now enter into a negotiation phase to determine the pricing. The customer can now make payment to the producer either as a prepaid mechanism or post-paid mechanism. Smart meters are linked with the e-wallets of their users. Once the user authenticates the payment, cryptocurrency is transferred from the consumer's wallet to the producer's wallet.

After receiving the payment from the consumer, the producer authenticates the energy transaction from producer to consumer. The amount of energy transacted is recorded on both producer's and consumer's smart meters. Each energy and money transaction is verified by the peer nodes. The blockchain model prevents the double-spending problem from occurring with consumers, and it is Byzantine fault-tolerant.

4.11.2 Threats in a distributed energy model

4.11.2.1 Selfish mining attack

Mining is the process of the creation of blocks with a predefined hash value. Miners invest their computing power in computing the hash value for the given set of transactions to meet the predefined hash value (Sathya, Premalatha, Vani, Rajesh Kumar, & Celestine, 2022). Competing in generating the block, the miner who creates the block first by identifying the correct nonce number needed to generate the block is rewarded with cryptocurrency. Selfish mining is a threatening scenario in a P2P network where the attacker attempts to build the block faster than the miner. The attacker, after creating the block, deliberately hides it from other peers leading to the creation of forks in the blockchain. Other genuine miners keep spending their time and computing in the mining process. Attackers with a pool of miners now create a separate private branch from the public chain of blocks and keep mining. As the private blocks grow in length, genuine miners in the public chain then realize their computing power was wasted on creating duplicate blocks in the network. The selfish pool of miners has greater control of the network, and rewards are higher in the selfish pool. Genuine miners are now tempted to join the selfish pool to earn more rewards than being in the public network.

4.11.2.2 Sybil attack

Sybil attack is the most common threat in P2P blockchain networks. As the blockchain is completely decentralized, there is a high probability of a single node becoming suspicious and gaining control over multiple nodes in the network (Andoni et al., 2019; Organisation for Economic Co-operation and Development, 2012; Premalatha & Sathya, 2020; Rahmadika et al., 2019; Rajasekar et al., 2021). As every message and every transaction in the blockchain is transmitted throughout the network, malicious nodes in the network will disrupt the block creation and functionality of the network, as depicted in Fig. 4.7.

4.11.2.3 Eclipse attack

Eclipse attack in a P2P network is gaining momentum. It is similar in working to the man-in-the-middle attack. An attacker node will be able to gain control of the blockchain network by eclipsing a node in the network. The peer node that was eclipsed (hidden) continues to communicate the messages to the attacker, thinking it is sending messages to the peers of the blockchain network. The attacker is now

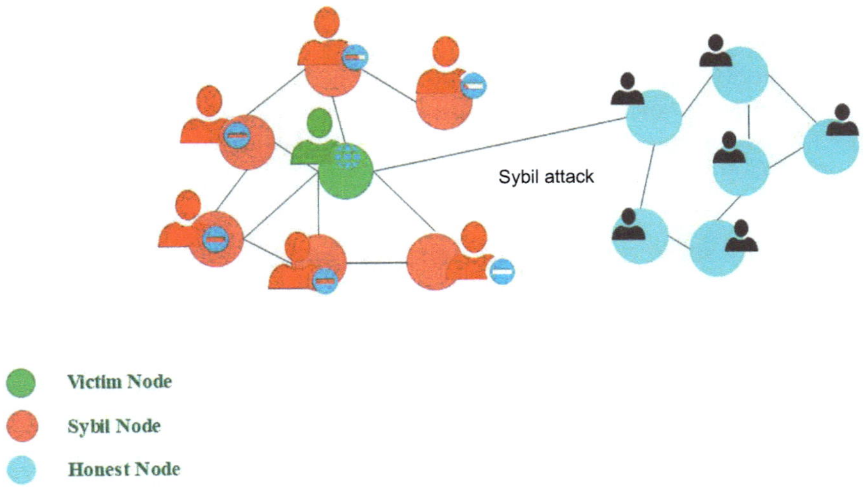

Figure 4.7 Sybil attack.

able to read all messages intended for the public network. The victim node is blinded from the public blockchain network, and the attacker has complete control over the connection of the victim with the network.

4.12 Blockchain-enabled energy industries

The blockchain touches the core of various sectors. The energy sectors can further be categorized into multiple domains based on the way of generation, utilization of energy, etc. This chapter explored the potentialities of blockchain that could dramatically improvise energy management, thus making the lives of consumers and producers easier. As a summary of this chapter, this section will conclude with the energy markets that could potentially benefit from the use of blockchain platforms.

4.12.1 Metering

Metering of energy in the energy market could gain benefits from a blockchain. Meters are connected in a decentralized network, and data are collected from the meters (Malathy, Sangeetha, Rajesh Kumar, & Vinayakumar, 2022). Smart meters can then be made a point of monitoring the grid to know the dynamic energy demand and consumption pattern.

4.12.2 Grid management

Electric grids are getting smarter with smart devices. In the smart electric grids, not only does electricity flow in the network but also the data related to grid management flows too. With the help of blockchain, the energy grids get dynamic data, are able to locate the point of failure, can predict future demands, and/or make payments for the electricity consumed by linking the digital wallets of users.

4.12.3 Decentralized generation

Decentralized generation of energy leads to the formation of P2P networks of varying sizes. Producers, consumers, and prosumers are interconnected to get involved in energy trading, and payments are made among themselves without the involvement of a central authority. A blockchain in such a decentralized network records every transaction of the network and makes the chain of blocks immutable. The exclusion of third-party stakeholders or central authority in the decentralized energy network makes the pricing of energy favorable to consumers and producers. The consumers get the choice of choosing the producer to buy their needed energy, and both can negotiate the pricing. The producers earn benefits from generating excess energy, thus, greatly reducing the carbon footprint.

4.12.4 Electric vehicle market

Electric vehicles (EVs) are an excellent alternative to reduce carbon emissions as mankind is facing the scarcity of fossil fuels. EVs were initially introduced as one of the electrical appliances consuming energy from the electricity grid. With the invention of smart grids, EVs are now smart objects capable of charging electricity from the grid, storing the energy in their large batteries, and returning the excessively stored energy back to the grid. The producers can now sell their energy to EVs by providing charging outlets. A blockchain at the backend keeps track of all records of energy transactions. EVs are charged by paying the producer in cryptocurrencies. When EVs return their stored energy to the electricity grid, the blockchain monitors all the transactions happening, and payments are carried out securely.

4.12.5 Internet of things

IoT is an interconnection of real-world objects into the existing Internet. Objects communicate with each other to perform their assigned tasks. A blockchain spun into the IoT makes the communication between objects tamper-proof and secure. The energy sector could deploy the IoT technology to save on energy by using smart appliances, having green buildings, energy maintenance, and so on. Security is a primary concern in IoT as the objects are the key players in the network, and objects can be compromised or made to act suspiciously to perform unintended tasks. A blockchain will protect these communications and make the IoT a better technology for a smarter energy market.

References

Alladi, T., Chamola, V., Rodrigues, J. J. P. C., & Kozlov, S. A. (2019). Blockchain in smart grids: A review on different use cases. *Sensors, 19*(22), 4862. Available from https://doi.org/10.3390/s19224862.

Andoni, M., Robu, V., Flynn, D., Abram, S., Geach, D., Jenkins, D., McCallum, P., & Peacock, A. (2019). Blockchain technology in the energy sector: A systematic review of challenges and opportunities. *Renewable and Sustainable Energy Reviews, 100*, 143−174. Available from https://doi.org/10.1016/j.rser.2018.10.014.

Basden, J., & Cottrell, M. (2017). How utilities are using blockchain to modernize the grid. *Harvard Business Review, 23*, 1−8.

Brilliantova, V., & Thurner, T. W. (2019). Blockchain and the future of energy. *Technology in Society, 57*, 38−45. Available from https://doi.org/10.1016/j.techsoc.2018.11.001.

Gai, K., Wu, Y., Zhu, L., Qiu, M., & Shen, M. (2019). Privacy-preserving energy trading using consortium blockchain in smart grid. *IEEE Transactions on Industrial Informatics, 15*(6), 3548−3558. Available from https://doi.org/10.1109/TII.2019.2893433.

Guan, Z., Lu, X., Yang, W., Wu, L., Wang, N., & Zhang, Z. (2021). Achieving efficient and privacy-preserving energy trading based on blockchain and ABE in smart grid. *Journal of Parallel and Distributed Computing, 147*, 34−45. Available from https://doi.org/10.1016/j.jpdc.2020.08.012.

Jamil, F., Iqbal, N., Imran., Ahmad, S., & Kim, D. (2021). Peer-to-peer energy trading mechanism based on blockchain and machine learning for sustainable electrical power supply in smart grid. *IEEE Access, 9*, 39193−39217. Available from https://doi.org/10.1109/ACCESS.2021.3060457.

Kumari, A., Gupta, R., Tanwar, S., Tyagi, S., & Kumar, N. (2020). When blockchain meets smart grid: Secure energy trading in demand response management. *IEEE Network, 34*(5), 299−305. Available from https://doi.org/10.1109/MNET.001.1900660.

Li, Y., Yang, W., He, P., Chen, C., & Wang, X. (2019). Design and management of a distributed hybrid energy system through smart contract and blockchain. *Applied Energy, 248*, 390−405. Available from https://doi.org/10.1016/j.apenergy.2019.04.132.

Malathy, S., Sangeetha, K., Rajesh Kumar, D., & Vinayakumar, R. (2022). Improved K-means based Q learning algorithm for optimal clustering and node balancing in WSN. *Wireless Personal Communications, 122*(3), 2745−2766. Available from https://doi.org/10.1007/s11277-021-09028-4.

Miglani, A., Kumar, N., Chamola, V., & Zeadally, S. (2020). Blockchain for internet of energy management: Review, solutions, and challenges. *Computer Communications, 151*, 395−418. Available from https://doi.org/10.1016/j.comcom.2020.01.014.

Organisation for Economic Co-operation and Development. (2012). *ICT applications for the smart grid: Opportunities and policy implications*. OECD Publishing.

Plaza, C., Gil, J., De Chezelles, F., & Strang, K. A. (2018). Distributed solar self-consumption and blockchain solar energy exchanges on the public grid within an energy community. In *Proceedings - 2018 IEEE International Conference on Environment and Electrical Engineering and 2018 IEEE Industrial and Commercial Power Systems Europe, EEEIC/I and CPS Europe 2018*. Institute of Electrical and Electronics Engineers Inc. Available from https://doi.org/10.1109/EEEIC.2018.8494534.

Pop, C., Cioara, T., Antal, M., Anghel, I., Salomie, I., & Bertoncini, M. (2018). Blockchain based decentralized management of demand response programs in smart energy grids. *Sensors, 18*(2), 162. Available from https://doi.org/10.3390/s18010162.

Premalatha, J., & Sathya, K. (2020). *Integration of blockchain into IoT* (pp. 121–134). Informa UK Limited. Available from https://doi.org/10.1201/9781003032441-8.

Rahmadika, S., Rauda Ramdania, D., & Harika, M. (2019). A blockchain approach for the future renewable energy transaction. *Journal of Physics: Conference Series*, *1175*(1), 012122. Available from https://doi.org/10.1088/1742-6596/1175/1/012122.

Rajasekar, V., Jayapaul, P., Krishnamoorthi, S., & Saračević, M. (2021). Secure remote user authentication scheme on health care, iot and cloud applications: A multilayer systematic survey. *Acta Polytechnica Hungarica*, *18*(3), 87–106. Available from https://doi.org/10.12700/APH.18.3.2021.3.5.

Sathya, K., Premalatha, J., Vani, R., Rajesh Kumar, D., & Celestine, I. (2022). A futuristic approach to generate random bit sequence using dynamic perturbed chaotic system. *Turkish Journal of Electrical Engineering and Computer Sciences*, *30*(1), 35–49. Available from https://doi.org/10.3906/elk-2010-137.

Wang, Q., Li, R., & Zhan, L. (2021). Blockchain technology in the energy sector: From basic research to real world applications. *Computer Science Review*, *39*, 100362. Available from https://doi.org/10.1016/j.cosrev.2021.100362.

Wireless communications in energy grid

5

C.N. Vanitha[1], *S. Malathy*[1], *S.A. Krishna*[2] *and K. Vanitha*[3]
[1]Department of Computer Science and Engineering, Kongu Engineering College, Erode, Tamil Nadu, India, [2]Department of Mechatronics, Kongu Engineering College, Erode, Tamil Nadu, India, [3]Department of Computer Science and Engineering, Faculty of Engineering and Technology, JAIN (Deemed-to-be-University), Bangalore, Karnataka, India

5.1 Introduction

The energy grid is a successful example of an interconnected network, an invention of the peer-to-peer network. This section discusses various ideas about the energy grid in wireless communication and its uses in this modern world. Nowadays, energy grids are used everywhere for transportation, weather forecasting, transmitting medical data (Vanitha & Malathy, 2021b), etc.

With the growing interest in remote communication, in the future, there may be more than 50 billion remote instruments in use (Mahmood et al., 2015). Thus, more Base Stations (BSs) are needed to coordinate the unlimited remote information traffic. The everyday energy cost for the cell network is a significant portion of the cost of operational use. Numerous cell administrators need to reduce their energy consumption by applying new arrangements; specifically, unique assignment plans have been made for various remote communication systems. The current arrangements illustrate two principal areas where energy savings may be made: energy-providing equipment and communication-related equipment of cellular networks.

On the question of energy provisioning, to improve the performance of distributed devices, the number of BSs needs to be increased. The energy spent by the cellular system adds a significant portion to the global "carbon footprint." An ecological and financial benefit is proposed by sustainable power sources from environment-friendly power-gathering instruments. Those instruments can provide cheap and green reusable energies (Panahi & Panahi, 2019). This is an overall solution to reducing the expenses of energy use or even a replacement for taking power from the main grid. Moreover, the cost of energy grid devices is lower than the cost of sustainable power-collecting instruments.

Since the reality is that in power fluctuations, the environment-friendly power is restricted from being exclusively beneficial, the sustainable power is equally difficult to store. It is unfeasible for BSs to depend solely on their individual power-creating ability; they need to connect with a power source to ensure a constant supply in the event that there is an inability to create their own energy.

Blockchain-Based Systems for the Modern Energy Grid. DOI: https://doi.org/10.1016/B978-0-323-91850-3.00015-9

Moreover, the power network is moving from a conventional grid to an adaptable network by using countless advanced intelligent measures which help in reducing energy costs (Yu et al., 2020). The primary distinction between a traditional framework and an intelligent grid is that a traditional framework just gets the power supply stream from a few large power plants and in a centralized manner, whereas the energy grid gets the power supply from many small plants and in a decentralized mode.

Fig. 5.1 illustrates a general flowchart for an energy trading and sharing model. In contrast to the customer power network, it has an intelligent grid that empowers two-way energy streams connecting power customers and the energy grid. Consequently, the intelligent measures can carry out the energy use monitoring, which enables the BS to take part in the intelligent grid.

5.2 Wireless communication

Remote communication includes the wireless transmission of data to a long distance from a sending to a receiving point without the assistance of wires, links, or other types of electrical devices. Fig. 5.2 illustrates that remote communication is an expansive term that joins all methods and types of sending and receiving data between at least two instruments utilizing a remote signal through remote communication advancements and instruments.

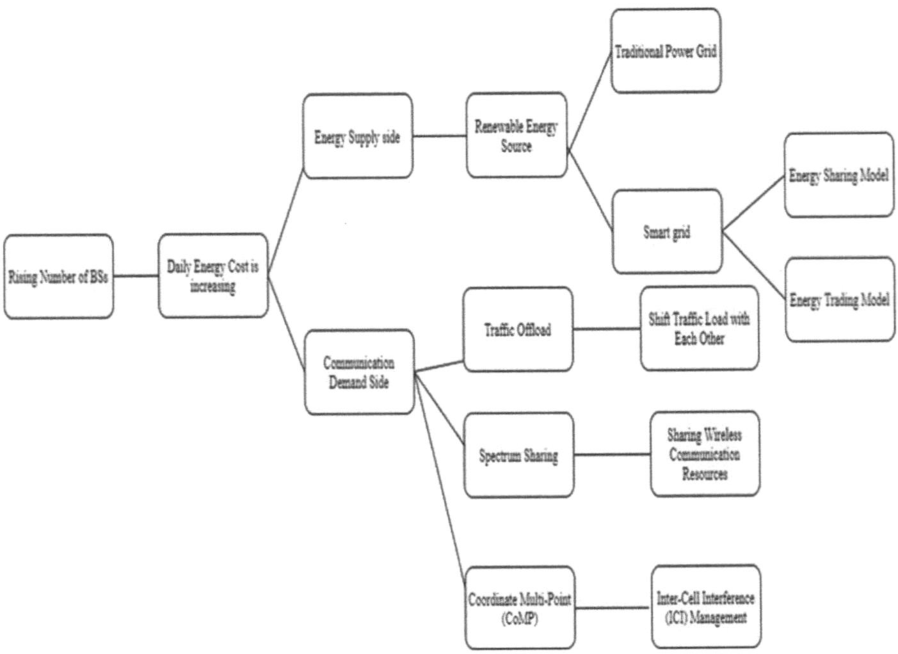

Figure 5.1 Flowchart for an energy trading and energy sharing model.

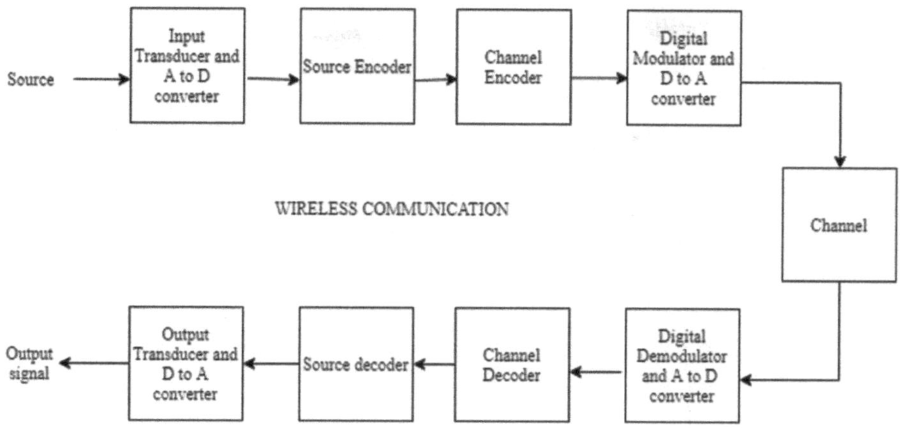

Figure 5.2 Block diagram for wireless communication.

5.2.1 Highlights of wireless communication

Wireless communications have many viable advantages. The growth of this remote communication invention has taken many giant strides, not the least of which is the spectacular distances covered by the wireless communications devices, which can be between a couple of meters (for instance, a TV controller) to many kilometers (for instance, radio communication).

Cellular communication is a major innovation of the remote communication methodology. Wireless communication is used for remote entry to the web, remote home systems administration, etc. Numerous instances of the need for better and improved communication have led to remote radio upgrades. These have been integrated into Global Positioning System (GPS) units, garage doorway openers, distant PC mice, control centers, headsets, headphones, radio channels, satellite TV, broadcast TV, and cordless telephones. Wireless communications are nowadays used for intelligent (Vanitha & Malathy, 2021a) communication for remote transmission to reduce energy use. Remote correspondence includes the movement of data with no actual association between the two ends. Remote communication has the following benefits:

5.2.1.1 Cost adequacy

Wired correspondence involves the utilization of association wires. In remote communications, there is no need for an actual framework or physical devices to be connected with wires, so the expense is minimum when the network must be increased.

Any organization which provides remote communication doesn't incur a great deal of expenses to add users to the network except for the cost of the user equipment. Thus, cheap and fast communication is assured for near communication as for long distance also.

5.2.1.2 *Adaptability*

Wireless communication adapts itself to various different network topologies, which have different considerations for efficient communications and cost. When the transmission happens through cross-layer protocols and architectures, the remote system empowers users to pass on and get messages in the most efficient and cost-effective ways.

Satellite devices depend on sensor communication to let users call their friends and family. Hence, wireless communication that depends on sensors for a self-correcting network by letting users always have a connection at the right time and the right place has improved overall governance by keeping neighboring sensors in contact for effective communication.

5.2.1.3 *Accommodation*

Wireless specialized instruments such as cell phones are very user-friendly and, in this way, allow anybody to use them, wherever they might be. To communicate, the users need to send and receive signals from the nearby base towers, which might not be their home station and so mobility is assured.

Model—Wireless communication (Chen et al., 2020) works successfully in administering internet innovations like Wi-Fi (wireless fidelity) which can connect multiple devices to the internet with just one router. With the development of the energy grid in wireless communications, we would now be able to associate with anybody, anyplace, whenever.

5.2.1.4 *Speed*

Wireless communication upgrades using energy grid result in high speed. Network availability or the OpenNESS in network significantly improves the exactness and speed.

Model—A remote, distant communication can work effectively quicker than a wired one. The remote-control switch of a machine that has stopped working will undoubtedly communicate that something is wrong by stopping working faster.

5.2.1.5 *Availability*

The remote communication innovation helps to communicate by opening up distant zones where ground lines can't be laid as expected. The availability of the signals is effectively associated with the network to connect anytime and anywhere.

Model—Instructors in various districts of our country present their online instruction through energy grid in wireless mode. Instructors presently don't have to make a trip to long distance to instruct.

5.2.1.6 *Steady network*

A steady network guarantees that the network can react to crises moderately and rapidly.

Model—A portable remote network can guarantee you a steady network, so you can move from one spot to another—while you travel—without worries of not having a connection with your handset because wireless communication moves with you.

5.3 Uses of wireless communication

In Fig. 5.3, the uses of wireless communication are denoted pictorially.

5.3.1 Settling on decisions

On a worldwide level, billions of individuals have subscriptions to cell communication. This framework works through bobbing radio waves being sent out from the cell towers and user equipment catching these signals so as to make and receive calls. This structure of the network communication works well for coordinating the radio waves. Hence, the call collectors function by handing over the caller who has moved from one cell to another by analyzing the cell location.

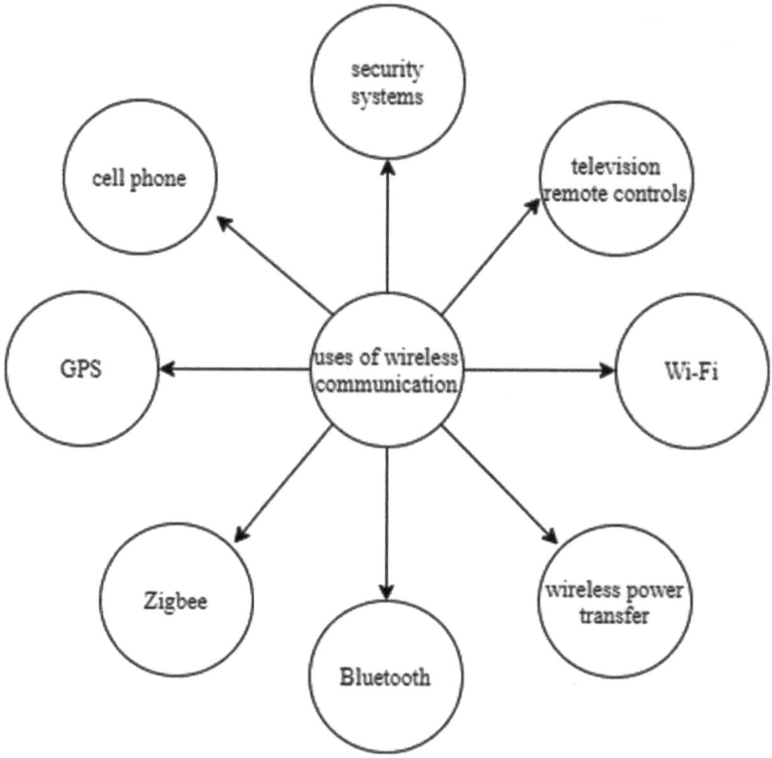

Figure 5.3 Uses of wireless communication.

In spite of the accessibility of the web, the network actually needs to settle on telephone decisions. It assists with client support and deals.

5.3.1.1 Associating devices

Bluetooth innovation permits individuals to manipulate different instruments. It does this through short-wave, low-energy radio signals. Bluetooth instruments pass through these waves to interface with and access other Bluetooth instruments.

For instance, individuals associate their cell phones with Bluetooth-empowered headsets. This is a choice for better safety for drivers who need to get directions or instructions.

In addition, Bluetooth permits the exchange of information starting with one instrument and then moving on to the next. It can be from a cell phone, a tablet, or a PC. Individuals who work in an office generally will print reports using this innovation.

IoT needs Bluetooth innovation to interface electronic instruments in the home. To interface indoor regulators, telephones, PCs, cameras, and light apparatuses, or to connect TVs and speakers, Bluetooth devices are indispensable.

Wellbeing aficionados use this innovation to combine wellness devices. Specialists likewise use it to get data from associated instruments like pacemakers.

5.3.1.2 Getting to the internet

Numerous individuals sign into Wi-Fi everyday without the knowledge that they are communicating using remote networks.

The communication networks use switches to send radio frequency (RF) signals to a different user equipment, like a router. The user equipment should have remote connectors to have the option to get the signals and access the web. This permits wireless communications to share the web in the workplace.

The lowest power networking technology is the Zigbee standard. It was grown unequivocally for ultra-low-power applications, including battery-controlled applications and minimal expense applications (Zhou et al., n.d.) like sensor networks for control. Wireless sensor communication is advanced to limit power usage and cost.

5.3.1.3 Upgrade security

The advancement of wireless communication improves security. For instance, walkie-talkies send and get radio signals. They don't have to face transmission issues like drop signals as do cell phones.

The security must be upgraded when the transmission crosses the jurisdictional and national boundaries. This is achieved by the energy gird itself. The energy grid doesn't allow communication beyond its boundaries.

5.3.1.4 For locating and tracking

In Fig. 5.4 A satellite communication permits satellites to interface with electronic devices on earth. In this methodology, the sensor devices are identified by their location and tracked when it is moving. Its most notable use is the GPS. In Fig. 5.5 the block diagram of satellite communication is demonstrated.

The wireless sensor communication permits the network to find and track their vehicles and their representatives.

5.3.1.5 Energy cooperation

The fundamental difference between the conventional energy network and the energy grid matrix is that the conventional network simply grants the expected power to the framework where the power is perceived as a one-way energy stream. Unlike the conventional power framework, the intelligent energy network conveys a reasonable measure of user independence by empowering two-way energy streams between energy clients and the energy grid.

The intelligent energy grid devices need power assistance; thus, this implies that the BS exchange and share the power inside the intelligent network. The ordinary energy-saving arrangement is different in the cell network because of exploited energy supply and communication between the client request side and server response side. Thus, it's important to consolidate the oversupply or excess demand created in the framework by resetting goals on both the power supply and communicated request sides. Considering the scenario discussed, we think that the safest option for the energy grid is to consolidate the energy exchanged and

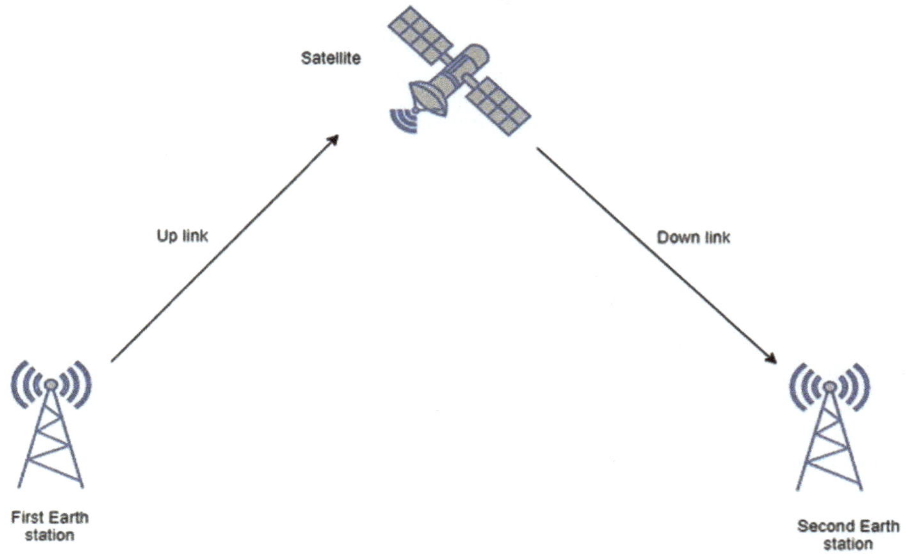

Figure 5.4 Satellite communication.

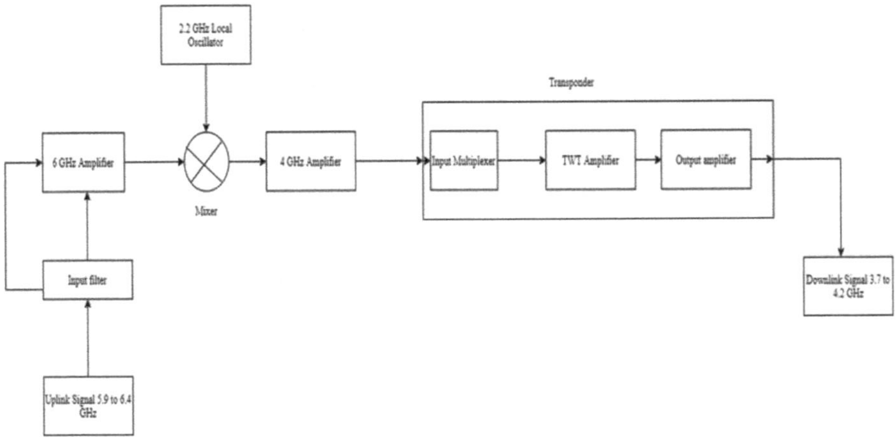

Figure 5.5 Block diagram for Satellite communication.

communicated by using a sensor network architecture by means of a power harvesting instrument in each central station.

Energy participation depicts the base station's technique of sharing and distributing energy to avoid the time-consuming process of transmission. Without affecting the ecosystem's power supply, the cost-cutting strategies are applied on both the client and server-side. While sharing the energy using wireless communication, if the number of hops in the middle between source and destination increases, the power also diminishes rapidly.

Fig. 5.6 represents a sensor communication network with two BSs. The functionalities performed for collecting power from the energy grid are highly recommended for wireless communication instruments.

5.3.1.6 Energy trading in aggregator

In the power exchange mode, the base station observes and assembles power to the power lattice by accumulation to execute the power exchanging process. In this methodology, the base station has sufficient and abundant gathered energy (Usha & Vanitha, 2016), which is collected from the energy grid.

Alternately, if the base station needs to collect the power, the architecture can help the base station to obtain the power from the energy framework. The architecture can help procure benefits to the energy prosumer or the consumer by unequivocally picking the power dispenser or spender, respectively, between the BSs, besides the power grid matrix.

In the power distribution method, the power is distributed via a transmission system to the consumers. The power distribution process allows the predetermined number of central stations to be used in a social event, with the extra get-together power being distributed to additional BSs with a smaller amount of energy. From or to the group, BSs can obtain or generate.

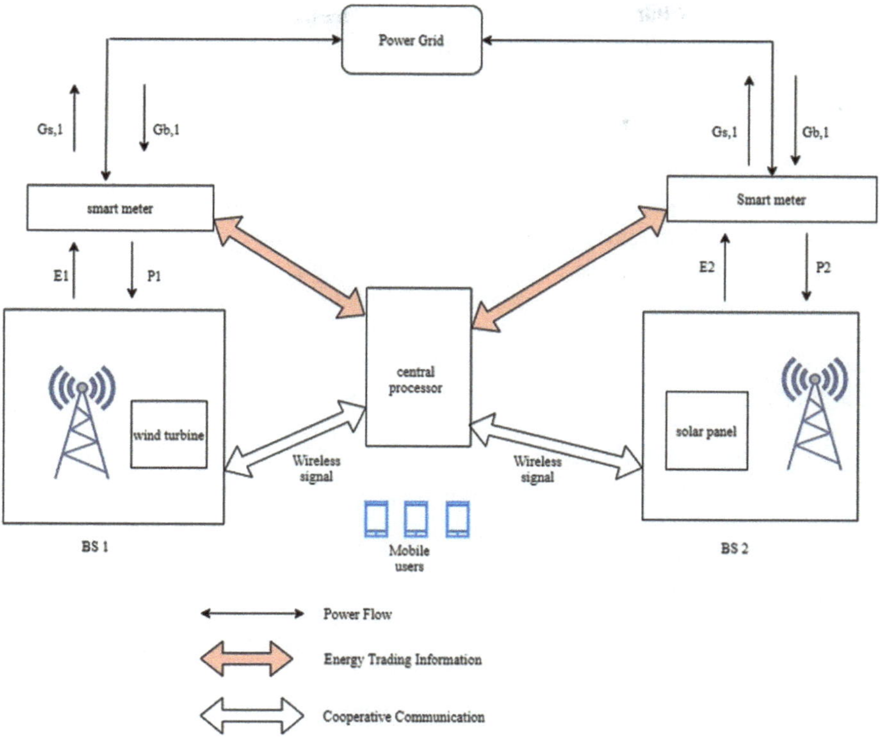

Figure 5.6 CoMP framework.

Table 5.1 Energy charge for energy trading and sharing in an aggregator.

	BS1 harvesting energy	BS2 harvesting energy	BS1 energy cost	BS2 energy cost	Total energy cost
Energy cooperation via energy trading in aggregator	4.14	8.36	4.14	18.28	10.51
Energy cooperation via energy sharing in aggregator	4.14	8.36	4.14	18.28	10.03

Table 5.1 gives details about energy charges for energy trading and sharing using an aggregator. To execute, the BSs must have harvested energy, and the energy cost has to be calculated. The energy trading and sharing are based on power generators, energy suppliers, and transmitters. The energy is generated by the energy generator, transmitted over the transmitter, and supplied to the consumers.

The energy which is harvested by the base station is transmitted over the transmission lines or hops in between the BS and the destination.

The energy grid process transmits high voltage energy, that is current to local central stations through the transmission lines. The high voltage current is then transformed into a low voltage current to distribute among the receiver points from the local central stations. Synchronization among energy grids is equally important to deliver communication through wireless methods.

The transmission of energy connects the power generator, efficient markets for electricity, distributing devices, suppliers, and consumers. The well-known truth is that security is always needed for any communication. Even when using the energy grid, security should be enabled to avoid suspicious and malicious activities. Nowadays, many types of security attacks are happening. In order to avoid the threat, the smart energy grid should be highly protected.

5.3.1.6.1 Energy gird utilities

A metering infrastructure is introduced as an effective solution for various industries that are searching for effective communication and proper connectivity to deliver the smart architecture. The energy grid communication is improved greatly because of the advanced metering concept. There are two types of spectrum signals available. One is publicly available, and another one is privately licensed. The public network is a mesh network, whereas the private network is a single-point to multipoint architecture.

When selecting among the above said two approaches, the following parameters should be considered:

- Type of Signal
- Architecture
- Benefits
- Features
- Regional challenges
- Wireless utilities
- Cost
- Reliability
- Latency
- Response time
- Interference

Each spectrum signal type differs and is distinct from each other, and it should be designed and selected according to the need. Even though both the architecture provide an advantage, there are key differences between those approaches.

5.3.1.6.2 Wireless grid architectonics

In the wireless grid architecture approach, the multipoint, mid-point, single-point, and mesh networks are used. These networks are two-way communication radio networks. The radio points on the network communicate with each other, and it can store, receive and redirect the signals or data to other points. Most of the private users get a private license for the spectrum signal. These separate signal bandwidths can be used only by the leaseholder and in only one particular region. The interference in that spectrum signal bandwidth is handled by the government.

The mesh network architecture produces high interference compared with a private network. As it is the default network, such that too much communication in one place makes the conversion worst. Likewise, the voice fidelity will not be clear most of the time in the mesh network. In a point-to-point network with a private separate spectrum signal, the communication would be naturally good. The interference ratio is also low compared with the public network.

The public unlicensed network signal range is very short because of the output wattage. For given that it's a short signal range, using this network, the communication can be done only for a short duration. If the data or energy needs to be transmitted over a longer distance, then many intermediate nodes have to be used to communicate point-to-point. This architecture makes the signal weak and noneffective.

In the case of latency, as discussed previously, when using more intermediate transmission devices increases, the latency of transmission also increases. In the licensed spectrum, the utilities used for communication allows energy and data in a metered network to improve so that the transmission is faster and the delay is lower.

The energy grid is a two-way communication electric system that is a highly secure cyber technology. In wireless communication, the energy grid system is computationally intelligent. It is heterogeneous in nature and also compatible with the usage of integrated components and applications. The advantage of using this system is that of mainly its being adaptable to the distributed new technology sensors, which makes it greatly helpful for secure communication.

5.4 Conclusion

In recent days, the usage of electricity has been high due to the emergence of many electronic devices in all fields. It is highly urgent to satisfy the customers and manage the remote transmission traffic. Numerous cell administrators desire to limit the power fee and exploit the quality of service of wireless communication. The energy grid helps in energy production, conditioning, and distribution. The energy grid in wireless communication ensures efficiency, quality, loss management, security, and its being economically cost-effective. Moreover, the energy grid engages new products, monitors intelligently, communicates effectively, controls, and provides services to the users of its products. The energy grid adapts to self-healing technologies and blooms in day-to-day life.

References

Chen, M., Poor, H. V., Saad, W., & Cui, S. (2020). Wireless communications for collaborative federated learning. *IEEE Communications Magazine*, *58*(12), 48–54. Available from https://doi.org/10.1109/MCOM.001.2000397.

Mahmood, A., Javaid, N., & Razzaq, S. (2015). A review of wireless communications for smart grid. *Renewable and Sustainable Energy Reviews*, *41*, 248–260. Available from https://doi.org/10.1016/j.rser.2014.08.036.

Panahi, F. H., & Panahi, F. H. (2019). *Smart grids and green wireless communications. Demand response application in smart grids: Operation issues* (pp. 1–35). Springer International Publishing - Volume 2. Available from https://doi.org/10.1007/978-3-030-32104-8_1.

Usha, M., & Vanitha, C. N. (2016). Pruning route modifiers in wireless sensor networks. *Wireless Personal Communications*, *89*(1), 27–43. Available from https://doi.org/10.1007/s11277-016-3249-1.

Vanitha, C. N., & Malathy, S. (2021a). A multi-syndrome pathology for breast cancer through intelligent learning. *IOP Conference Series: Materials Science and Engineering*, *1055*(1), 012073. Available from https://doi.org/10.1088/1757-899x/1055/1/012073.

Vanitha, C. N., & Malathy, S. (2021b). Optimizing wireless multimedia sensor networks path selection using resource levelling technique in transmitting endoscopy biomedical data. *IOP Conference Series: Materials Science and Engineering*, *1055*(1), 012071. Available from https://doi.org/10.1088/1757-899x/1055/1/012071.

Yu, X., Xu, D., Sun, Y., Ng, D. W. K., & Schober, R. (2020). Robust and secure wireless communications via intelligent reflecting surfaces. *IEEE Journal on Selected Areas in Communications*, *38*(11), 2637–2652. Available from https://doi.org/10.1109/JSAC.2020.3007043.

Zhou, Z., Wang, W., Dong, F., Jiang, J., & Kavehrad, M. (n.d.). Application of optical wireless communications for the 5G enforced smart grids. In 2020 IEEE 4th Conference on Energy Internet and Energy System Integration (EI2) (pp. 756–760). Available from https://doi.org/10.1109/EI250167.2020.9347307.

Blockchain in internet of entities - issues and challenges

6

Dahlia Sam[1], *K. Jayanthi*[2], *Rishabh Tiwari*[3], *A. Shamila Ebenezer*[4], *S. Deepa Kanmani*[5] *and Adlin Sheeba*[6]

[1]Department of Information Technology, Dr. M.G.R. Educational and Research Institute, Chennai, Tamil Nadu, India, [2]Department of Computer Science, KGiSL Institute of Information Management, Coimbatore, Tamil Nadu, India, [3]Department of CSE, Dr. M.G.R. Educational and Research Institute, Chennai, Tamil Nadu, India, [4]Department of CSE, Karunya Institute of Technology and Sciences, Coimbatore, Tamil Nadu, India, [5]Department of IT, Sri Krishna College of Engineering and Technology, Coimbatore, Tamil Nadu, India, [6]Department of CSE, St. Joseph's Institute of Technology, Chennai, Tamil Nadu, India

6.1 Introduction

The Internet of Entities, or IoE, denotes billions of physical devices around any geographical location that are now connected to the internet, all collecting and sharing data. The sharing of information and the authentication of data are only carried out through a central server, but this can lead to data security and data privacy issues. To avoid the high chances of device spoofing and false authentication rife with IoE, the concept of a central server to be used for signing in to one's data—where third-party management of cloud servers and data blockages are not unknown—is being dislodged in every sector including healthcare, finance, e-commerce, and supply chain management. Instead, Blockchain technology is being adopted widely as part of IoE.

Many business organizations are on their way to adopting Blockchain technology, but they face several challenges. Firstly, there are some critical problems associated with the adoption of the Blockchain, including scalability, interoperability, energy consumption, lack of talent, and lack of standardization. Secondly, given the Bitcoin protocol dictates a limit on the size of blocks which makes small applications better at handling Bitcoin transactions, which means the Blockchain network can process less than 10 transactions per second, it follows that Bitcoin is not suitable for larger applications with its voluminous data-crunching capabilities. Similarly, most blockchains work in independent storage towers which cannot communicate easily with other peer networks, which means they cannot send or receive information from other Blockchain-based systems.

Blockchain mechanism requires high computational power to solve complex problems in networks, which by itself is a challenge. Thus, fourthly, the overall cost,

Blockchain-Based Systems for the Modern Energy Grid. DOI: https://doi.org/10.1016/B978-0-323-91850-3.00012-3

data privacy, and data security concerns are some other factors challenging the implementation of Blockchain technology in a real-life scenario. Fifthly, there is no universal standard for Blockchain, which means it can lead to issues related to interoperability, high costs, and problematic mechanisms making mass adoption very difficult. Further, the lack of regulations in the Blockchain industry makes it an easy target for manipulators in the technological market. Lastly, to add to these issues, there is a severe lack of Blockchain professionals and developers. Some of these issues cannot be overlooked and have to be addressed while adopting Blockchain technology in various domains. In this chapter, we will elaborate on the different issues and challenges of Blockchain in IoE.

6.2 History of blockchain

Blockchain is an emerging technology introduced by the famous scientists Stuart Haber and W. Scott Stornetta in 1991. Blockchain works well with cryptocurrencies because cryptocurrencies are not backed by physical assets but by a consensus between the network who agree to approve future cryptocurrency transactions (records or blocks) belonging to a chain of transactions (records) that are transparent to the network, accountable by the network, and owned by the network, that is decentralized without a trusted third-party taking care of the data/money.

Digital time-stamping—where a block is stamped with the date and time of the transaction before being added to the chain of blocks—is a way to create a sequential chain of records so that it is immutable because even though the block is not unchangeable, the history of transactions is. This means no record is lost once created, which enhances the security of transactions from accidental deletes and data thefts which helps to secure the Blockchain transaction and link the current block to the previous block.

In 1992, the Blockchain became more powerful due to the integration of a new concept called Merkle Trees with the design of blockchains. This design integration presented a more efficient Blockchain while securing the chain of blocks, for instead of one block per transaction, the Merkle Trees allowed for more transactions/records to be collected in a single block: storing a series of data records that are part of one transaction in a block with each data record connected to the previous one. Thus, the newest record in the chain will contain the history of the whole chain. Unfortunately, this technology went unused. A simple blockchain is shown in Fig. 6.1 below.

Later in 2004, the cryptographic activist and researcher Hal Finney enhanced this Blockchain technology with the Reusable Proof of Work or the RPoW System. Because of this enhanced version of Blockchain technology with RPoW, the Blockchain obtains a non-exchangeable Hashcash-based Proof of Work token, and it creates RSA(Rivest, Shamir, Adleman algorithm)-based signed tokens also. It facilitates more convenience in the person-to-person transfer of records. Double-spending, a major problem in the banking and financial world, can also be resolved with the registered token ownership on a trusted server. It permits users from

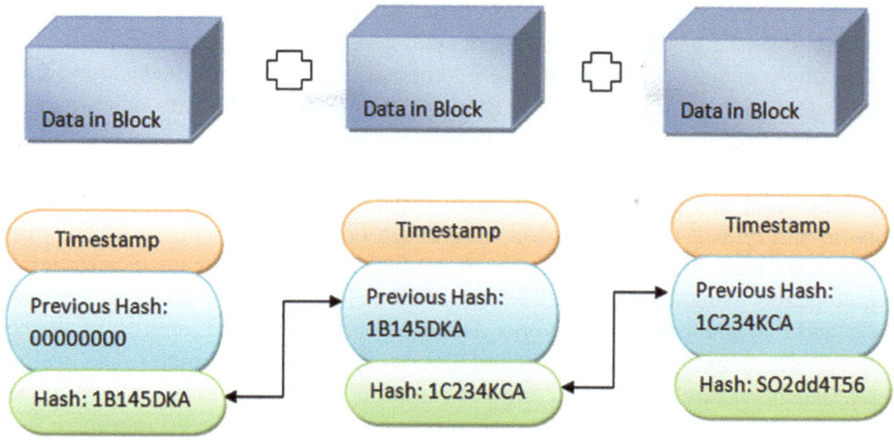

Figure 6.1 Simple blockchain.

anywhere in the world to confirm its integrity and accuracy in real-time. RPoW played an important role as a primary prototype and a major step in the progression of cryptocurrencies.

In Bitcoin-based blockchains, the problem of double-spending is protected against each and all transactions by distributed or decentralized peer-to-peer protocol, which actually tracks and confirms double-spending didn't take place.

Generally, Bitcoins are mined by distinct miners and later confirmed by the decentralized nodes in any network. The Bitcoin concept came into existence when 50 Bitcoins were mined by Satoshi Nakamoto in January 2009. The first Bitcoin transaction was received by Hal Finney in January 2009 when he received 10 bitcoins in the world's first transaction of bitcoins.

The evolution of blockchains has ever since been steady and quite promising. The words block and chain were initially used separately by Satoshi Nakamoto, but later around 2016, the two words became popular as a single word Blockchain. Currently, the file size of cryptocurrency blockchains has drastically increased from 20 GB to 100 GB.

6.3 Issues and challenges faced by blockchain in internet of entities

A Blockchain provides us with good solutions for many problems, that is, the need for transparency, lack of decentralization, integrity, need of immutability, and need for security that a trusted centralized third-party cannot bring for human error, corruption, etc. because as soon as we move from a trusted third-party in a centralized network to a trustless, decentralized network, we are gaining ownership over our assets and a transparent system evolves where changes to records are secure, visible, and permanent with a timestamp that cannot be deleted without breaking the chain. However, there are a few more challenges to be faced before we realize the complete potential of the Blockchain-IoE symbiosis for different applications.

The single biggest challenge, unlike interoperability issues, revolves around standards and around latency issues. The Blockchain is probably the only interorganizational technology that needs to bring experts together.

From the introduction of IoE, we have observed that every entity is involved in data collection, like traffic lights on main roads receive data from sensors in order to avoid and reduce road congestion, or garbage bins send data to the urban waste management team to intimate its status of getting filled, etc., so that IoE effortlessly communicates using assorted devices with various functionalities in centralized networks and machine-oriented networks. Due to the strength of IoE entities in all sectors, the substantial number of linked devices generates enormous data traffic, which can be a blockage in meeting the essential quality and service level agreements, mainly due to the storage management and bandwidth-related issues.

In this analytical field, recently, Blockchain technology has progressed and tries to avoid the essential of trust among entities. A significant remedy for eliminating trust from the traditional way of storage or online mediators is blockchain technology (Agrawal et al., 2020).

Blockchain technology and IoE collectively reduce operative costs, manage distributed resources like IoT devices, and are impervious against malicious attacks and vulnerability threats.

When IoE and Blockchain technology become one, it overcomes substantial challenges of implementing the IoE platform and storing the collected data in the future connected data-driven world. Blockchain technology was first developed for cryptocurrencies, a decentralized append-only ledger technology as a public data source, for example, Bitcoin. Later in 2008, Satoshi Nakamoto advanced Blockchain technology to another level which has fascinated emerging peer-to-peer technology for decentralized and distributed computation and sharing of data.

Many business organizations are planning to adopt Blockchain technology in the face of several challenges. Given the implementation of Blockchain technology along with cryptoanalysis technology, and where there is no unified central data systems or a central data storage, the system can be safeguarded from virus attack outbreaks that want to take control of the IoE-based system. Now, a unique feature of the blockchain has attracted many data scientists in 2013 to the Ethereum. Because of this powerful and feature-laden Ethereum, the Blockchain is becoming practically possible in many sectors across cryptocurrencies. It is powerful because of its transaction-based, state-machine, the privacy of transactions, security, the immutability of data, auditability, the veracity of data, endorsement, transparency of a system, and fault tolerance.

6.3.1 Resiliency against combined attacks

Internet public service improvements and the rise of the IoT socio-technical systems have regularly produced more insecure and unreliable protocols to accelerate the rapid intercommunication among centralized servers and an IoT device (Scalable Blockchain for the Internet of Things IOT, n.d.).

Numerous security solutions for Blockchain-based IoE have been projected, each of which is intended to deal with diverse security issues and aggressive models. The major difficulty here is to find how to propose a security solution that can resist pooled communication attacks and consider the practicability of the execution plan, especially when resources are insufficient, restricting IoT devices as secure against the full range of security threats.

Around 2019, ransomware attacks increased by 62% around the globe. Everyone is aware of the SolarWinds Sunburst malware that infected SolarWinds Orion systems, where more than 18,000 organizations downloaded the Orion software that was infected. Extra vulnerabilities that were unrelated to these Sunburst attacks were spotted in the Serv-U FTP software. The fallacy and general impact of these attacks have made the private organizations examine the cyber resilience of their own organizations and also among their supply chains.

In the recent past, data has been under constant attack from precocious adversaries. Threats have been evolving at a higher pace than the traditional security methods are able to protect. Government and private-sector organizations are together partnering to ensure the success of improved cyber resiliency across countries.

Some common vital steps that can be taken to protect sensitive data include a renewed attention to database security, anticipating a breach, collaborating across public or private sectors, enabling continuous monitoring and testing of applications, and training and retaining cybersecurity experts.

6.3.2 Dynamic adaptive security framework

A variety of devices are deployed in IoE networks, from lower-end power devices to high-end servers. Therefore, due to the different numbers of Blockchain-based IoE architecture resources, we cannot implement a one-size-fits-all security solution. Therefore, security solutions must first get used to the array of resources and decide which security services to offer to meet up the security requirements of end-users from a minimalist viewpoint. Therefore, a challenge worth paying more attention to is how to design a Blockchain-based, dynamic, adaptive, and secure IoE architecture framework in future implementations.

Sun Microsystems first coined the term Adaptive Security in the year 2008. This adaptive security architecture will be able to predict and respond to threats. It also tries to reduce threat amplification, the attack surface, the speed, and also retrieval time. This architectural prototype at the microscopic level counterfeits the biological auto-immune system and the environmental systems at a macroscopic level.

Dynamic adaptive or environmental systems usually respond to and then try to adapt to unpredicted conditions. The auto-immune system responds to threats in a dynamic way using the inbuilt and voluntary response. Generally, ecological systems are diverse and varied while being independent of each and every entity for their survival. But these systems are dependent on their reactions and feedback to improve their response to threats and vulnerabilities. This security framework is imitated dynamically and autonomously by the developers of adaptive security architecture designs.

Adaptive security plays an important role in the faster detection of security compromises and its reaction over the system whenever such malicious events occur. Some of the advantages of dynamic adaptive security architecture include data theft prevention, a service interruption avoidance, a proactive threat covering when it occurs, decreasing downtime or reducing the duration of threats in the system, identifying the current security breaks, evading the environment that is more similar to the same kind of culture systems.

Adaptive security is a multi-level system rather than a single system with an all-time monitoring system that progresses as and when cyber threats are found in a system, or it becomes more cultured and strenuous against more new multifaceted threats coming across the system.

6.3.3 General data protection regulation compliance

The European Union's new data privacy and security regulations went into effect on May 25, 2018. It imposed strict laws on organizations around the world as long as they collected data about people in the European Union. In 2008, end-users of various internet-based goods and services, like online banking, had new powers over personal data like name and email addresses, location and gender information, etc., and data controllers who are either the owner of an organization or an employee of one who handles data to have assumed new obligations for data privacy and security. Data processing makes general data protection regulation (GDPR) implementation difficult since it contains the premise that data should be kept to a bare minimum and only processed for purposes that have been specified in advance. Yet, these rules can be difficult to apply to blockchain technology.

In the public chain, all data is copied and shared across all machines on the network, which anybody can see and read. This problem can only be solved by combining reliable hardware solutions. Some preliminary attempts have been made, but this is a very promising area to research.

As IoT is a concern, the main functionality of IoT networks often is to collect enormous quantities of data from clients and consumers, and it's the responsibility of the associations collecting data to ensure that Blockchain technology to secure an IoT network is in accordance with the GDPR, which is a real live issue within an organization.

The GDPR is a legitimate framework that brings strategies and rules for gathering and processing individuals' personal data, which apply to people residing in the European Union. However, the regulation applies to any organization even if it is not in the European Union, that is all locations worldwide. The GDPR makes it mandatory for few disclosures of data. The regulations were approved and adopted in April 2016 after passage in the European Parliament and came into effect on May 25, 2018.

6.3.4 Blockchain-specific infrastructure

IoE, with Blockchain-specific infrastructure storage limitations, may not be able to store large blockchains that grow as blocks rise. Furthermore, it is generally believed

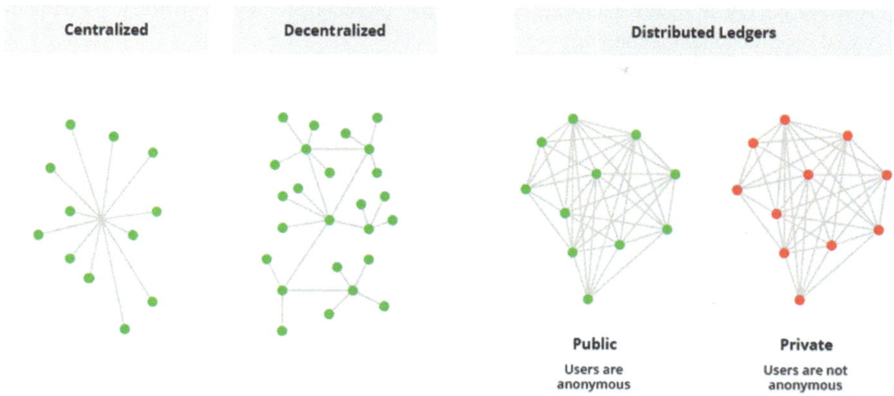

Figure 6.2 Blockchain structures.

that the Blockchain data stored by IoT devices is not even used except exclusively for their own transactions. Therefore, especially for the devices that support the distributed storage of big data on the Blockchain, the large and small blockchains have become a challenging problem. Furthermore, address administration and the essential communication protocols play an important role in Blockchain communications. Most importantly, teams with rich computing resources must be built on top of the Blockchain infrastructure. Lastly, the API should be easy to use.

Blockchain structures generally fall into these two groups, as shown in Fig. 6.2.

1. The public Blockchain architecture—The data and the admission to the system is available online for whosoever is willing to join. Some known examples are Bitcoin, Ethereum, or Litecoin Blockchain systems.
2. The private Blockchain architecture—the data and the admission to the systems are controlled explicitly by an organization, so one can participate with authorized invitation only.

6.3.5 Advertisement dissemination in vehicular cloud

Smart cities have become renowned in contemporary metropolises due to the development of entrenched and connected smart devices, techniques, and technologies. They have supported the connectivity of every "thing" to the worldwide web. It is expected that the IoT and the internet of vehicles (IoV) will play an important role in newly emerged smart cities.

In order to increase safety driving and other support, advertising dissemination in automobiles is a communication technology in which sensors connect to each other to transmit information, based on the decentralized model of the blockchain. People have proposed various unidentified solutions to cover the identity on the IoV. Proliferation aside, the key security issues are as follows:

1. How to propose a single-attribute contact control protocol based on Blockchain technology to protect the confidentiality of transaction information and severely punish the spread of advertising in the cloud about the car?

2. How to design a secret exchange scheme that protects privacy to identify participants based on Blockchain technology?
3. How to design low convolution verification between RSU and Peer Drainage Vehicle and use Blockchain technology in advertising and communication?

6.3.6 Handling skyline queries

In the IoE environments like smart grids, smart homes, and smart cities, data with multifaced and tedious attributes are generated in copious amounts regularly. These data make it hard to provide interoperability with a substantially low assortment. Maintenance and management of such assorted data carry difficulties in data management and value mining. These skyline-related issues can be sorted out by a unified collaboration of the IoE and Blockchain as blockchain of things.

Querying Databases with needed preferences or likings is an important aspect of every database or data storage-related research problem. There are many approaches being used for database querying: querying with partials or preferences, where skyline querying is the most widespread in this category. The major problem with the skyline framework is that all attributes are of equal rank in skyline preference relations. Thus, the magnitude of the results of skyline queries may produce exponentially voluminous results based on the number of skyline attributes.

Skyline queries have turned into an important topic in database research, such as centralized databases, distributed databases, and similarity searches. The series under examination has not yet explored the likelihood of using skyline queries in combination with Blockchain.

6.3.7 Mining of energy efficiency

The Blockchain mechanism requires high computational power to solve complex problems in networks, which by itself is a challenge. Mining involves operating Blockchain algorithms that detect fraud, such as the Proof of Work (PoW) consensus algorithm. In addition, the Blockchain grows as users accumulate their transactions. Therefore, extra controlling miners are needed to hold the Blockchain compromise protocol. Numerous energy-saving consensus algorithms store the most recently recommended Blockchain transactions, such as Proof Space Delegated, Proof of Space, Proof of Stake, and mini-block chains.

However, IoE devices with limited resources and power aren't always enough to fulfill the needs of huge computing and power usage like blockchain technology. As a result, an architecture capable of managing these precious resources is essential to manage such confined situations. Therefore, designing a power-saving consensus protocol is one of the most important research challenges in blockchain IoT technology.

6.3.8 Social networks and trust management (Social IoT)

The lack of regulation in the Blockchain industry makes it an easy target for data breaches as technology vulnerability manipulators in the technological market can take over the network for malicious reasons. Security choices must consider that fake news could be part of a cyber-attack, with the spread of rumors causing serious social and economic losses. For organizations or countries, especially those using online social networks, this is a particularly nagging issue. Blockchain can be a means to limit the spread of rumors, as described before, by securing social networks that support Blockchain (The Convergence of IoE and Blockchain: Security Challenges, n.d.). Social Internet of Things (Social IoT) is the idea of integrating the social network concepts with the IoE, in which each entity in the network is capable of determining social interactions with each other. In these networks, to conserve data privacy and security, the access to IoT networks can be carried out based on social responsibility assessment of the devices.

Building trust among the Blockchain and IoE ecosystem aims to include identity authentication, transaction model design, and process design, which improves trust among social media networks. All these attributes can be compressed as smart agreements or limits so that completing these operations by invoking agreements is accomplished quite easily. The operations and design of agreements can be done using the parameters and values given in Table 6.1. Authentication used for identification aims to issue identity for edges of the network in a concurrent way so that diverse edges can authenticate each other effortlessly.

Every transaction contains parameters as described in the above table, and it is notable that the owner field ID which belongs to the transaction has to reach an agreement, while Pubkey belonging to the ID represents the Public key that gives access to the Blockchain network, while the communication is signed by its existing owner. In simple words, the parameters ID, owner, and signature will denote the current owner or immediate owner, or the next upcoming owner, respectively, and the three can be dissimilar. Generally, other IoE-related transactions are verified with its Public Key (Pubkey parameter).

6.3.9 "51%" attack in blockchain

A Blockchain is a chain of blocks, which are a sequence of data that record every single transaction during a particular cycle. When a transaction is carried out on a Blockchain, like Bitcoin or any other cryptocurrency, it is usually put in a

Table 6.1 Message of identity.

Value	Parameter
Unique identity	ID
Operation type	Type
Edge owner	Owner
Public key of the edge	Pubkey
Validity period	Validity
Description of the device	Description
Signature of who invoked the contract	Signature

collection of unsubstantiated transactions. Miners can choose transactions from the pool to form a block of transactions. A "51% attack" arises when a particular miner or mining party takes majority power of a PoW-based Blockchain and when they are ready to pay double the amount of its coin (Uddin et al., 2021).

To add a transaction into a Blockchain, a miner needs to find a perfect solution for a puzzle; that is, they need to solve some complex mathematical puzzles by using hashpower or computational power.

There is no such situation of 51% attack that has been supported by Bitcoin so far; however, it has occurred in networks of altcoins with less hashpower and bad network security. A 51% attack is purely theoretical. To broadcast an exact solution to a puzzle to supplementary miners, almost all transactions in a block must be valid and timestamped, corresponding to the present record on a Blockchain; fraudulent miners, on the other hand, in trying to transact a suspicious block, are left with little time to carry out an attack before the network is alerted to the situation. Given a suspicious transaction is not timestamped with respect to the history of transactions happening before, it is invalid; the miner with an invalid record must break the immutable chain of records, and to gain time to carry out an attack, he won't broadcast answers to complex puzzles to the remainder of the network.

A 51% attack is perhaps a feasible problem that the majority are afraid of in the entire Blockchain engineering. In the majority of situations, however, 51% attacks either fail or don't succeed for long.

The 51% attack problem, as shown in Fig. 6.3, is one of the possible restrictions or drawbacks of Blockchain. Given Blockchain works through an agreement that says if 51% of the managing power in a particular network combines to change a contract, then that change will be recognized and established as valid change (Uddin et al., 2021).

Notably, the amount of power needed for mining hardware companies has accumulated in recent business transactions. Professional companies like ASIC mining company have improved their mining computer hardware, making them tremendously influential. This rising mining hardware setup is the major concern of many individuals and other groups. Such an increasingly influential setup would pre-empt

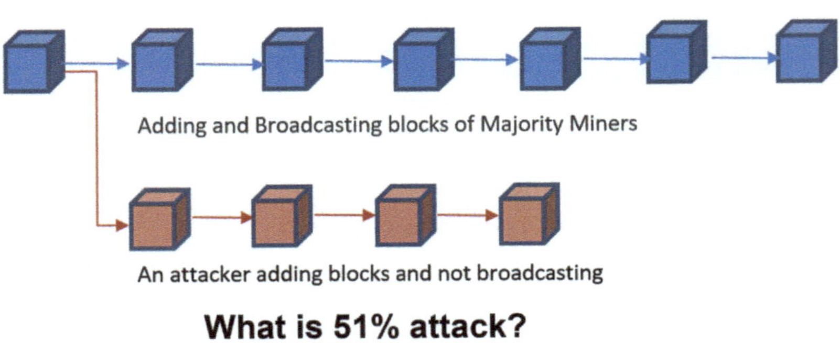

Figure 6.3 51% attack.

the control of the Blockchain network, and given they are very highly talented and strongly influential with commendable hashing power to create 51% attack whenever they wish to create, they are an opponent to contend with before it becomes too late.

6.3.10 Scalability of blockchain

When a Blockchain expands, data accumulates and becomes larger and larger, and the loading and computing create the blockages in the Blockchain network that will also get tougher and harder to access, leave alone operate. Because a distributed data/ledger is stored in many locations, the system will have to match the data multiple times. As the data grows, scalability will become a huge concern. Scalability with respect to a Blockchain procedure implies its capacity to support high transactional quantity and immediate progress.

Proof of work-based authorization-less blockchains currently deal a transaction data of about 10 transactions per second. Whereas visa can deal with almost 1700 transactions per second, which is ordinary. But with simple technical corrections, the speed at which blockchains function could possibly be increased to about 100 transactions per second without weakening the security provisions of an open and distributed network. The opportunity for implementing Blockchain technology in various fields is there, but scalability is the present bottleneck of this technology.

Scalability issues mean that as usage increases and acceptance of Blockchain technology accelerates, the execution of a scalable Blockchain will not suffer in adoption as of now. Blockchains that make inadequate provisions for scaling out as mass adoption rises are said to lack scalability. The Blockchain trilemma shows that better scalability is possible, but security, decentralization, or both will tolerate as likely. Scalability is the prominent way for Blockchain networks to rationally battle with legacy, centralized platforms at this stage. Even though many platforms like Bitcoin have established decentralization and security in blockchains, accomplishing scalability is the key challenge for today's prominent decentralized systems. Whenever the Blockchain network was scaled with a greater number of nodes, there was a possibility of having broadcast traffic, and it was tough to manage the overheads of a network.

Simultaneously, an issue with processing emerges when it is connected to low bandwidth IoE devices. Signaling and transferring data between devices use a lot of power and consume a lot of CPU processing power. You need minimal battery depletion and low power usage with all of this connectivity.

6.3.11 Lack of adoption

A Blockchain is scattered in many locations and linked to form a ledger that is visible to all and functions as distributed systems. It may look reliable, secure, special, and autonomous, but it is not exactly true. The concept of transparency in the Blockchain is the strength of this technology, from which strength it is thought to scale the Blockchain without draining the sources; at the same time, it has no regulations to be complied with because there are no universal laws as regards blockchains. Records and files collected within a Blockchain are considered not significant in any law court.

There are decent motives that convince us to be optimistic that the adoption of Blockchain will improve soon. Business associations are gradually approaching the technology together and have started to form collaborative Blockchain working groups to face common discomfort points and develop resolutions that will promote the technology in the eyes of everyone without exposing patented information (Agrawal et al., 2020).

The integration of Blockchain technology along with IoE recently became popular among researchers since data collection, and high computational storage from various nodes are quite easy. The method of collection of data and its processing is steady and easy to debug whenever the system malfunctions at a single point.

Blockchain technology, along with IoE, has some restricted capabilities in order to provide innovative new business ideas for the execution of new roles and facilities in innumerable domains (Agrawal et al., 2020).

For example, during the COVID-19 pandemic, various drug manufacturing business groups came forward to create the Blockchain for Clinical Supply Chain Industry to help it work no matter where in the world.

6.3.12 Skills gap

The Blockchain is considerably evolving as a technology, but the skills needed to be built for Blockchain technology to progress are in short supply. In a major survey, 49% of research contributors indicated that these skills gaps might be reason enough and a primary task to be carried out in earnest is to fill the skills gap. Blockchain skills are very competitive and have been in the marketplace for a few years. The demand for Blockchain technology was determined by looking at the skills of people.

The investment in blockchain can be utilized fruitfully, but the shortage of talent in this field adds to the organizations' concerns; while adopting Blockchain and incorporating it with the traditional heritage systems is a need, talent acquisition must be speeded up too.

Another way to neutralize the skills gap in this technology is to make use of Blockchain as a service, or "Business as a service" (BaaS). Through this, an organization can extract the advantages of Blockchain. When BaaS advances, the organization need not code in-house and develop bots because they can check for abundant product-making companies who have the knowledge base to implement Robotic process automation and convert it for every organization's requirement. Blockchain has the potential to completely transform the financial and commercial industry, amongst others, and so the necessity for talent must be faced, and talent encouraged.

However, to do this, there must be reinvigorated attention on the education system and on the practical application of book learning, putting the onus of the responsibility on the numerous Blockchain and crypto courses offered in global academies to bring into use, and expand, a set, standardized curriculum. Additionally, business organizations and the academic world must be ready to work together to spread awareness of Blockchain as influencing society. The skills gap is no longer an issue when providing openings for new talent to understand and realize the skills to implement Blockchain.

To channel the considerable gap between talent aquisition challenge and talent supply, it's mandatory for companies with vast resources and technical talent to train in-house, providing the prospect for recognized developers in other areas to move into a growing field (Agrawal et al., 2020; Uddin et al., 2021). Companies with enormous resources and technical skills are now beginning to train in-house to bridge the significant gap between talent need and supply, allowing veteran developers in other fields to transfer into a developing profession.

6.3.13 Integrated cost problem

The planning and development of Blockchain will include a lot of cost, time, and money to change the existing central system, particularly when it's an established and accepted infrastructure that people have become used to. This cutting-edge technology must not only generate economic gains by satisfying the needs of management and administration, but it must also interact with traditional organizations, and it must always deal with the challenges of present internal organization (The Convergence of IoE and Blockchain: Security Challenges, n.d.; Scalable Blockchain for Internet of Things (IOT), n.d.).

The incorporation of new technologies like IoE, cloud computing, and Blockchain has proved to be more valuable when it is incorporated rather than an individual technology. Computer society acknowledged the huge capacity of modernizing IoE with the integrated Blockchain technology with decentralized data storage structure and high computational power.

Because of Blockchain's having trustworthy data division service and security, the information is so consistent and can be unique from other services. Security plays a key role here, for security breaches can be detected at any point in time, and the data is left unchanged [22] because of the integrated technology.

During pandemic situations like COVID, these integrated and enhanced instruments in these fields would increase the feeling of safety among people, disburse information among sponsors, and reduce the time spent searching for the information.

Nowadays, Blockchain, along with IoE, is being used in other areas such as smart cars and smart cities for steadfast data communication while supporting the addon of new contributors to the network.

The integration of Blockchain with IoE has started to be established everywhere and Blockchain technology is recognized as the key to solving scalability, privacy, energy efficiency, handling queries, and trust problems related to the IoT and IoE paradigm.

6.3.14 *Blockchain and internet of entities technology distribution*

When compared to previous similar technologies, getting the computer science community, users, and academics to accept new and integrated technologies like IoE and Blockchain is a significant problem.

6.3.14.1 *Internet of entities technology spread*

Even though it is likely to produce a new network of devices that can function with the anticipated IoE paradigm, we can considerably reduce this problematic issue by fitting the IoE network into the pre-existing IoT, that is a wireless-based one like a mobile network with IoT facility.

Adopting this strategy would permit us to increase the IoE capability. Some of the simple adoptable strategies are as follows.

1. Manipulating non-complex and crystal-clear procedure of incorporation of the needed IoE functionalities in the pre-existing IoE tracker devices as a service. In this case, the entity like tracker can be implemented in smartphones, tablets, etc., with periodic and simple well-documented software upgrade process or by constructing accessible applications.
2. Distribution or spreading of technology awareness by making effective drives of information dispersals expected to highlight the rewards for each user who joins the Blockchain and IoE network. The new user can understand the added opportunity to exchange a valuable and huge amount of information among a large community of Blockchain users with lower computational power.
3. Generally, the integration with IoE and Blockchain benefits apart, the joining user to the IoE network can benefit from trackers, and the joined system can participate in the distributed-ledger registration tasks and detection of an entity. The integration benefits comprise of few free uses of service related to IoE network, such as remote data storage.

6.4 Discussion

There is no doubt that there will be more and more business giants and cyber organizations adopting Blockchain technology along with IoE integration in the near future. However, as we have discussed in this chapter, there are several challenges that are the reasons for the concerted adoption of this technology.

The critical issues associated with Blockchain and IoE include its scalability difficulties, interoperability lacunas, energy consumption overheads, lack of talent, and lack of standardization. An example of a Blockchain application called Bitcoin has a significantly slow processing time since it can carry out roughly less than 10 transactions per second, which makes Blockchain technology not feasible for large-scale real-time applications. Interoperability is the next big issue that is the reason behind big organizations and financial industries not approving this technology till now. Most countries blocked the Bitcoin concept because of no proper standardization in the technology. The blockchains work with the concept called silos which makes communication with other peer networks difficult, as they cannot send or receive information from another Blockchain-based system easily. Similarly, Blockchain technology uses the PoW mechanism to validate each transaction for data security. This PoW mechanism would work for solving multifaceted or compounded mathematical puzzles and needs a huge amount of computational power to process the data, verify it, and secure the entire network.

Along with the above-mentioned challenges, the demand for professionals with good knowledge of Blockchain technology is increasing each day. As of 2021, the statistics say, the global demand for Blockchain developers is above 600% over the preceding year, and the discovery of talent is very challenging. Though Blockchain developers are in high demand, there is a sharp shortage of experts in this field. Also, talking about standardization, there is no universal standard yet followed by Blockchain technology. This lack of standardization gives rise to issues like hassles of interoperability, high cost, tough mechanisms among new participants, etc. This acts as a hindrance for the budding new developers and the businessmen too. Apart from cost, security issues, data privacy, etc., there are other challenges mainly inhering in an inability to cause the implementation of blockchains to scale out.

However, the single biggest challenge is not technology-related because lots of smart people are working to solve those challenges, like interoperability, especially around standards, and around latency issues, etc. The Blockchain is probably the only interorganizational technology that needs to bring multiple experts of diverse domains to function together.

Now, talking about Blockchain in IoE, it is a known fact that Blockchain provides us with good solutions for many problems like the need for transparency, decentralization, integrity, immutability, and security without the need of any trusted third-party. There are still a few challenges remaining to be solved before realizing the complete potential of the Blockchain-IoE symbiotic existence for different applications.

6.5 Conclusion

Blockchain technology is a novelty, but it's quite complicated with many challenges in its implementation. Its integration into any domain requires the appropriate building blocks to converge to achieve the final desired objectives. It is quite promising to see Blockchain technology along with IoE result in desirable outcomes. The integration of these two technologies can be addressed with proper study and communication standards for faster adoption. The approval from appropriate regulators is the most important key factor for the inclusion of Blockchain and IoE as part of governmental communications also. This adoption would definitely speed up the collaboration between the common man, global businessmen, governments, and organizations.

On the whole, it can be concluded that Blockchain has become a revolutionary technology in the last decade. There is no doubt that many business organizations are on their way to adopting Blockchain technology. Several senior executives think that Blockchain technology will eventually achieve mainstream adoption. However, the adoption of Blockchain within IoE is still a challenge due to the several issues that were discussed in the chapter. If these issues are properly addressed, then Blockchain integrated with IoE will be a promising technology for progress to take its first tentative steps on the road of the common good and, thus, mass adoption to take shape as reality.

References

Agrawal, R., Paprzycki, M., & Gupta, N. (2020). Big data, IoT, and machine learning: Tools and applications.

Scalable Blockchain for Internet of Things (IOT). (n.d.). https://research.qut.edu.au/trusted-networks/projects/scalable-blockchain/.

The Convergence of IoE and Blockchain: Security Challenges. (n.d.). https://ieeexplore.ieee.org/abstract/document/8832287/metrics.

Uddin, M. A., Gondal, I., & Balasubramanian, V. (2021). Blockchain: Research and applications. In Blockchain: Research and applications *2*(2). https://doi.org/10.1016/j.bcra.2021.100006.

Further reading

A New Approach to Bridging the Blockchain Skills Gap. (n.d.). *IEEE innovation at work*. https://innovationatwork.ieee.org/a-new-approach-to-bridging-the-blockchain-skills-gap/.

Adaptive Security Framework for the blockchain on IOT. (2019). International Journal of Innovative Technology and Exploring Engineering.

Agarwal, S., Makkar, S., & Tran, D.-T. (2020). Privacy vulnerabilities and data security challenges in the IoT.

Blockchain for Internet of Energy management: Review, solutions, and challenges. (n.d.). https://www.researchgate.net/publication/338537673_Blockchain_for_Internet_of_Energy_management_Review_solutions_and_challenges.

Blockchain security Attacks, Challenges, and solutions for the future Distributed IoT Network. (2021). IEEE access.

Blockchain Technologies for the Internet of Things: Research Issues and challenges. (2019). IEEE Internet of Things Journal.

Blockchain Technology: A Step Towards Sustainable Development. (n.d.). International Journal of Innovative Technology and Exploring Engineering.

IEEE XPLORE. (2020). Blockchain-based trust management in social internet of things. https://doi.org/10.1109/DASC-PICom-CBDCom-CyberSciTech49142.2020.00024.

Internet of Everything (IoE): Security and Privacy Paradigm. (n.d.). https://www.routledge.com/Internet-of-Everything-IoE/book-series/CRCIOESPP.

Jabbar, R., Kharbeche, M., Al-Khalifa, K., Krichen, M., & Barkaoui, K. (2020). Blockchain for the internet of vehicles: A decentralized IoT solution for vehicles communication using ethereum. https://doi.org/10.3390/s20143928.

Panda, S. K., Elngar, A. A., Balas, V. E., & Kayed, M. (2020). Bitcoin and blockchain: History and current applications.

Ren, Y., Zhu, F., Sharma, P. K., Wang, T., Wang, J., Alfarraj, O., et al. (2020). Data query mechanism based on hash computing power of blockchain in internet of things. https://doi.org/10.3390/s20010207.

Reyna, A., Martín, C., Chen, J., Soler, E., & Díaz, M. (2018). On blockchain and its integration with IoT challenges and opportunities. *Future Generation Computer Systems*, *88*, 173–190. Available from https://doi.org/10.1016/j.future.2018.05.046.

Rogers, R., Apeh, E., & Richardson, C. J. (2017). Resilience of the internet of things (IoT) from an information assurance (IA) perspective. *IEEE Xplore*. Available from https://doi.org/10.1109/SKIMA0.2016.7916206.

Sharma, S. K., Bhushan, B., & Unhelkar, B. (2020). *Security and trust issues in internet of things: Blockchain to the rescue*. CRC Press, Published December 3, 2020.

The Blockchain Technology for Secure and Smart Applications across Industry Verticals. (2021). In N. K. S. Aggarwal, & P. Raj (Eds.), *Advances in computers* (1st ed.). Elsevier.

Blockchain-based security for internet of everything

7

Michel Rwibasira and R. Suchithra
Jain University, Bengaluru, Karnataka, India

7.1 Introduction

The Internet has revolutionized the way people share and receive information. Before discussing internet of everything (IoE), we will review the internet of things (IoT), which connects and monitors all physical devices via the Internet. Whenever a user wants to do something or implements a change to the processes performed by an object, they can communicate via cyberspace. Every change to the device sensor alerts its owner to take action. This is a one-way communication performed by IoT. IoT sensors gather information from the surrounding environment and send it to an actuator to communicate through the Internet to the owner of the device, who may take effective measures. For instance, when the fridge runs out of food items, the sensors identify that a given item is no longer accessible. This information is turned into digital form, compressed to fit the network bandwidth, and sent as an update alert to the owner.

All devices would communicate with other devices and the owner of the network using their unique identity, and this system of communication has been used for decades. In other words, it is one-way communication between a device and human beings. Unfortunately, the complete conveyor is available anytime, and we communicate with a device as well. Hence, there is a need for a two-way communication system via TCP/IP because we have to communicate with things, and they, with us. This whole communication cycle is the foundation of a network device.

Every network is now truly a TCP/IP network device. Around 99% of such networks can manufacture a wireless network card with the entire TCP/IP setup built-in with a format as small as a fingernail. The TCP/IP version 4 is now being widely used, and as we march into the future, version 6 will be available.

When we talk about devices, they can now not only send data, but they can also receive the data and do things to the data. Even though everything is dealt with by a TCP/IP network, it is essentially bidirectional communication. IoE records information, sends that information to the computer and the server; and then it is possible for the workstation or host to transmit it directly to the device in order to perform some action. Currently, we perform the system based on autonomy that moves the network completely based on programming without human involvement.

Blockchain-Based Systems for the Modern Energy Grid. DOI: https://doi.org/10.1016/B978-0-323-91850-3.00005-6

The Internet can now be used to control operations like the air conditioning of a home because the thermostat of the air conditioner can now take commands from sensors—with sensors sending and receiving commands from you, and each other—and convey information to a server where data is processed, and the sensors commanded to relay the processed command to the thermostat to switch on. If we look at a restaurant—for analogy—where you have entered to dine, the restaurant network comprises the frontend sensors—the waiters; the backend support staff—the cooks; and the controller—the manager; and each will work seamlessly to convert commands from you to a dish for your taste. Similarly, a sensor(waiter) takes the order from you and relays it to the thermostat (support staff) about what to do via the server's (controller) knowledge, who controls the processing of every command. So, with the world of things on the Internet, a thermostat on the wall is another thing attached to the IoT; just as your iPad is attached to the home IoT so that when you are about to return home, you can plug into the IoT with the iPad and convey the command to the sensor that you will be home in half an hour and want the air conditioner to be switched on in that time. However, with the IoE, matters are speeded up as your iPad/mobile phone attached to the IoE can be programmed to answer with "yard" to the car's queries about where you want to travel, in which interaction the cellphone in your coat pocket senses that you are in the car and tells the car where to go, the car identifies that you're on your journey back home and would instantly communicate the information to the heating, envitilation, air conditioning system at home, and the air conditioner tells the network to set the temperature that you prefer so that it happens automatically, and without you lifting a finger. A family—husband, wife, and kids—will have different temperature setting requirements so that even if you come home and the other members might still be away, your house might stay at the temperature you prefer. When your wife is on her way home, the information will be communicated to the air conditioning system to automatically adjust the temperature as soon as she comes home so that it is set to her preferred temperature.

So, we are no longer talking about a very simple, very basic, essentially one-way communication, but a dynamic two-way communications that makes just about any kind of thing possible. That is the kind of stuff that IoE powers. Why is IoE important for all of us? When we talk about the Internet, everything is all about servers, cellphones, and cars for the time being. But in the coming decades, everything will be operating through IoE. Mainly, now, Nike has started to produce shoes that use the IoE. And that is the reason why the TCP/IP version 6 protocols are so important. TCP/IP address version 4 can only find around 4.7 billion addresses. In the coming days, there will be 50 billion pieces of equipment. That is why there will be a mass migration to TCP/IP version 6, which offers a larger address pool to accommodate the numerous universe of objects. As a result, all devices connecting to the Internet will cooperate automatically through different IP addresses without the need for translating an IP address to a unique device ID using an interface. Given data sent to these devices will have commensurate Internet IP addresses, all these devices will have their own external IP address through the TCP/IP version 6 with its ability to handle a larger array of objects, which is important to IoE in the technology world.

7.2 A brief history of the internet of everything

The IoE was chosen as the leader among the new technologies of 2015 by Gartner. Various differences exist between IoE and IoT. To be clear, let's begin with the definition, practices, and details of IoE theory. IoE brings people, processes, information, and objects together. It creates a much larger network link than that which exists in operations. It adds capabilities, power, and expertise by providing an exceptional economic opportunity for businesses, individuals, and nations. In other words, the IoE is a practice of linking people, processes, data, and objects through the Internet. Besides, it details a self-sufficient world where each object connects to the Internet via the smallest device called a sensor. A sensor collects data from surrounding objects and sends a data signal in the form of system control to compress data in the digital format. Therefore, it uses the public network to send and receive data from/of the private network by applying proper standards and protocols.

Building blocks of the IoE:

People: Ties up people in collecting data for use in suitable ways.

Data: Reshaping data to arrive at a better solution.

Process: Transforming data to an authorized user at the right-exact time.

Things: A device connecting via the Internet for participating in smart decision-making.

IoT is a network of physical objects which are interconnected and operated by the Internet. All objects are designed in a way that they might interact with the inner and external functioning of the device. In other words, they are designed to give meaning and communicate by interacting with changing information to make a final decision.

7.2.1 The gap between internet of everything and internet of things

The IoE has four pillars: persons, processes, data, and objects built on the IoT with a single point. IoE helps businesses in improving their operations and gaining profits. Moreover, IoE changes people's lives by complementing the achievement of the IoT (Dave Evan, Chief futurist Cisco Consulting Service).

Next, the IoE will train techno-industries on the three pillars: business process, business model, and business time. "On the spot, this digital technology outlines our harvest, services, and processes that our customers have vital experiences with, as well as how we accomplish things in our organisation in collaboration with our partnership," said Hug Le Hong.

Whenever the organization follows a digital process in production, a new business approach for industries is formed. The IoE will construct millions and millions of devices with sensors that process real-time information. "Data is money," said Nick Jones, Vice President of Information Security at PRC in Longwood, Florida, United States. Therefore, it needs to have a big data and storage platform to gather all the information generated by millions of objects through the IoE that facilitates information analysis and storage. In addition, to change data into money, businesses and service providers need to decide if they have enough time and resources to invest in IoE. Often, they might not have the ability to make the right decisions as information, analysis, and storage require high processing power on the part of the organization. As a result, these days, digital technology is embedded in most objects used in daily life, and everything requires following a digital strategy.

7.3 The concept of internet of everything

Internet of everything is a broad term used to refer to Internet-connected sensors and smart products with enhanced digital capabilities. It is a theory that the destiny of technology is decided by many different sorts of products, devices, and products—all of which have access to the worldwide network.

7.3.1 People

Internet of everything plays a vital role in every network between people and devices via the Internet. The Internet is the best channel of communication whenever people need to exchange data between themselves or to any other device. IoE adopts the Internet as the communication channel for people and devices to exchange information or vice versa. IoE allows a device to communicate with a human being to make a decision based on their real-time needs. For example, when a prisoner wears a tracker on his body, a device can alert the prison guard through the Internet to verify if the rules and orders of the prison are being followed. The real-time data exchanges between the device and the security person help to arrive at the right decision of maintaining discipline in prison.

7.3.2 Process

The process is defined as all the necessary steps and procedures followed to deliver information at the right time. The IoE process uses small sensors as a tool for collecting data about its environment. The data is sent to an actuator in the form of electrical signals, and the actuator converts it into digital format. An actuator compresses digital information into a size suitable for the entire system as most electronic devices use digital data. The process steps are based on real-time data exchanged between people and a machine or a machine and a machine. The process helps deliver data to a dedicated recipient within the correct timeframe (Fig. 7.1).

A general phrase used to describe Internet-connected sensors and smart objects with improved digital capabilities is "the internet of things." People say that many

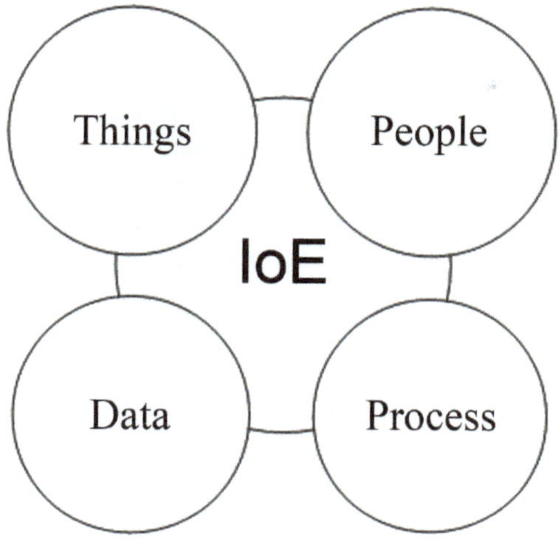

Figure 7.1 The concept of internet of everything.

Table 7.1 Connections of IoE Pillars.

People to People	Machine to People	Machine to Machine

different kinds of items, gadgets, and products that can connect to the global network will shape the future of technology (Table 7.1).

7.3.3 Thing

To understand the IoE, one must understand the idea of a thing. A thing is anything present in our surroundings, i.e., those that we use in our everyday life, such as machines, goods, and animals. In today's market, the security camera uses Wi-Fi to link up its operation of surveillance to other cameras via the Internet, and every camera on the Internet has a unique identification to recognize any other camera anywhere. IoE uses IPv6 addressing mode, which offers a larger pool of IP addresses to accommodate the vast array of objects on the Internet. These objects are controlled by the people via the Internet through sensors placed on the body of the objects that obtain information from our surroundings and forward data to an actuator, which compresses the data to form digital facts and, finally, sends that knowledge to the computer through the Internet.

7.3.4 Data

IoT devices collect "data" and send it over the Internet to a centralized point where it is analyzed and digested. Things connected to the Internet will grow more intelligent as their capabilities improve, synthesizing data into more relevant knowledge. Connected devices will soon communicate higher-level knowledge directly to devices, computers, and humans for further examination and decision-making, rather than merely reporting original data. In IoT, this transformation from data to knowledge is vital because it will help people to make faster, more informed decisions and better govern their surroundings.

7.4 Protection and isolation problems in internet of everything

Internet of everything is a new-age technology where a device performs an assigned task supported by an Internet connection. The security of information exchange between different objects across the Internet is of high importance. Each company takes up security and confidentiality measures for its data based on its area of operation and data values. Therefore, to date, there is no consistent approach available to define the security and privacy of the IoE.

7.4.1 User confidentiality and data shelter in the internet of everything

The confidentiality of user data is the most important aspect of the IoE. People, processes, things, and data are the four pillars of the IoE, while data gathering, data exchange, data storage, and data security are the IoE processes related to data. Additionally, as data is vulnerable to threats when transmitted via the Internet, user privacy for every object is an essential requirement of the IoE.

7.4.2 Authentication and identification

The IoE is a new-age technology that implements authentication and identification requirements of data to protect it from either an unauthorized person or a device. For instance, when two objects need to exchange information on the IoE communication channel, each device authorizes its counterpart to ensure that each other is the dedicated point. Each device also has the task of protecting its credentials.

7.4.3 Authorization and access control of blockchain-based security for internet of everything

Blockchain technology plays a vital role in access control and ensuring user privacy in the IoE. Incorporating the blockchain technology into IoE protects and verifies each node via a digital contractor. It secures data from attackers through the Internet. Chain storage decentralizes transparently to every node, infers every activity on a blockchain network, and updates any changes that might occur on a chain of information to all devices on the same network as a security method. In addition, all nodes work together based on a consensus algorithm instead of being a standalone object. This process protects distributed blockchain data from threats and unauthorized users to tamper on a secured blockchain network to a comparatively high level of security.

7.4.4 End-to-end protection

Cryptography helps secure the IoE data communication channel and data sharing between any object via the Internet. A security key, used in ciphering, is most valuable in data communication of the IoE through an Internet connection as it reduces data alteration through malicious attacks and upholds the security of all objects of IoE. Therefore, cryptography key pairs create a comprehensive network communication across the Internet.

7.4.5 Attack unyielding

The IoE is a set of different types of objects and people connected to create a bidirectional system via the Internet. In such a scenario, a device is prone to threats and attacks that cause damage to an entire system storing sensitive information. Data can be accessible to an unauthorized person who can cause malicious attacks like denial of services, man-in-the-middle, and phishing through the Internet.

7.5 Internet of everything based on blockchain with distributed ledgers

Because of the large size and wide diffusion of IoE connections, securing and protecting personal information on the IoE would be a difficult task. IoE makes it easier to monitor the exchange of information between people and devices via the Internet. It is a new technology used by companies and industries to exchange data through the Internet and make business decisions based on shared data. IoE uses sensors to detect and record information and forwards it to an actuator to compress data to be sent over the limited bandwidth of the Internet to a server to investigate its use and accuracy. For example, Toyota driverless cars use IoE by connecting it with a mobile phone to aid the exchange of data between the car, its owner, and the data service provider to achieve the following purposes:

- Monitor the spare parts of the car
- Trace tires that indicate low pressure
- Alert when it is time to service the car
- Point out faults with the functioning of the car lights

- Watch over the car battery life and alert when it runs low on charge
- Sense obstacles and traffic signs and signals while driving

In IoE, the idea of a thing is anything and everything in our surroundings, and these include all things used in daily life, such as machines, goods, and animals. An electrical gadget (electricity counter) connects to the Wi-Fi and calculates how much electricity has been consumed at home or a corporation. Every object communicating via IoE has a unique identity that makes it easy to recognize them anywhere. IoE works via a private IPv6 address, which offers a larger IP address pool. Objects using IoE are controlled at the initial state by sensors to gather information. The sensors then forward the information to an actuator that compresses it to the desired digital format compatible with a computer through the Internet. Adopting blockchain technology with distributed ledgers enhances IoE data security and prevents malicious attacks which cannot tamper with data, given data exchanged over the Internet is in the unencrypted form called plain text and can be tampered with otherwise.

Blockchain technology was first developed in 1991 by IT professionals as a sequence of blocks that contained classified information for making digital timestamps on documents. The blockchain is irreversible or tamper-proof but was not used widely. It was being used even though it was acquired by Santoshi Nakamoto in 2009 to establish bitcoins, the first electronic money. Whenever information has been stored inside this blockchain, it is difficult to manipulate it since the records have been encrypted using a secret key, which makes it almost impossible to replace.

Cryptography is the technique of exchanging a secret message between people on a network that no one else can read it apart from the sender and the receiver. It is a way of converting plain text into ciphertext and vice versa. Cryptography is used to secure messages between different receivers at different locations when communicating through untrusted channels. Cryptography uses keys and algorithms to encrypt data and also to decrypt them back into plain text. For example, supposing individual A wishes to communicate with individual B through the Internet. Both A and B are at different locations. Individual A's message is in plain text format that everyone can read. Individual A's message is then encrypted and converted to a ciphertext and sent to individual B, who applies decryption to read the message received. Such a process of encryption and decryption avoids access to the shared message through any communication channel as asymmetric keys are used.

Different keys are used in asymmetry key encryption: secret key and shared key are two different types of cryptography. Let's look at an example to better understand the distinction between these two keys. Individual A wants to send a message to the air conditioner in his home to set up a particular temperature. Individual A encrypts the message to be sent using his public key, and the air conditioner decrypts it using its private key. This means anyone should have the ability to cipher a document by using another shared key and decode that by using their own secret key to view the content in cleartext.

When a block is created within a blockchain, its hash is calculated. Each block in a chain contains data. As new data keeps getting added to each block, every biometric data block evolves, and it is no longer the same throughout the whole series of blocks. The same hashes of the preceding block are included in the current block, and indeed, the link that is generated as a result is trustworthy. For example, Let's say there are several blocks, each with a unique hash code upon it. The hash value of block 1 would be directed towards the hashes of piece 2, as the hashing of frame 2 is assigned to the ciphertext of unit 3. The block header is the first frame in a chain of blocks that starts with a pinpoint block and progresses through the chain.

The keys to IoE, as well as asymmetric cryptography, involve the most important information because, without these, no transaction is possible. When asymmetric cryptography is adopted in IoE to secure data through the Internet, nobody can read the transactions or decrypt them unless secret keys are used. In asymmetric cryptography, everybody knows your public key and algorithm but does not know the other's private key. The chance of guessing the private key is close to zero. That's the reason bitcoin uses this technology to make transactions.

For example, if Mike wants to send a document to Sutanu, he will send it by using his public key. The public key is not only used for encryption but also used as a safety lockbox, and everyone can deposit it in the safety box by using the public key. Even if no one ever knows your private key, that document will be safe in the network forever. Therefore, the public key acts as a lock.

Let's say that Mike wants to pay his fees to Jain University. In this case, Mike will use the private key and, with the help of cryptography, generate a hash that can be used to encrypt the message. In other words, Mike takes the transaction and encrypts it with the private key to create the hash. Remember that Mike's public key is at Jain University's account to decrypt the message. So, you may have different private keys to encrypt every message you want to exchange with others through the network and one public key to decrypt the message.

In fact, you can visit bitaddress.org, where you can create one private key-public key for yourself. This site will generate one public key to share with others on the network and keep generating private keys for every message to be transferred. Then, how are malicious acts and entry of unauthorized persons into the network stopped? Let's assume that Mike is a malicious node and needs to share information with Sutanu. Mike will protect data with Sutanu's shared key as well as use his private key, and the message will be sent to Sutanu as the destination. Unfortunately, Sutanu won't have the capacity to decipher the communication since it is encrypted because he has no access to Mike's public key into his node.

Asymmetric key cryptography and blockchain technology enables the use of smart contracts. The smart contract can be programmed to sell stuff and receiver funds when it meets the required conditions. For example, Mike is running a funded project through smart contracts. The contributors of the estimate will forward the digital currency through the crypto contracts. If such a project receives its full funding, the funds will be sent automatically to the proposal's founders' bank accounts. In the event that the initiative does not accomplish its objectives, all the money will be returned to the supporters. As the smart contracts are stored inside the blockchain technology, everything is completely and evenly distributed within the ledgers. With this technique, no one is controlling the money. When it comes to smart contractual arrangements, why else would consumers have faith in them? To secure transactions on the ledger, agreements on the ledger use an RSA, having intriguing features that are irreversible and spread is a great advantage. Being unbreakable implies that once something has been made, it cannot be modified; as a result, no one would go ahead and change it. Being distributed signifies that the contract-based ledger is legitimate for all network participants. In fact, tampering the smart contracts is almost impossible. Contracts would be used to safeguard operations such as the automated reimbursement of credit facilities, the processing of damage claims, and the execution of postal services.

How does this work? Let's cross-check the block. Each block contains data to be transferred, hashes of the block for security purposes on the new block, and hashes of the previous blocks. For example, Mike wants to send money to Peter. The bitcoin block contains the specifics of each transfer of funds, such as the fact that Mike paid money to Peter as well as the cash Mike sent. Hash values are also associated with a block of data. If you have a thumbprint, you may examine it for the hashes. Each block, as well as its constituent pieces, are still distinct, much like biometrics. Whenever a block is formed, its hash is determined, and modifying anything within the block could trigger the hash of any and all subsequent blocks to alter as a result of the change. Whether or not any signature within the block compromises, the biometrics within the block are no longer the same as they were previously.

The cipher of the preceding block serves as the third piece of the block structure. This basically forms a public blockchain, and the distributed ledger used for this is protected. Suppose I have three boxes with their own hashes, and each box has a connection to the hashes of the next box. The bridge from block I to block II is indeed the ciphertext of the next package, and so on until block III holds the reference from block II. There is a particular block at the beginning of the chain, which we refer to as an outset block. Consider the scenario in which you changed the hash of its first block. It does have an effect on the 2nd and 3rd blocks as well as the first.

On the contrary, any change on a single block causes the others to be invalidated, and they will no longer have the same hash as the previous blocks. But using a hash is not a reliable technique to protect against data tampering. Furthermore, in order to successfully modify a block, several of the hashes of all previous blocks must be computed in order to keep the ledger legitimate. As a result, in order to prevent hashes from being manipulated, computations are carried out by means of the application of a procedure known as power computation. In other words, it is a technique that delays the process of adding a new block to the chain. Estimates indicate that it will take around 10 minutes to compute the requisite proof-of-work and update the block of transactions. A block tampering attempt—that results in the computation which includes all verification for all other blocks, regardless of their order in the public ledger—is difficult because of this method. As a result, the creation of hashes and proof-of-work methods contributes to the trustworthiness of the ledger.

Blockchain security uses distributed ledgers instead of using the centralized management chain. Blockchain uses peer-to-peer networks, and every object is allowed to join. Then a new node joins the community, receives a complete copy of the data, and begins mining, which uses the mining method to check if everything is in order. If anyone generates a new element and distributes it to all of the community, it is known as a "broadcast," and each node verifies the block to make sure that no tampering has occurred when each node adds it to its blockchain. Furthermore, all of the stations in this cluster work together to develop a consensus protocol for validating a transaction. If it is not verified, that block is considered tampered with and will be rejected by others on the network. In addition, successful tampering requires modification of all blocks and acceptance of tampered blocks by everyone on the blockchain network—this is almost impossible to manipulate as the blockchain is adapted to control the denial-of-service attack of the peer-to-peer network connection.

Blockchain on the IoE protects information between devices when people connect on the same platform from different locations. Next, it increases user trust and fights malicious attacks on the IoE network. Blockchain technology is still leading in the security of everything that passes via the Internet. It has the most approached feature of storing data in different storage. Next, decades before adopting blockchain technology into various communication channels of data exchange through the Internet, data have been stored in a single node called a server. Once a server has any security issues, the entire system fails. In particular, if an attacker has the chance to compromise the central node on the IoE, like denial of the service, the whole system collapses. However, by applying blockchain with distributed ledgers, every object has its storage. Therefore, attacks may take place in fewer percentages, and the communication would still be possible due to the strong security of the blockchain and because it is based on maintaining decentralized ledgers.

7.6 Application of blockchain-based security for internet of everything

As technology advances, all data are digitized, and hence data security has become a critical consideration. Globally, data sharing via the Internet is the most popular means of exchanging information between people and a device or any other object that has a unique number as the IP address. Earlier, people used to exchange data via radio frequency signals and microwaves, and choosing the best mode would depend on the availability of supporting infrastructure, adequate data security measures, and reasonable costs that match the user capability.

Blockchain-based security for IoE facilitates the protection of people's data from malicious attacks via the Internet while exchanging information. Blockchain with distributed ledgers ensures data is secured, and all data available on the same network but stored in different locations diverge to an existing one that is centralized. Every object on the blockchain network has a uniform pattern of storage that is visible to all other nodes but invisible to outsiders. This method of storing data in multiple databases enhances data security and protects them from various threats via the Internet. The traditional approach of a centralized collection of data is still used in some sectors of our routine life.

1. Military operations and defense systems
   a. It is used in inland communication and air force operations and in radar and missile systems.
2. Banks and business industries
   a. It has been adopted in the banking processes of money transfer and withdrawal.
   b. It is used in shipping goods and tracking their locations.
3. Driverless cars and manufacturing technology
   a. At this point, it is used in driverless car technology to monitor and report every part of the car's movement to the right person.
4. Firefighting and disaster recovery
   a. Fire detectors monitor for any emergency situation and exchange data regarding it through the Internet with the person in charge.
5. Airspace and civil aviation
   a. The communication channels between the ground stations and flights are critical and sensitive. They need reliable upkeep as any slight error can be catastrophic.
6. Petrol and natural gas people who work in the petroleum and natural gas industries must be able to share real-time data with their objects in a safe manner through the Internet and have complete details in order to choose the best option to make the right decision.
7. Weather and land surveillance
   a. Internet is widely used by weather satellites and devices used for gathering data related to land surveillance.
8. Voting system
   a. Votes cast by citizens via a digital voting system are sensitive data that need security in every aspect through their transmission via the Internet.
9. Geology
   a. During the mining of natural resources, objects need to exchange information with operators via the Internet, and adopting a high-security channel assures confidentiality, integrity, and availability of data.
10. Medical services and health centers
    a. As medical technology grows, people do not need to consult their doctors in person. They can use monitored objects to gain access to medical care and advice. Such sensitive patient data needs complete protection.

7.7 Summary

The IoE technology shares data from different locations via a secure Internet connection to make the right decision. Many attackers take advantage of this shared data being transmitted over the Internet by exploiting those data that are not securely transmitted via the Internet. The adoption of blockchain-based security for the IoE provides many advantages when compared with other kinds of data security methods. Blockchain technology with distributed ledgers stores data in different locations, and thus every object online has its unique storage space. IoE data that are stored by following blockchain security protocols are confidential, assure integrity, and provide easy sharing of data between users and the devices they use via the Internet. Using blockchain technology, users are protecting system networks from a variety of security threats, including certain man-in-the-middle assaults and DoS attacks, not to mention hackers, SQL injection threats, and network phishing attacks (Alsaafin et al., 2020; Bera et al., 2021; Dorri et al., 2017; Douglas et al., 2020; Gagneja & Kiefer, 2020; Huang et al., 2018; Karthiban & Raj, 2019; Miaz & Ali, 2020; Mohanty, 2020; Priyadharshini & Canessane, 2021; Qiu et al., 2020; Saia, 2018; Saini et al., 2021; Thakker et al., 2020; Wei et al., 2019; What Is Blockchain Technology? A Step-by-Step Guide for Beginners, 2016; What Is Blockchain Technology? How Does It Work? Built In, 2021).

References

Alsaafin, A., Talib, M. A., Qasse, I. A., & Nasir, Q. (2020). Lightweight blockchain-based system for internet of things security. In *Conference Proceedings - IEEE SOUTHEASTCON* (Vol. 2). Institute of Electrical and Electronics Engineers Inc. Available from https://doi.org/10.1109/SoutheastCon44009.2020.9368272.

Bera, B., Das, A. K., Obaidat, M. S., Vijayakumar, P., Hsiao, K. F., & Park, Y. (2021). AI-enabled blockchain-based access control for malicious attacks detection and mitigation in IoE. *IEEE Consumer Electronics Magazine*, *10*(5), 82–92. Available from https://doi.org/10.1109/MCE.2020.3040541.

Dorri, A., Kanhere, S. S., Jurdak, R., & Gauravaram, P. (2017). Blockchain for IoT security and privacy: The case study of a smart home. In *2017 IEEE International Conference on Pervasive Computing and Communications Workshops, PerCom Workshops 2017* (pp. 618–623). Institute of Electrical and Electronics Engineers Inc. Available from https://doi.org/10.1109/PERCOMW.2017.7917634.

Douglas, A., Holloway, R., Lohr, J., Morgan, E., & Harfoush, K. (2020). Blockchains for constrained edge devices. *Blockchain: Research and Applications*, *1*(1–2), 100004. Available from https://doi.org/10.1016/j.bcra.2020.100004.

Gagneja, K., & Kiefer, R. (2020). Security protocol for internet of things (IoT): Blockchain-based implementation and analysis. In *2020 6th International Conference on Mobile and Secure Services*, MOBISECSERV 2020. Institute of Electrical and Electronics Engineers Inc. Available from https://doi.org/10.1109/MobiSecServ48690.2020.9042948

Huang, X., Xu, C., Wang, P., & Liu, H. (2018). LNSC: A security model for electric vehicle and charging pile management based on blockchain ecosystem. *IEEE Access*, *6*, 13565–13574. Available from https://doi.org/10.1109/ACCESS.2018.2812176.

Karthiban, M. K., & Raj, J. S. (2019). Big data analytics for developing secure internet of everything. *Journal of ISMAC*, *1*, 129–136.

Miaz, M. H., & Ali, M. (2020). Integration of blockchain and IoT: An enhanced security perspective. *Annals of Emerging Technologies in Computing*, *4*(4), 52–63. Available from https://doi.org/10.33166/AETiC.2020.04.006.

Mohanty, S. P. (2020). Security and privacy by design is key in the internet of everything (IoE) Era. *IEEE Consumer Electronics Magazine*, *9*(2), 4–5. Available from https://doi.org/10.1109/mce.2019.2954959.

Priyadharshini, K., & Canessane, R. A. (2021). Blockchain-based security algorithm on IoT framework for shielded communication in smart cities. In *Proceedings of the 3rd*

International Conference on Intelligent Communication Technologies and Virtual Mobile Networks, ICICV 2021 (pp. 320–327). Institute of Electrical and Electronics Engineers Inc. Available from https://doi.org/10.1109/ICICV50876.2021.9388497

Qiu, S., Wang, D., Xu, G., & Kumari, S. (2020). Practical and provably secure three-factor authentication protocol based on extended chaotic-maps for mobile lightweight devices. *IEEE Transactions on Dependable and Secure Computing*. Available from https://doi.org/10.1109/TDSC.2020.3022797.

Saia, R. (2018). Internet of entities (IoE): A blockchain-based distributed paradigm to security. *arXiv*. Available from https://arxiv.org.

Saini, S., Maithani, A., Dhiman, D., Rohilla, A., Chaube, N., & Bisht, A. (2021). Blockchain technology: A smart and efficient way for securing iot communication. In *Proceedings of 2021 2nd International Conference on Intelligent Engineering and Management, ICIEM 2021* (pp. 567–571). Institute of Electrical and Electronics Engineers Inc. Available from https://doi.org/10.1109/ICIEM51511.2021.9445292.

Thakker, J., Chang, I., & Park, Y. (2020). Secure data management in internet-of-things based on blockchain. In *Digest of Technical Papers - IEEE International Conference on Consumer Electronics* (Vols. 2020). Institute of Electrical and Electronics Engineers Inc. Available from https://doi.org/10.1109/ICCE46568.2020.9042998.

Wei, L., Wu, J., Long, C., & Lin, Y. B. (2019). The convergence of IoE and blockchain: Security challenges. *IT Professional*, *21*(5), 26–32. Available from https://doi.org/10.1109/MITP.2019.2923602.

What Is Blockchain Technology? (2016). A step-by-step guide for beginners.

What Is Blockchain Technology? (2021). How does it work? Built In.

Blockchain utility in renewable energy

8

Vani Rajasekar and K. Sathya
Department of Computer Science and Engineering, Kongu Engineering College, Erode, Tamil Nadu, India

8.1 Overview of renewable energy and internet of energy

Fossil fuels are a nonrenewable natural resource that has fueled our economy by providing electricity for industries and homes. Our dependency on electric energy grows as our civilization gets more digitally savvy and sophisticated. On the other hand, renewable energy uses dynamically regenerating but fluid sources, like water, to produce hydropower, for example (Afzal et al., 2020). These resources have an essentially endless storehouse and can be produced for an infinite duration though they still have a constrained power supply per given period. India's renewable energy business is the world's fourth most appealing renewable energy market. As of 2020, India has been rated fifth in wind energy, fifth in solar energy, and fourth in renewable energy generation capacity. Fossil fuels, often known as nonrenewable energy, are restricted in supply, and at present consumption rates, they are expected to run out during the early 22nd century. This shortfall, as well as the known environmental difficulties associated with carbon emissions which is an inevitable fallout of burning coal to produce thermal power, has prompted a greater focus on alternative sources of energy, particularly renewable energies such as wind and solar power. Thus, society has turned to IoE for transforming the energy sector in several ways during the last decade. Firstly, the IoE intends to create a conscience intelligent energy infrastructure for information processing and use, resulting in reduced energy waste. Secondly, IoE has revolutionized the production, distribution, and transmission system by combining sensors, the Web, smart grid, renewable sources of energy, AMI, and autonomous techniques that has raised the efficiency of the system and cut costs while improving the performance of the electricity generation and distribution infrastructure. Thirdly, electricity producers have brought electricity market paradigm shifts by helping distribution operators grappling with the rise of distributed generation units concomitant with problems such as power volatility, demand-supply mismatch, and so forth, to adapt to the idea of local energy markets to efficiently reward and dispatch shared resources; while the steady transition toward dispersed renewable resources meant that power generation is becoming more weather-dependent and that this random nature of power generation necessitates new demand-supply management approaches among providers. Thus, on a warm day, there would

Blockchain-Based Systems for the Modern Energy Grid. DOI: https://doi.org/10.1016/B978-0-323-91850-3.00010-X

be too much electricity produced, but there may be a shortage of energy on a cloudy day (Alladi et al., 2019). In such instances, the most important necessity is to maintain a balance between electricity generation and consumption. The regional electricity market is a small-scale subset of the electricity market, including renewable power trading among the channel's various utilities. However, the implementation of the local energy market necessitates a modern communication structure as well as a unique payment system. An intelligent planning and monitoring system is necessary to establish a coordinated, efficient, and sophisticated energy market that includes distributed energy resources, flexibility, and consumer visibility while using renewable sources of energy. This new technology will deliver a power system that is efficient, accessible, flexible, and autonomous. A smart grid system preserves stability by balancing electricity output and utilization among utilities. Microgrid ecosystems are commonly used to combine the energy infrastructure with distributed generation units.

Furthermore, electric vehicles, which are a key component of smart grid cyber-physical systems, have grown well in recent times as a result of their ability to provide a green, environmentally safe atmosphere. When electric vehicles are combined with wind and solar power plants, they provide superior storage and power stabilization services, which can help alleviate the fossil fuel problem (Apergis & Payne, 2012). However, the absence of charging facilities in the electricity sector limits the use of electric vehicles. The main characteristics of IoE are shown in Fig. 8.1.

8.1.1 Issues in internet of energy

Internet of energy technology could provide sophisticated communication among diverse electric utilities to create a single system. As the IoE market grows, so will

Figure 8.1 Characteristics of internet of energy.

the number of connections, communication instances, infrastructure complexity, and malware programs. As a result, there are numerous challenges in integrating various energy generation and use methods. Also, the primary purpose of this chapter is to take a look at the places where the combination of blockchain and IoE are strongest, where challenges are overcome in developing IoE from making a transition from centralized architectures to decentralized power markets.

8.1.1.1 Payment mechanism

A power grid spanning across nodes such as electric vehicles, microgrids, smart grids, and smart houses in an IoE smoothens the process of payment of power usage bills, which is then utilized for services such as load forecasting, flexible pricing prediction, and optimized energy usage planning. The price for electricity trading services in the power grid or decentralized energy supplies should be agreed upon between electricity companies engaged in the purchase process (Chen, 2018). Traditional payment options, such as debit cards, credit cards, and PayPal, require a trusted third-party bank or financier to exchange the user's banking details with merchants. Furthermore, exchanging bank details among nodes may raise privacy concerns and divulge electricity usage patterns, let alone pose a danger to confidential personal information, like name, email id, and location. Several solutions based on anecdotal payment mechanisms have been proposed to address the issue of online privacy loss. Although these solutions protect privacy, they are dependent on centralized structures and frequently need payment by a third-party auditor. The use of a pseudonym for every new transaction, on the other hand, can make it difficult to identify malevolent attackers.

8.1.1.2 Energy trading mechanism

In the IoT, peer-to-peer energy trading is critical but still poses some security and privacy concerns. Microgrids interconnect smart houses and streets, whereas current power infrastructure employs a big centralized grid that generates energy and transmits it over distances via transmission lines. Electric utilities, suppliers, and market agencies work together in transmission and distribution networks (Chitchyan & Murkin, 2018). Trading between smart buildings, residences with wind and solar power stations, electric cars and electrical grids to manage load peaks, hybrid vehicles and electric vehicles, renewable sources to grids, V2G and G2V are examples of P2P electricity market possibilities for IoE. Automated verification by third-party auditors such as brokers, traders, trading agencies, and banks is common in the current power trading sector, leading to high computational complexity and lower efficiency. As a result, several nodes with excess energy refuse to participate in the energy sale process for privacy reasons. Obviously, as the population grows, the IoE scenario will become more complicated. When a centralized organization fails under the weight of a tremendous workload, it might result in high operational costs. Furthermore, traditional energy supply companies provide utilities at set prices, making them less expensive.

8.1.1.3 Demand management

Since a huge quantity of energy is needed to recharge each electric car and renewable energies are unreliable, the transition to electric vehicles and renewable energy sources creates issues of resource scheduling and load variations for grids. Whenever the network nodes grow, so does the amount of transactional data, making substantial demand management more complex. Load management control is critical for the successful implementation of IoE technologies. Its goal is to produce an optimal load by balancing power usage activities (Ghosh et al., 2018). Some of the benefits of load management control include increased asset utilization efficiency, improved penetration and consistency for green energy supplies, bill savings, capacity expansion, volatility of stock reduction, administrative cost reduction, and controlling peak-hour loads. Energy consumption management, supply-side management, and changing economic response are used to describe predictive control management. The absence of an energy storage system is yet another reason to create a load management control system. In this setting, direct load management and price optimization are two extensively used demand response mechanisms.

8.1.1.4 Security threats

Cyber-attacks are one of the specific dangers to the IoE, as they jeopardize data's validity, availability, security, and integrity. Users and vendors are frequently concerned with the usefulness and affordability of IoE while ignoring cyber security. Because of the lack of privacy standards and procedures, the ever-increasing information flow obtained at edge devices is obviously at risk. As IoE technology advances, the number of nodes, buildings, and devices it connects to grows, increasing the attack vector for different cyber-attacks. Current security solutions that use public key encryption require verifiable other source authentication, posing a privacy risk. Furthermore, the usage of numerous communication techniques and communication systems exposes the power system to hackers by giving them access to personal details of electricity users like user identification and personal data. Malware, disseminated denial-of-service (DDoS), jamming attacks, hacking, and bogus data injection are all examples of cyber-attacks against the power system. Poor measurements and misleading data injection assaults during state estimation are other potential cyber-attack. Condition estimation is a technique for tracking and determining the current state of an energy system to ensure its long-term viability and efficiency. Dependency analysis is performed by the control center using the information from state predictions.

8.1.1.5 Pricing mechanism

Introducing dynamic pricing, in which power rates fluctuate dynamically over a period, is one technique to manage the real-time responses of users to market demand-supply conditions. The goal of dynamic pricing is to change the current cost of electricity based on market demand. This dynamically evolving pricing structure encourages the entire network to alter market forces as supply adjusts to demand in maintaining a market equilibrium in which the price is the tool in the hands of a conscientious electricity consumer. Thus, prices are low when there is little consumption and increased production, but they are high when there is high demand and limited production. It adjusts energy usage during peak hours, and customers may be rewarded for reducing peak-hour electricity usage. Customers are forced to adjust their usage of power to off-peak time as a result of high prices during peak hours. Smart meters can aid in this regard by evaluating the user's power consumption trends and sending him alerts of price hikes when usage has to be cut down and price cuts when usage can be accommodated. Genuine pricing, peak power rebates, and peak power pricing are examples of dynamic pricing schemes.

8.1.2 Blockchain preliminaries

The blockchain is a decentralized process that helps people to execute transactions without a central authority (e.g., instead of a bank looking after your paper securities, there are digital assets that have no actual assets backing them up but only a consensus in the network). It's a decentralized cryptocurrency that's duplicated across a network of peers, with each peer keeping a local record of transactions. Everybody can read that ledger given the public key (which is a hash pointer, a variable containing the address of the previous block, that also keeps a hash of the data stored in that address which makes the data in the chain read-only), but to write, the owner of a block needs the private key of the next block (that requires solving a hash function connecting the private key to the public key, but only the miner who finds a solution first according to what the consensus mechanism dictates will have found the private key) such that changing the record of valid transactions, even minutely, needs a mathematically astute mind to keep the change secret from others making the hacker's work not only impossible to do without informing all in the chain but also computationally difficult. People may use the same blockchain for a variety of purposes. For example, in the finance industry, a blockchain can be used as insurance in the shipping business. It can also be used to replace currencies or prepaid cards in digital money payments or to follow the distribution network (Goranovic et al., 2017). The strongest inventions in history have all been motivated by a desire to fill a void, and in the case of cryptocurrency, that void is one of trust. In economic interactions, the overall importance is linked to the necessity for two parties to trust one another. A third-party authority is necessary in the absence of a blockchain. In fact, by using a decentralized system, two parties can agree without the need for any intermediary, making commercial operations easier. Another difficulty we wish to avoid is centralization. Although the blockchain is naturally decentralized, there are instances where centralization emerges, resulting in a more vulnerable network. When centralization is achieved, an attacker will be able to exert control over the channel's status, selecting which payments are validated and which are dismissed. The basic properties of blockchain are shown in Fig. 8.2.

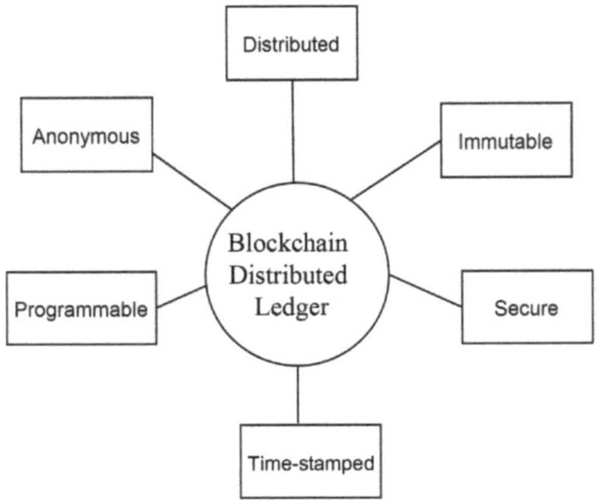

Figure 8.2 Properties of blockchain.

8.1.2.1 Block

A block made up of "data" and "header" is the most elementary component of the blockchain. It contains a portion of blocks which are consistent. The goal of hashing blocks is to store cryptographic keys—public key(a variable, called a hash pointer, pointing to the previous node and with a hash of the data stored in that node) and private key(this is the output of a hash function connecting the public key with the private key and with a consensus mechanism telling the miners what to look for)—in each new block's header that makes it difficult for a hacker to guess the private key of the block from the public key which only the miners, or nodes in the network, can solve by agreeing on all the information they can take as given about the blockchain and match to get a complete dataset to link with the previous block. Thus, we build a cryptographic chain. The enormous use of the cryptographic algorithm behind the consensus mechanism is what makes up the chain. Blocks are scrambled, and then each hashed string of digits is incorporated in the next block provided there is consensus, thereby maintaining the security of the network. The first block, known as a block header, has no preceding block, but all succeeding blocks have a progenitor.

8.1.2.2 Smart contract

Smart contracts, which are a part of business logic and can be included in the database, are computer programs that can be deployed by the blockchain and executed by the nodes when certain criteria or pre-defined terms are satisfied and are maintained on a blockchain. Smart contracts are often used to streamline the implementation of a contract so that all parties of the smart contract are sure of the output right away, without the need for any intermediaries or wastage of time. They can also automate a workflow, starting with the terms coded with "if/when...then..." conditions to make smart contracts run automatically when the conditions are met by the blockchain's nodes. When preset circumstances are satisfied and validated, the activities are carried out by a computer network. These activities could include transferring payments to the proper parties, licensing a vehicle, providing alerts, or delivering a ticket. When the contract has been concluded, the blockchain is modified. The advantages of a smart contract are

1. Speed, efficiency, and accuracy
2. Trust and transparency
3. Security
4. Savings.

8.1.2.3 Consensus mechanism

Blockchain solutions do not require a third-party organization that, contrary to the blockchain's network owners, holds power to sanctify a transaction depending on assets. Rather, blockchain uses cryptocurrencies that are not backed by physical assets but by digital assets and follows a decentralized consensus method to ensure the data and transactions are reliable and consistent. There are four basic consensus processes in extant blockchain systems. Firstly, the Proof-of-Work (PoW) technique is used by the two most common blockchain systems, Bitcoin and Ethereum (Imbault et al., 2017). The PoW technique is also used by Ethereum, and several other cryptocurrencies, including Peer Coin, Shadow Cash, and others, employ the PoS methodology. The PoW procedure is also referred to as mining, and miners are the nodes that tackle difficult mathematical riddles that necessitate a lot of computing power. Miners use a variety of mining technologies for resolving the transaction based on consensus, including CPU mining, Gpus mining, FPGA extraction, resource extraction pools, ASIC mining, etc. Secondly, the PoS, on the other hand, doesn't resort to resolving mathematical puzzles but determines who receives an opportunity to produce the next block through a randomized procedure. For the purpose of becoming a validator, blockchain individuals can start locking up their tokens for a set period of time. Users who have earned the status of validator will be able to create blocks. Validators could also be chosen based on the bitcoin's structure. Thirdly, individuals can stake their coins and choose a specific number of delegates in the Delegated Proof of Stake procedure. The importance of a participant's vote is determined by their stake based on, for example, his size of holdings. Lastly, responses to complex mathematical riddles are recorded in digital storage like hard disks using the Proof of Capacity approach. Proof of Time Elapses is the time process that chooses the creator of a new block randomly and fairly depending on the amount of time they have waited. A user's private key is compared to an authorized identity in Proof of Identity. Proof of Identification is a mathematical piece of evidence for a user's encryption key that is cryptographically connected to a specific transaction.

8.1.3 Blockchain in internet of energy

The three kinds of blockchain technology are permissioned, permissionless, and consortium. IoE is a possible use case for permissioned blockchain. The devices in the IoT continuously collect environmental data in terms of balancing peak demands and keeping the lights on. As a result, only minimal permissions must be allowed to certain sensor devices. As a result, several ideas for monitoring the effectiveness of blockchain in IoE have chosen authenticated and consortium type blockchains. Therefore, to develop a structure in the energy market that is free of disputes, legal laws and guidelines must be properly specified. All IoE nodes operate randomly without the need for a central authority, similar to a blockchain-based network, responsible for security and resilience versus attacks. Transactions are recorded in a decentralized manner on each node, ensuring autonomy from third-party authority. The elimination of third-party interference minimizes transaction expenses as well. Unlike a non-sequential information layout, however, blockchain technology allows for a sequential arrangement in which data elements are kept in a chain structure and grow in sequence. Furthermore, the connectivity between public blockchain members ensures that any energy utility may achieve peer-to-peer sharing of information, which facilitates automatic planning. Blockchain technology's basic principles, that is mining, linking, and creating keys, enable legal confirmation of the many sources of power transactions to happen between utilities. When a blockchain is used in the energy sector, delivery to clients of the data of electricity costs and other data may be automated. Self-participation is encouraged by blockchain, which enables transparency of power generation and distribution, as well as adaptable demand-side management systems and real-time price comparison (Mengelkamp et al., 2018). The transparency provided by blockchain technology opens up prospects for prosumers, allowing smart houses to not only produce but also trade energy generated from renewable resources. This approach also allows customers to be more flexible, resulting in a more customer-focused experience. Blockchain technology also ensures anonymity because users are identified by their cryptographic keys instead of their true names. The use of zero-knowledge evidence can also enable anonymity in the blockchain. Aside from the energy markets, the blockchain has applications in the smart grid meter removal process, as shown in Fig. 8.3.

8.1.3.1 Security management

Traditional credit and debit card-based systems necessitate the sharing of credit card details with a third-party, like a bank. The bitcoin revolution promotes decentralization and offers an alternative to dealing with third-party issues. All financial transactions are securely maintained on a global ledger with a cryptocurrency infrastructure. Because ledgers are accessible to all network members, each transaction's validity and consistency can be confirmed easily. Several blockchain-based project plans have been developed to address the challenge of secured amount exchange for energy transactions. The majority of them make use of cryptocurrency to make payment transactions. All payments to be exchanged are verified by the sender with the paying account's secret key. A new electronic cryptocurrency called Energycoin has been developed for power trading. NRGCoin and SolarCoin were built specifically for renewable power exchange in the smart grid system. The node

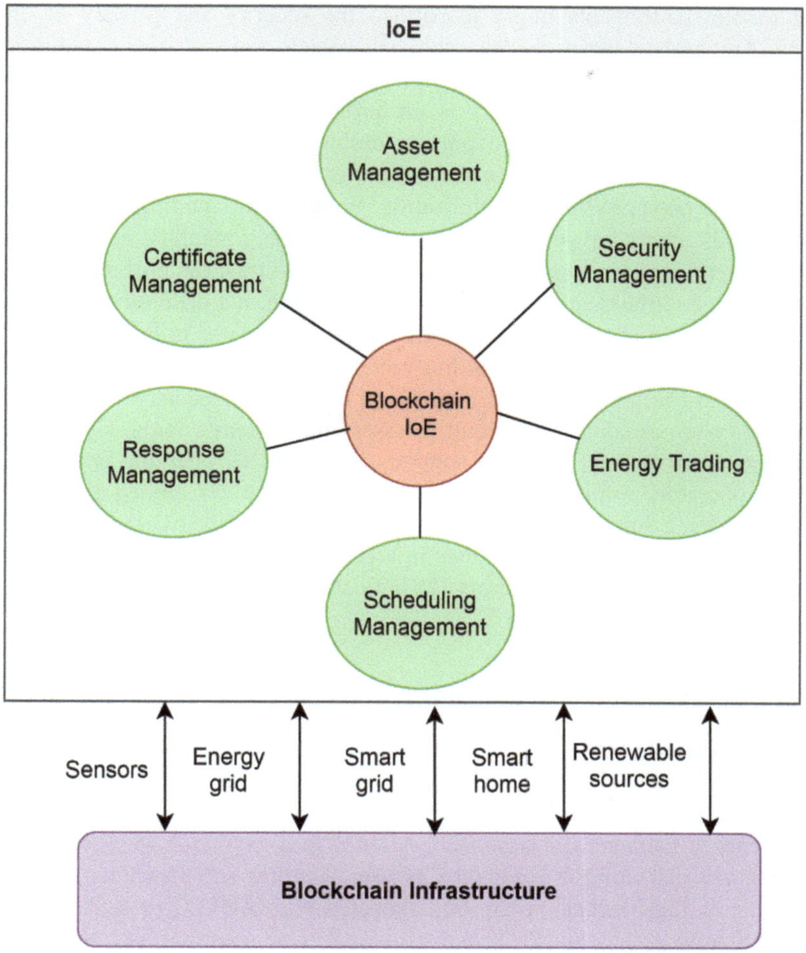

Figure 8.3 Integration of blockchain in internet of energy.

that provides renewable energy technology receives cryptocoins according to the amount of energy produced in NRGCoin and SolarCoin. Secure information sharing, privacy protection, audits, and dependable payments are all achieved through this technique. This plan, though, is not completely decentralized because it includes a provision for the services for auditing and recording the actual identity of peers.

8.1.3.2 Energy trading

Smart grids, microgrids, electric vehicles, payment centers, and authentication centers are all involved in charging or discharging processes. By transmitting bogus data, impersonating others, and charging exorbitant electricity charges, unscrupulous energy market participants might jeopardize the security and privacy of electric mobility. Automobile sensitive data, including vehicle and recharging station identities, charging and discharging times, and vehicle type, must be protected. Another danger for electric vehicle mobility is the promotion of bogus charging services. Three types of attackers may be present during the charging or discharging of electricity-powered vehicles. Large-scale electrical vehicles recharging and discharging can cause power system destabilization. Rather than charging electric vehicles from the grids, charging points give the vehicles the possibility of exchanging a depleted battery for a rechargeable one. It's difficult to trust the grid in terms of battery interchange fairness and respect in such a situation. It's difficult to persuade the grid to be truthful and equitable about the electricity transfer in this situation (Peck & Wagman, 2017). Knowledge regarding battery interchange battery brands, remaining energy, and charging current is maintained in a centralized server in the existing centralized system, making it difficult to assess the system's fairness. In a permissioned blockchain system, a private contract theory for EV charging is suggested to maximize operators' utility. To construct efficient contracts among providers and electric vehicles, a smart contract model is relevant. Specifically, the chosen electric motors and the grid can audit activities in a permissioned blockchain system, and a decentralized byzantine fault tolerant (DBFT) consensus process can be utilized to obtain a shared accord.

8.1.3.3 Scheduling management

The optimal advance scheduling of load strategy is suggested in an electrical power distribution system in order to reduce the power distribution in a system with distributed generation units, which was a major attempt to use the blockchain in the efficient energy management approach. ADMM is a technique for breaking down a complex optimal control issue into smaller, simpler sub-problems. During the fusion process, the blockchain can be connected with ADMM to enable all members of the network to validate progression results of solutions. As a result of its convergent characteristic, ADMM is suited for blockchain implementation.

8.1.3.4 Response management

According to, expected to witness considerable growth is anticipated to increase by more than 25% by 2040, resulting in a supply-demand imbalance. Demand-supply disparity is caused by a lack of finance, social implications, and the long time it takes to produce new electricity. Certain nodes could use more energy than others due to varied energy demands, whereas others encounter power shortages. As a result, it's critical to strike a balance between the generation of electricity and its consumption among various consumers. Response management tries to balance energy requirements with production by planning electric vehicles or dispersed power sources to keep off-peak periods as well as peak periods consistent. Despite delivering significant benefits, demand response techniques have a number of security and compatibility problems. Different protocols have been created by organizations such as NIST and CENELEC to address interoperability difficulties. Blockchain technology, for instance, can address privacy and security concerns in request-response mechanisms. Smart contracts are used to calculate incentives or consequences, identify imbalances, and validate the load management agreement, allowing for an autonomous and flexible system.

8.1.3.5 Certificate management

Blockchains have recently been utilized to construct green energy certifications. Solar panels enable prosumers to produce electricity at home and sell it to power companies. A green certificate verifies that energy was generated using renewable energy sources. The advanced metering system sends the physical record to the centralized agency rather than forwarding it to the centralized agency. Several marketplaces are installed in a decentralized environment to interchange energy and allow consumers and transmission operators to plan for market-based mobility. The certificate of standardization for renewable energy resources improves energy transparency and predictability and gives plant owners more choices. It provides renewable resources, identification, type, and location of energy generation. GaO allows for consistent pricing when selling and acquiring energy. Customers who are ready to profit from renewable energy providers can buy tokenized GaO on a simulated blockchain utilizing Ethereum and smart contracts, where tokens in GaOs are automatically sent among renewable energy providers and customers via blockchain. Commonly, other possible uses of blockchain are to provide protection for authenticated records, which are utilized in the provision of energy certificates to companies.

8.1.4 Energy networking

Due to the scarcity of fossil fuels and the resulting environmental challenges, people are beginning to create renewable energy from renewable sources like solar and wind energy (Rajasekar, Jayapaul et al., 2021). Furthermore, there will still be numerous issues to be resolved in the field of renewable power production. Since these energy supplies are frequently acquired and used in a dispersed rather than a centralized manner, management and trade efficiency are very often weak, and unethical trading is common. A new type of power grid system, the energy network, is used to regulate renewable energy production and optimize the electricity generation system, given it is dependent on power storage computers and the Internet. There are numerous issues in the energy system right now, including generation cost and various security issues induced by centralized administration. Furthermore, as many energy situations shift to a dispersed structure, the previous centralized management paradigm is obviously not possible. The introduction of blockchain technology offers answers to these challenges. The electric power Internet has grown significantly as a result of the advent of distributed ledger technology. It is feasible to create an energy Internet that is truly efficient. The administration efficiency of the new decentralized energy network has been substantially enhanced owing to the deployment of blockchain technology in the areas of management, trade, tracking, as well as other connections in the energy sector. However, as a large number of producers and consumers develop, how and where to manage them becomes a challenge we must address. For instance, when the number of participants in the electricity market grows, it becomes more challenging to keep their huge data. Many issues will arise as a result of storing vast volumes of data in a centrally managed organization. The volume of information will grow as the number of respondents grows, and the accompanying storage cost will rise as well. It will be impossible for other parties to obtain past data if all data is held in one organization, and openness will be tough to enforce. Furthermore, if central organizations are attacked or hacked, all data is at risk of being exposed or deleted. Voltage regulation, on the other hand, is critical in a decentralized power grid since it affects the system's operational stability. However, it appears that dealing with these issues will be challenging for the typical centralized administration technique.

8.1.5 Peer-to-Peer energy supply in grid

The purchases and sales of energy between two or more grid-connected entities are known as peer-to-peer energy trading. Any surplus energy, which is usually in the form of solar energy, could be exchanged and sold to other clients via a secure website. Consumers can choose who they buy energy from and whom they offer it to due to peer-to-peer commodities trading. Surplus solar energy is being transmitted to the grid for a minor feed-in tariff. Nevertheless, as more individuals desire flexibility and control over how their goods are dispersed, this approach is becoming outmoded. Energy is traded on a secure network, which frequently employs blockchain technology. Blockchain is a network technology for processing and storing data, including asset transactions. Renewable energy certificates, which may be exchanged through the blockchain's network, are one example of these assets. Anyone without solar cells can still buy renewable energy from their peers at a fair price, and those who sell their surplus power can sell it for a higher price than they would earn as a grid-connected tariff from their supplier. Energy transmission costs are reduced because energy would not have to be delivered from centrally situated power facilities. According to Genesis Energy, 41.1% of your electricity bill is spent on managing and measuring the poles and cables that carry power from producers to consumers' homes (Rajasekar et al., 2022). Renewable energy can be used to generate electricity, which offers a number of advantages to prosumers, as mentioned above. Power Ledger is a smart contract peer-to-peer sustainable energy trading platform established by an Australian technology startup. The website enables users with solar cells to exchange their excess solar energy with their neighbors in real-time, allowing them to buy and sell renewable-generated energy instantaneously.

8.1.6 Blockchain-based energy trading

Citizens are being pushed to adopt the new notion of peer-to-peer electricity markets as solar energy and energy storage systems become more prevalent. Consumers can gain from using excess power generated by someone using this approach. People can exchange power generated without the use of a middleman if they agree to the terms and conditions (Rajasekar, Premalatha et al., 2021). The transparent transaction benefits both the provider and the recipient. However, it is essential to find unified and efficient peer-to-peer energy trade utilizing the consortium blockchain technology. It avoids the use of a third-party mediator. For fast and efficient energy markets, this peer-to-peer commodities trading model proposes a safe, efficient resource payment approach. Electricity vendors are nodes that have excess energy to sell, whereas energy buyers are units that require energy. The electricity buyer's request will be submitted to a neighboring energy aggregator's network. The server will keep track of energy requests and broadcast them throughout the network in order to locate energy sellers. The energy aggregator works as a broker, in this case, arranging trade between the two nodes. Transmission lines or wireless transmission transfers are used to convey energy from energy suppliers to appropriate energy consumers. The energy coins are sent from the wallet account to the buyer node. This suggested paradigm could encourage the use of renewable energy storage infrastructure. A large-scale decentralized energy trade can be carried out using a trusted, secure approach, though with all participants worried about privacy. However, this collaborative blockchain network is a publicly available digital ledger that verifies and immutably records all energy and financial transactions, even recording the minutest change. All authorized nodes are responsible for maintaining a heterogeneous distributed database.

8.1.7 Green certificate and energy trading

Cars make travel more comfortable, but they also bring with them a slew of issues. Firstly, cars not only speed up the usage of fossil fuels, but they also produce a considerable amount of greenhouse gases, causing environmental issues. But, the widespread adoption of electric vehicles has lowered not only carbon dioxide emissions but also the users' gasoline costs. In addition, it encourages the advancement of renewable energy technology (Velliangiri et al., 2021). However, electric vehicles are not without their share of problems: the rise in the number of electric vehicles has attendant complications, including the installation of charging infrastructure, mode selection, and safety concerns during the charging process. The buildings where charging infrastructure will be installed are quite expensive if centralized administration is used. It's also impractical to expect a single department to oversee such a vast number of electric vehicles. As a result, individuals are exploring techniques to handle distributed EV charging, the majority of which are built on blockchain applications. However, using a distributed charging system necessitates taking into account pricing mechanisms as well as privacy protection during the power transaction process. The disclosure of sensitive private information, such as EV drivers' routes and preferred charging stations, will cause problems for their owners. Electric automobiles may be recharged in a variety of methods, including direct battery replacements at supercharger stations, without having to stop for significant periods of time. However, utilizing this strategy may result in a problem with transaction impartiality. The transaction's fairness will be affected by the different battery models and wear levels. Given this information would be housed on a centralized computer in a traditional centrally administered facility, it follows that fair trading would become more difficult to accommodate. Carbon emission trading is a commercial method for reducing global carbon emissions and encouraging the reduction of greenhouse gas emissions, given its main goal is to reduce the consequences of climate change (Svetec et al., 2019). Companies are assigned greenhouse gas emission quotas by government authorities because various companies have varying levels of ability to manage carbon emissions due to a variety of circumstances. Some businesses struggle to keep their carbon emissions under control, and some have plenty. Underneath the carbon emission exchange mechanism, organizations that do not fulfill the reduction of rated emissions can acquire carbon emission credits from various sources that can overcome the issue of non-fulfillment of objectives. The allotment of carbon emission quotas permits the identification of sources of real emission. The establishment of trade laws and pricing, including regulation, are all issues in the carbon emission trading community right now. Because of the enormous number of enterprises, certifying each energy producer's carbon emission quota would be a huge amount of work in the public sector. Thus, a green certificate, which is a digital certificate issued by a government body to businesses that generate power using renewable energy, can be sold by these businesses, which lightens the burden on the government. The use of blockchain technology in an economical operational system ensures that energy asset transactions are traceable. Producers and consumers can subsidize sustainable energy producers and promote the development of renewable energy by exchanging GoOs. GoOs is a type of green certificate as well.

8.1.8 Characteristics of blockchain in energy trading

Blockchain technology is mostly utilized in distributed management settings to make the preservation of users' data possible, as well as to improve data security and privacy. It can also be used to manage the grid's decentralized infrastructure. In peer-to-peer trade, blockchain can help price, control supply and demand, and assure accountability, security, and user privacy (Teufel et al., 2019). When it comes to EVs, blockchain is primarily employed for paying charging prices. The price can sometimes be determined through bidding and other times by a consensus mechanism that reacts to the market situation automatically. Blockchain technology is primarily employed in carbon emission pricing and green certificate exchange to provide accountability and data confidentiality during the trading system. Because it fits the needs of a decentralized power system with the following qualities, blockchain has widespread use in the energy industry. The characteristics of using blockchain in energy trading are given below:

1. Transparency

   The blockchain's data is synchronized across all nodes, allowing any node to access all transaction information, maintaining transaction transparency. This will enhance transparency in decentralized energy transactions and increase energy trading transparency.

2. Anonymity
   When customers engage in decentralized energy exchange, this feature can assist in preserving their privacy.
3. Decentralization
   This feature of blockchain complies with the criteria of a distributed energy system in terms of decentralized user management and data cloud processing, reducing management costs, and increasing data security.
4. Security
   The decentralized storage design of blockchain makes it harder to have a point of failure because each node might have a comprehensive copy of the entire database. As a result, an attacker cannot take control of the entire database by assaulting or managing a single node. As a result, network security is improved. When used with decentralized energy networks, it also improves the security of consumer information.
5. Democracy
   Through the use of consensus mechanisms in blockchains, transaction choices are more transparent. Because each participating node becomes a manager under the blockchain consensus method, there is no need to be concerned about corruption. Energy management becomes more democratic and effective as a result. Because of the various qualities of the blockchain, like secrecy, decentralization, and anonymity combined with the requirement of the distributed electricity market, its use in the energy field will grow more and more widespread, and the essential technologies will mature. However, there are several issues or hurdles that we must address when applying them to the energy sector.

   Issues such as government interference and legal oversight will arise as a result of adopting blockchain technology for the energy industry. There are more cases where blockchain technologies in the energy sector are important to examine, in addition to potential solutions to the above concerns. The right consensus mechanism should be chosen and improved to prevent high costs in the implementation of the consensus mechanism. Real-time information updates based on the blockchain will aid in the exchange of information between the different nodes and drastically reduce supply system wastage. There are numerous roles engaged in the process of transportation systems, and there are activities and contests with them in the field of power transportation.

8.1.9 Efficient blockchain model for energy utility

Utilities supply energy to customers based on their needs. Residents also generate energy, which they and other residents consume. The data of utilities and consumers is safely stored on a blockchain. Authorized users can analyze and access the data (Tsao & Thanh, 2021). The efficient blockchain framework for energy utility based on smart grid is shown in Fig. 8.4. Smart Grid covers power generation, distribution, and administration. Cloud-based multi-dimensional delivering services, systems, and blockchain-based distributed smart contract payment systems are being used. The main function of this entire framework is a rapid storage and verification infrastructure. Our main goal is to improve smart contracts so that dynamic price per unit calculations can be made. Energy users and producers are included in the SG area, and data is exchanged among SG and blockchain. The cloud is being used to store data indefinitely. Hyperledger's elements include smart contracts,

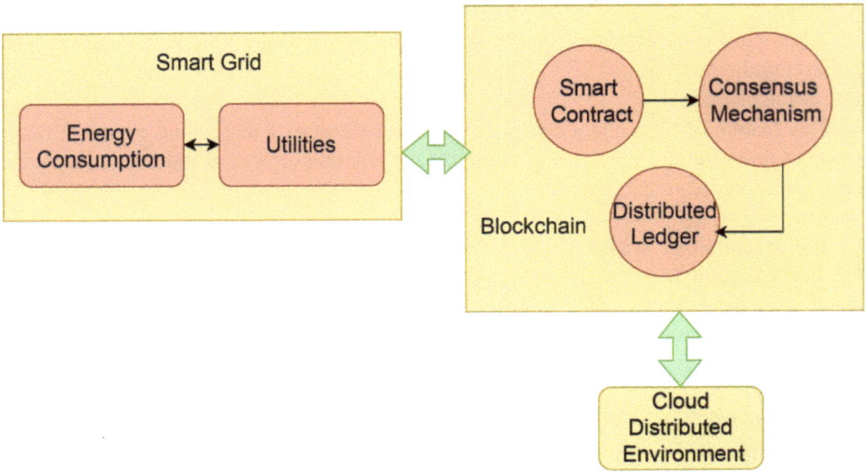

Figure 8.4 Smart grid energy utility model.

consensus, and a ledger (Wang & Su, 2020). A smart contract ensures that the transactions between the producer and the consumer are legal and that the transactional data is recorded in the ledger. The utility creates and distributes electricity to its customers. The bank is used to exchange money for coins. Every citizen is a consumer, and only a handful is both consumers and producers. Residents who are producers generate RER and sell it to other neighbors. When a household produces an extra unit of power, the information is updated. The total amount of electricity consumed changes every time. When a resident receives electricity in return for coins, the account of coins is also monitored and updated.

A single utility cannot meet the needs of all citizens. As a result, producers must sell any excess electricity to other users. If the price of power is set, and manufacturers can only offer this for the same price with which it was produced, producers will get greedy and store all of the electricity generated for them. If the producer chooses a high selling price for power in a bidding procedure, it becomes simple for the customer to purchase electricity from the utilities rather than from other residents at a high cost. Utility demand is increasing, yet producers are not driven to supply electricity. Hyperledger is a project that aims to enhance blockchain technology (Wu & Tran, 2018). In December 2015, the Linux Foundation established the platform. It is the result of collaboration between thirty foundations members, including professionals from finance, banking, IoT, supply chains, production, and innovation. Hyperledger Composer is a set of tools for constructing smart blockchain networks, including smart contract and blockchain-based creation, management, and testing. Participants/parties who are directly involved with the exchanges are interconnected on a hyperledger-based infrastructure. The ledger is modified and updated when electricity and money are exchanged. As a result, privacy and secrecy are preserved. Our challenge is solved using the Hyperledger Fabric framework. It offers assistance for distributed ledgers. It has elements that help with consensus and membership operations. While the participant's enrollment is validated by the membership facilities, the transaction's validity is ensured by the consensus mechanism. As a result, the participant's legitimacy is preserved, and security is provided. Concurrency control is used by Hyperledger Fabric to boost throughput by assistance in completing in parallel. Fabric's features make it a massively scalable platform, particularly for blockchain systems. The programmable logic that produces new regulations and facts that are recorded in the ledger is defined by the smart contract. Smart contracts are computer programs that give an output from pre-defined terms and conditions, or inputs, that are determined by both customers and producers. Those contracts outlined a corporate strategy. These business paradigms manage relationships between consumers and producers.

8.1.10 Benefits of using blockchain in the energy sector

The energy sector plays an important role in its being transformed by distributed ledger technology or blockchain. Innovations such as rooftop solar, electric cars, and smart metering have continually stimulated the energy market. With its smart contracts and network interoperability, the Enterprise Ethereum blockchain currently promotes itself as the next key technology to stimulate growth in the energy industry. By monitoring the chain of custody for system components, the distributed ledger technique offers a great ability to increase efficiencies for utility companies (Zhang et al., 2018). Blockchain provides special possibilities for renewable energy distribution in addition to provenance monitoring. Enterprise Ethereum solutions have the potential to improve traditional energy industries like oil and gas. Blockchain technology has the potential to improve complex systems with various actors. The following are the primary advantages of blockchain in the electricity sector:

1. Cost savings
2. Ecological sustainability
3. Higher visibility for stakeholders.

Consumers can benefit from improved efficiency and production over their energy sources thanks to blockchain technology. A safe and real-time updating of energy consumption data is also provided by an immutable ledger. Market prices, variable costs, energy law compliance, and fuel costs are all examples of different sorts of energy statistics. Electric power companies are vast, sophisticated businesses that generate electricity from power plants, solar farms, and other sources. In comparison to financial institutions or the banking business, utility companies do not interact with others. The capacity of a blockchain technology framework to trace and verify supply chains could prevent massive waste. Furthermore, a platform that makes use of intelligent legal contracts can save time, power, and money for all of the energy businesses involved. An immutable database can aid in the management and tracking of data required at all phases of energy production (Zhao et al., 2021). By ensuring that nothing will be forgotten or destroyed, the technology helps save time and money. The new, disruptive electrical network is evolving at the same time as blockchain-based networks, and the two technologies have symbiotic relationships: blockchain may be a significant facilitator of the electricity network, allowing distributed electricity generation and an operating network having a local decentralized system. The combination of these two technologies might open up a slew of profitable prospects for companies working to solve these energy problems and increase renewable energy consumption.

8.1.11 Chapter summary

Blockchain could provide a secure, low-cost mechanism to record and confirm operating and financial transactions along a decentralized system with no central authority. Sensors and modern communications are used by customers and suppliers to collect the required data and forecast demand and supply. The present energy sector began changing towards distributed and decentralized approaches with the utilization of smart grid energy utilities like electric vehicles, smart grid, and vehicle-to-grid technologies. The capacity of a blockchain technology framework to trace and verify supply chains could prevent massive waste. Furthermore, a platform that makes use of intelligent legal contracts can save time, energy, and money for all of the energy businesses involved. An immutable database can aid in the management and tracking of data required at all phases of energy production. The combination of these two technologies might open up a slew of profitable prospects for companies working to solve these energy problems and increase renewable energy consumption.

References

Afzal, M., Huang, Q., Amin, W., Umer, K., Raza, A., & Naeem, M. (2020). Blockchain enabled distributed demand side management in community energy system with smart homes. *IEEE Access*, *8*, 37428–37439. Available from https://doi.org/10.1109/ACCESS.2020.2975233.

Alladi, T., Chamola, V., Rodrigues, J. J. P. C., & Kozlov, S. A. (2019). Blockchain in smart grids: A review on different use cases. *Sensors*, *19*(22), 4862. Available from https://doi.org/10.3390/s19224862.

Apergis, N., & Payne, J. E. (2012). Renewable and non-renewable energy consumption-growth nexus: Evidence from a panel error correction model. *Energy Economics*, *34*(3), 733–738. Available from https://doi.org/10.1016/j.eneco.2011.04.007.

Chen, D. (2018). Utility of the blockchain for climate mitigation. *The Journal of the British Blockchain Association*, 1–9. Available from https://doi.org/10.31585/jbba-1-1-(6)2018.

Chitchyan, R., & Murkin, J. (2018). Review of blockchain technology and its expectations: Case of the energy sector.

Ghosh, A., Aggarwal, V., & Wan, H. (2018). Exchange of renewable energy among prosumers using blockchain with dynamic pricing.

Goranovic, A., Meisel, M., Fotiadis, L., Wilker, S., Treytl, A., & Sauter, T. (2017). Blockchain applications in microgrids: An overview of current projects and concepts. In *Proceedings of the fourty-third annual conference of the IEEE industrial electronics society, IECON 2017* - (Vols. 2017, pp. 6153–6158). Institute of Electrical and Electronics Engineers Inc. <https://doi.org/10.1109/IECON.2017.8217069>.

Imbault, F., Swiatek, M., De Beaufort, R., & Plana, R. (2017). The green blockchain: Managing decentralized energy production and consumption. In Conference Proceedings - 2017 seventeenth IEEE international conference on environment and electrical engineering and 2017 first IEEE industrial and commercial power systems Europe, EEEIC/I and CPS Europe 2017. Institute of Electrical and Electronics Engineers Inc. <https://doi.org/10.1109/EEEIC.2017.7977613>.

Mengelkamp, E., Notheisen, B., Beer, C., Dauer, D., & Weinhardt, C. (2018). *A blockchain-based smart grid: Towards sustainable local energy markets. Computer science - Research and development* (33, pp. 207–214). Springer Verlag. Issues 1–2. Available from https://doi.org/10.1007/s00450-017-0360-9.

Peck, M. E., & Wagman, D. (2017). Energy trading for fun and profit buy your neighbor's rooftop solar power or sell your own-it'll all be on a blockchain. *IEEE Spectrum, 54* (10), 56–61. Available from https://doi.org/10.1109/MSPEC.2017.8048842.

Rajasekar, V., Jayapaul, P., Krishnamoorthi, S., & Saračević, M. (2021). Secure remote user authentication scheme on health care, iot and cloud applications: A multilayer systematic survey. *Acta Polytechnica Hungarica, 18*(3), 87–106. Available from https://doi.org/10.12700/APH.18.3.2021.3.5.

Rajasekar, V., Premalatha, J., & Saracevic, M. (2022). *Cybersecurity in 5G and IoT Networks. EAI/springer innovations in communication and computing* (pp. 29–46). Springer Science and Business Media Deutschland GmbH. Available from https://doi.org/10.1007/978-3-030-79766-9_3.

Rajasekar, V., Premalatha, J., & Sathya, K. (2021). Cancelable Iris template for secure authentication based on random projection and double random phase encoding. *Peer-to-Peer Networking and Applications, 14*(2), 747–762. Available from https://doi.org/10.1007/s12083-020-01046-6.

Velliangiri, V., Manoharn, R., Ramachandran, S., & Rajasekar, V. R. (2021). Blockchain based privacy preserving framework for emerging 6G wireless communications. *IEEE Transactions on Industrial Informatics*. Available from https://doi.org/10.1109/TII.2021.3107556.

Svetec, E., Nad, L., Pasicko, R., & Pavlin, B. (2019). Blockchain application in renewable energy microgrids: An overview of existing technology towards creating climate-resilient and energy independent communities. In Proceedings of the international conference on the European energy market, EEM (Vols. 2019). IEEE Computer Society. <https://doi.org/10.1109/EEM.2019.8916292>.

Teufel, B., Sentic, A., & Barmet, M. (2019). Blockchain energy: Blockchain in future energy systems. *Journal of Electronic Science and Technology, 17*(4), 100011. Available from https://doi.org/10.1016/j.jnlest.2020.100011.

Tsao, Y. C., & Thanh, V. V. (2021). Toward blockchain-based renewable energy microgrid design considering default risk and demand uncertainty. *Renewable Energy, 163*, 870–881. Available from https://doi.org/10.1016/j.renene.2020.09.016.

Wang, Q., & Su, M. (2020). Integrating blockchain technology into the energy sector—From theory of blockchain to research and application of energy blockchain. *Computer Science Review, 37*, 100275. Available from https://doi.org/10.1016/j.cosrev.2020.100275.

Wu, J., & Tran, N. (2018). Application of blockchain technology in sustainable energy systems: An overview. *Sustainability, 10*(9), 3067. Available from https://doi.org/10.3390/su10093067.

Zhang, T., Pota, H., Chu, C. C., & Gadh, R. (2018). Real-time renewable energy incentive system for electric vehicles using prioritization and cryptocurrency. *Applied Energy*, *226*, 582–594. Available from https://doi.org/10.1016/j.apenergy.2018.06.025.

Zhao, J., He, C., Peng, C., & Zhang, X. (2021). Blockchain for effective renewable energy management in the intelligent transportation system. *Journal of Interconnection Networks*, 2141009. Available from https://doi.org/10.1142/s0219265921410097.

Impact of Blockchain-IoE on economy

9

S. Malathy[1], C.N. Vanitha[1], Rajesh Kumar Dhanaraj[2] and C. Kotteswari[3]
[1]Department of Computer Science and Engineering, Kongu Engineering College, Erode, Tamil Nadu, India, [2]Computer Science, and Engineering, School of Computing Science & Engineering, Galgotias University, Greater Noida, Uttar Pradesh, India, [3]Department of Computer Science and Engineering, Velalar College of Engineering and Technology, Erode, Tamil Nadu, India

9.1 Introduction to blockchain in supply chain management

Blockchain is a tamper-proof distributed ledger system (DLS) that is immutable and traceable. It is used in a common, synchronized platform where the entire transactions are approved by users. It provides a decentralized platform in which all network users can securely interact without requiring the assistance of a trusted authority. As a result, it eradicates the necessity of a centralized organization by validating and maintaining all transactions with a distributed consensus (Shala et al., 2019).

9.1.1 *Workflow of a blockchain*

Blockchain is made up of a sequence of interconnected blocks that can be easily tracked back to former blocks, making the technology crystal clear and reliable. Every block possesses its own unique identifier and the previous block's hash, which ensures transaction security. Users in that network approve and record all transactions; they are also time-stamped and put in a certain order (either in alphabetical order or in increasing numerical order) linked to the block before them and permanent after they have been put into the network. The overall blockchain structure classifies it as a trustworthy technology, and its workflow is listed in Fig. 9.1.

The so-called "consensus mechanism" is one of the most significant features that make the blockchain trustworthy, safe, and transparent. Records are stored in blocks that are connected by hash values, which has a significant impact on adding a new block to the system based on consensus. Any change to a current block requires approval from the whole set of users to build a lengthy branch, which aids DLT in maintaining historical data protection through collaboration mechanisms.

Blockchain-Based Systems for the Modern Energy Grid. DOI: https://doi.org/10.1016/B978-0-323-91850-3.00007-X

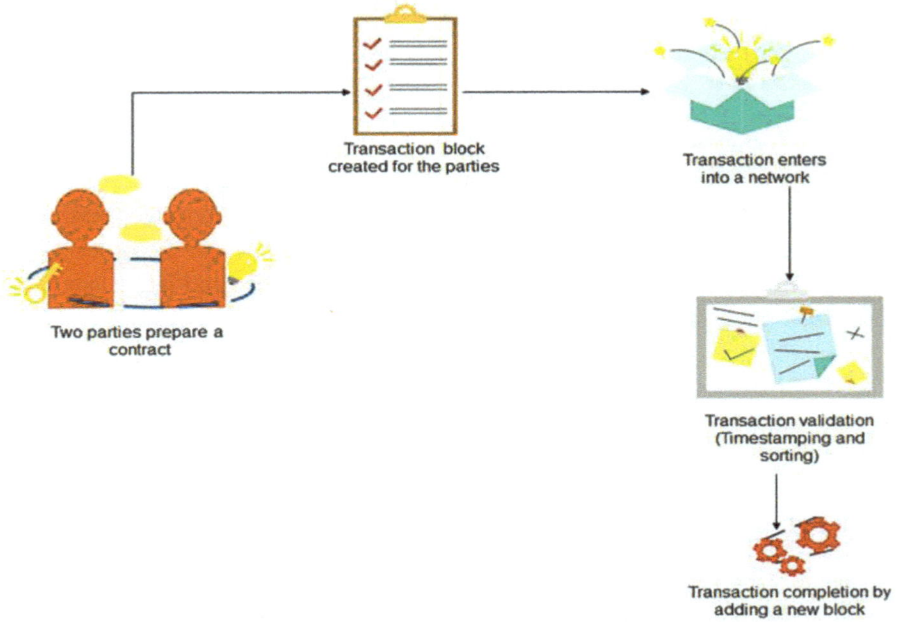

Figure 9.1 Workflow of blockchain.

9.1.2 Features of blockchain

Some of the characteristics of the blockchain (Dutta, Choi, Somani, & Butala, 2020) that make it distinct and interesting in the future for industrial purposes are listed in Fig. 9.2.

1. With network consensus, data is stored and saved in the network. The users can view the data and trace the data throughout the data having come into existence.
2. The system's data may be viewed, evaluated, stored, and updated across other systems, making it decentralized.
3. Blockchain provides timestamps and controls to ensure immutability of history.
4. The transactions on the blockchain are irreversible in nature. Each blockchain keeps a precise and verifiable record of every transaction ever made.
5. Each node existing in the blockchain may safely store and update data without the involvement of a third party, making it autonomous.
6. Everyone in the network has open-source access to the blockchain.
7. Every document transferred on the blockchain has a unique hash code that records who owns it.
8. Every product in the blockchain has a digital record document that proves its legitimacy and provenance.
9. Contract automation.
10. It is a small automated application that assists with contract execution.
11. It eliminates the requirement for a traditional contract while improving data protection and minimizing the cost of each transaction.

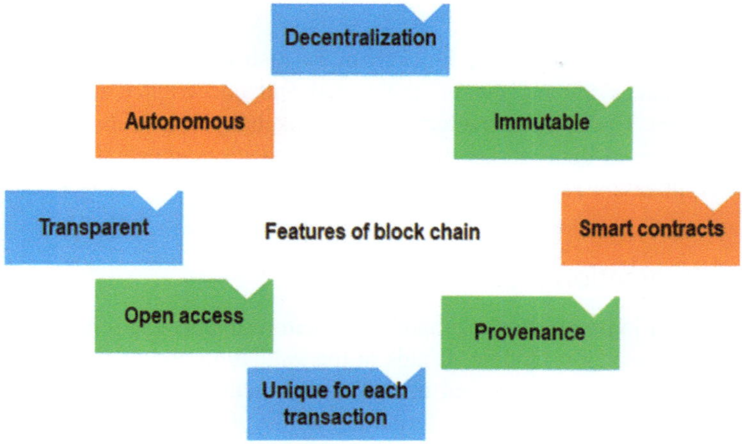

Figure 9.2 Features of blockchain.

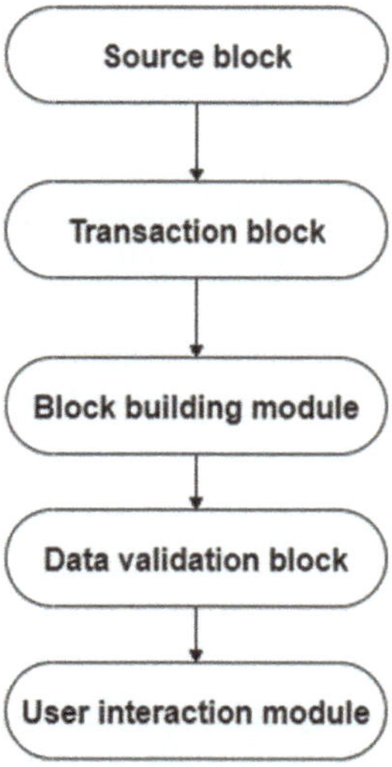

Figure 9.3 The architectural design of blockchain.

12. Smart contracts are typically designed with rules, fines, and actions that are applicable to all parties participating in the transaction.
13. Smart contracting (SC) allows for swift response operations in the event of supply chains.

9.1.3 The architectural design of blockchain

There are five essential blocks present in the architectural design of the blockchain, which are illustrated in Fig. 9.3.

9.1.3.1 Source block

This module aids in creating the blockchain by distributing and sharing the databases. It guarantees the process of unchanged and uncorrupted data retrieval by blockchain users. The main characteristics of blockchain include data unchangeability, timestamp-based data storage in any form, and a common information ledger via a data Application Programming Interface.

9.1.3.2 Transaction block

The "transaction journey on blockchain" is monitored, managed, enabled, and supported by the transaction module. It aids in the validation and facilitation of blockchain additions. Data is exchanged using SC transaction gates. The blockchain is responsible for the movement of information across the SC, as well as the shared visibility of transactions. Transactions are grouped and transmitted in the form of a block to each node. It's worth noting that in blockchains, transactions are nearly impossible to undo or reverse once they have been completed.

9.1.3.3 Block building module

Blocks can be thought of as various data arrangements that miners have generated. They store message and transaction information that is duplicated among the network's nodes. The block creation module allows you to add fresh blocks to an existing blockchain by giving the previous block's hash values and linking it with the new block. Transaction sequences are recorded in alphabetical order, and blocks containing false transactions may be easily discovered and traced.

9.1.3.4 Data validation block

Proof of work and state techniques are utilized to check and validate all transactions to avoid data corruption. The distributed network's data consistency is ensured by cautiously developed consensus algorithms. Distributed consensus aids in the confirmation of transaction authenticity as well as the construction of links between blocks present in the blockchain network.

9.1.3.5 User interaction module

It keeps track of transactions and aids in the provision of real-time information on intelligent and smart contracts. This module incorporates all of the blockchain applications with its data collection and validation platforms, algorithm, and procedures. Depending on the application, numerous shared ledger platforms are available with consensus algorithms that could be made accessible in the market, regardless of whether the blockchain is unrestricted, restricted, available after obtaining permission, or nonpermittable.

9.1.4 Supply chain management and blockchain

Supply chain management (SCM) is a massive industry that serves as the backbone of all others. Traditional SC systems, on the other hand, are not extensible or transparent enough to meet future needs and means of the economy, resulting in significant impacts in terms of bugs and error fixing, expenses, governance, and fake data management (Casado-Vara et al., 2018). The significant advantages of using blockchain in SCM (Azzi et al., 2019) are illustrated in Fig. 9.4.

9.1.4.1 Data collection and administration

1. Blockchain allows for the calibration of data from various SC.
2. It enhances the safety of data stored.
3. All information is captured in real-time.

9.1.4.2 Improved transparency

1. Aids in the tracking of an item state.
2. Content analysis tasks are automated.
3. End-to-end transparency via a hierarchy of permission levels.

9.1.5 Improved response speed

1. Creates a dynamic, real-time supply chain that makes greater use of its resources.

9.1.6 Management of smart contracts

1. Individualized and customized contracts are developed for each and every block, which in turn should be coordinated with one another, and have also paved the way for effective design of business processes.
2. Enhances visibility and eliminates the need for a middleman.

Figure 9.4 Advantages of using blockchain in supply chain management.

9.1.7 Increased efficiency in operations

1. Improves the speed of the SC process from start to finish.
2. To make the process more robust, it identifies flaws and issues early on.

9.1.8 Disintermediation

1. This results in a continuous flow of transactions.
2. Integration of blockchain with SCM boosts the speed of the process.

9.1.9 Immutability

1. Builds trust among process stakeholders.
2. For all modifications, there is a consensus mechanism.
3. Ensures that all transactions are secure.

As a result, deploying blockchain in the supply chain is not only cost-efficient, given it lowers costs of operations but also enhances relationships between all the stakeholders. It helps to build trust and streamline corporate processes. There have been successful blockchain implementations with SC in the wood and transportation industries. Blockchain has been deployed in the wood SC to enable end-to-end traceability from the time the wood is chopped to the time it is transformed into usable commodities. Blockchain incorporation changes the design of the supply chain by facilitating cooperation among all stakeholders built on the usage of digitization and intelligent contracts (Sandner et al., 2020).

9.2 Utilization of blockchain in smart contracting functionalities

Fig. 9.5 illustrates some of the primary services that can be modified with blockchain technology that include supply chain reengineering, safety, flexibility, data origin, process, blockchain, and product management.

9.2.1 Provenance

The use of traceability restrictions on the Ethereum blockchain helps in determining the unchangeability of physical items that are created and delivered through complicated, local, or international organizations prevailing in the supply chain (Longo et al., 2019). SC provenance is ensured by providing proper certificates, tracing and verifying rules of product details, as well as a guarantee of origin, data authentication, and integrity across the whole supply chain.

For diamond supply chains, it has been the fundamental role of the blockchain. The "Hyperledger Sawtooth" project, in particular, has used blockchain and IoT technology. A blockchain integrated SC provenance framework has been developed

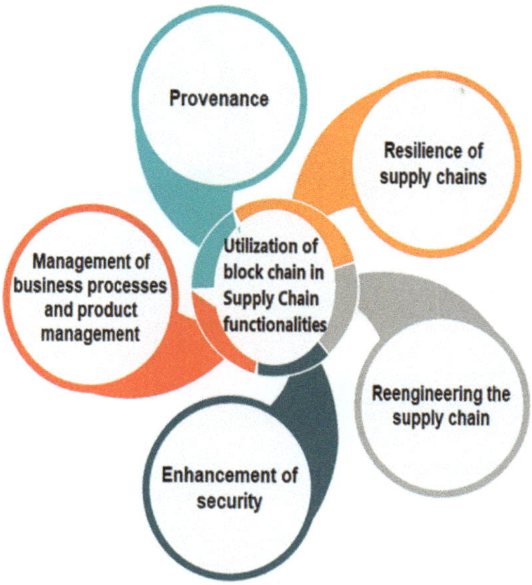

Figure 9.5 Utilization of blockchain in smart contracting functionalities.

for storing and retrieving all the essential information and using secure encryption techniques for data protection.

9.2.2 The resilience of supply chains

SC resilience is enabled by blockchain technology, which reduces the impact of interruptions, employs an obstructive and retroactive risk management approach, and provides protection on the multiple layers of the supply chain network (Chen et al., n.d.). Blockchain's structure aids in capturing both network and organizational risks related to any supply chain.

9.2.3 Reengineering the supply chain

Through various approaches such as data privacy and traceability, blockchains improve the visibleness of the supply chain and the transparency of transactions, where the automation process enables real-time tracking without the interference of the middleman, all of which are cornerstones of SC reengineering. In all areas of business, a well-reengineered SC may accomplish synchronization of tracking information. Furthermore, the use of smart contracts minimizes the money spent and the time required for the supply chain reengineering process.

9.2.4 Enhancement of security

Substantiation, secrecy, access code control, information and resource provenance, service integrity, and privacy guarantee are all possible with the blockchain. It also

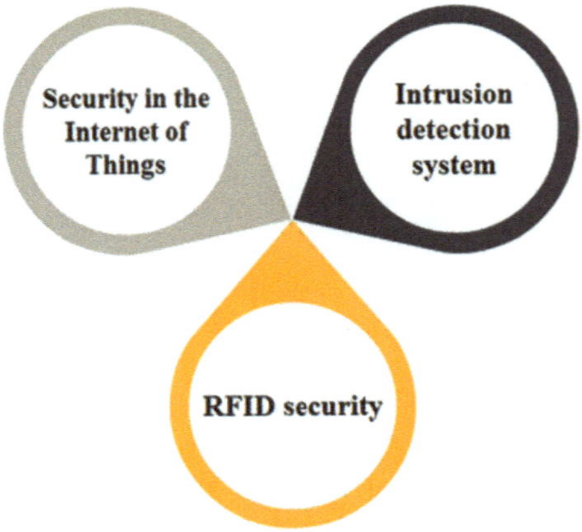

Figure 9.6 Enhancement of security.

allows for the creation of a risk analysis platform to investigate the relationship between business, content, and technology, as well as the gathering of an analytics viewpoint on digitization in SC.

Due to its capacity to scale up as needed, a blockchain can provide comprehensive risk management even in a large production organization. Due to its superior cybersecurity and performance, blockchain is far more secure than conventional IoT systems or data protection services. When blockchain technology is combined with the following systems, it can result in a more secure SC, which is shown in Fig. 9.6.

9.2.5 Security in the internet of things

1. The typical IoT system is a centralized digital integration network.
2. By offering an agreement method for ever-changing data storage, guaranteed end-to-end transfer of data, and furnishing product assessment and visibility, blockchain technology guarantees the advanced safety of IoT systems (Vanitha, 2019).
3. Rules and regulations-based agreement methods all aid in improving the safety of IoT devices while also facilitating transaction throughputs.
4. Furthermore, because it stipulates a decentralized distributed platform for verifying the data and an unchangeable ledger, a blockchain is very little vulnerable to change of state and stronger in recognizing fraud.

9.2.6 Intrusion detection system

1. Blockchain aids in the development of cooperative invasion sensing systems in which commodity codes can communicate with one another and exchange information during their trip.
2. There is a danger connected with code meddling and safety, which can be mitigated by the adoption of blockchain methodology (Min, 2019), which ensures data storage integrity and transparency.

9.2.7 *RFID security*

1. A blockchain increases RFID transparency, data protection cover, dependability, and cost administration using an ultra-lightweight shared certification RFID protocol.
2. This protocol allows for easy communication with RFID devices and is compatible with decentralized databases.
3. RFID is mostly used in SC for product and service traceability, which can be greatly enhanced by combining it with blockchain.
4. Walmart and IBM used blockchain technology to track consumer products using RFID systems.

9.3 Management of business processes

By combining the activity flow and business concern logic of interorganizational enterprise processes, a blockchain enables effective enterprise process administration through intelligent contracts. Smart contracts can also be utilized with blockchain for hyperconnected logistics. Triggers enable these controls, which serve as a link between organizational applications and the blockchain.

There exists a data gap between several stakeholders, which leads to suboptimal enterprise performance. This leads to heated debates and an advanced level of uncertainty among the people involved. For any request for modification or payment, a well-implemented smart contract will automatically start a process flow based on necessary permissions, resulting in speed, trust, and increased business performance. Blockchain technology can also be used to increase the effectiveness, traceability, and perceptibility of orders through customer-order-process management (COM). The blockchain also facilitates asset management with its proof of concept algorithm, which ensures openness, dependability, and efficiency.

Pilot prototypes such as DAO (Decentralized Autonomous Organization) were put in place to substitute for conventional administration systems with machine-driven governance; however, they had to be suspended owing to a security issue. Blockchain technology acceptance and enforcement should be done in stages to achieve an effective combination with enterprise operations and decreased safety concerns. The use of blockchain in conjunction with fuzzy sets to register the greatest feasible ideas and make tactical SC strategy decisions are also on the anvil.

9.4 Management of products

Through its integration with SCs, blockchain increases quality, cycle time, and efficiency, thereby providing new business options in various industries such as agriculture, medicine, food, and automobile (Madumidha et al., n.d.). It allows for commodity goods differentiation in an efficient way. An extensive study has been done on how to efficiently delete products and manage prices using blockchain technology. The next few sections take a look at certain specifics, as illustrated in Fig. 9.7.

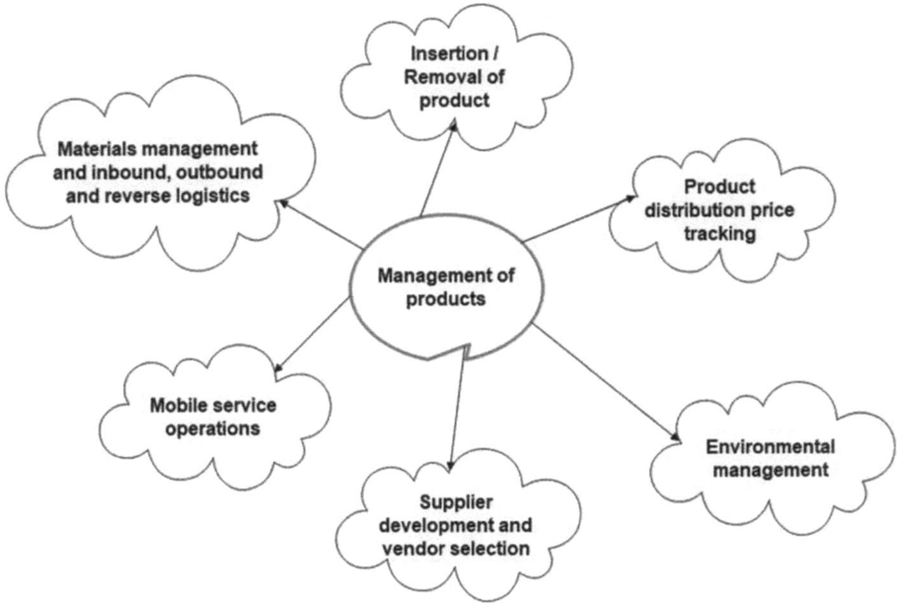

Figure 9.7 Management of product and its services.

9.4.1 Removal of a product

Product elimination necessitates recognition, analysis, and renewal, as well as review, decision-making, and implementation.

1. Real-time tracking
2. layered data administration
3. improved supply prediction
4. risk minimization, and machine-driven selection making

are all facilitated by the blockchain. The blockchain, circular economy, and removal of an item all have a connection. The data saved in the blockchain can be utilized to examine the company's plan of action on product omission and the circular economy.

9.4.2 Product distribution price tracking

The standard price tracking system is opaque and does not indicate end-to-end pricing fluctuations across the SC. During transaction completion, it only presents the final price of the item to the customer. However, thanks to the blockchain's end-to-end price tracking mechanism, consumers will be able to see perfect pricing from unprocessed materials through providers to suppliers, which will make all data available to the public. As a result, all stakeholders will be able to share accurate information, ensuring price transparency. On the Ethereum network, a trial implementation was carried out using smart contracts, which ensured price transparency.

9.4.3 The resilience of supply chains

1. Minimizes the effect of disruptions by using obstructive and retroactive risk management techniques and offering multiple-layer supply chain network protection.
2. The blockchain's hierarchical architecture aids in the capturing of both enterprise and network risks.

9.4.4 Provenance in the supply chain

1. Determining the unchangeability of physical items that are created and delivered through complicated, local, or international organizations prevailing in the supply chain.
2. Provides product information with proper certificates, and tracing and verifying rules.

9.4.5 Reengineering the supply chain

1. SC is more transparent and visible, and process automation is possible.
2. Traceability, privacy, and data management solutions remove middlemen and pave the way for real-time tracking.

9.4.6 Enhancement of security

1. Authentication, secrecy, privacy, and data control, as well as service integrity assurance.
2. Blockchain, when used with IoT and RFID, improves security, agreement mechanism for data storage, data clarity and security, reliability, and cost control.

9.4.7 Management of business processes

1. SC allows for efficient enterprise process management by assembling the activity flow and enterprise logic of interorganizational activities (Gong et al., n.d.).
2. Effectively used for managing the assets and order processing results in increased order efficiency, traceability, and visibility.

9.4.8 Management of products

Through its connection with SCs, the blockchain improves response time as well as enables product differentiation.

In end-to-end SCs, the blockchain aids in the removal of an item and pricing tracking.

9.4.9 Management of the environment

Blockchain is quite important for developing long-term supply chain operations. It has the potential to significantly improve the environment and increase sustainability. A blockchain aids in the creation of sustainable and environmentally friendly SCs in the following ways:

9.4.9.1 Supplier development and vendor selection

A blockchain enables suppliers to track their previous sustainability performance, execute supplier improvement, and eliminate middlemen. The environmental implications of the ingredients used by upstream suppliers can be tracked by the blockchain technique. This not only helps businesses to choose vendors but reduces the wastage of time.

9.4.9.2 Mobile operational services

Most local operational service activities have been interrupted by the COVID-19 pandemic. Service at doorstep operations and the conduct of service activities on a truck have been investigated. Blockchain utilization is also emphasized in order to create certainty and increase operational clarity.

9.4.9.3 Material logistics and inbound management

Blockchain enables the traceability of materials with sustainability values and product histories. Transportation effects on the environment may also be monitored using blockchain. The use of electronics parts based on blockchain SCs is being researched to improve traceability, long-term logistics while shipping, and security.

9.4.9.4 Production and internal operations

Blockchain aids in the management of required certifications, the transformation of the shop floor into an environmentally friendly surroundings, and the evaluation of sustainability performance as well as the continual development of all activities. In a carbon taxation regime, a solution based on the blockchain can be created to reduce both emissions and operating expenses for an ideal SC.

9.4.9.5 Outbound logistics and marketing

A blockchain allows for the certification of transportation's sustainability performance in outbound logistics, as well as the improvement of smart and traceable packaging using green marketing. As an example, blockchain could aid in the implementation of the "physical internet" for long-term logistics.

9.4.9.6 Reverse logistics

A blockchain allows for the tracking of goods and the authentication of all parties involved in the recycling process. This is important because it improves closed-loop SCM.

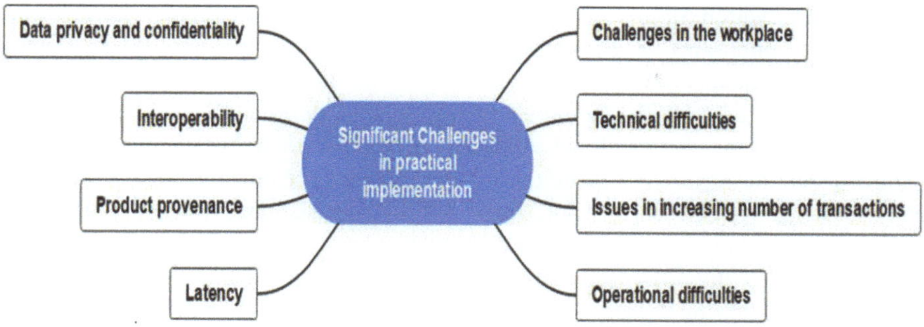

Figure 9.8 Significant challenges in practical implementation.

9.5 Significant challenges in practical implementation

Blockchain-based applications are rapidly increasing, yet interoperability is a challenge, leading to standards issues. The practical challenges in implementing a blockchain with the supply chain are shown in Fig. 9.8. SC's blockchain integration improves the entire value chain's visibility, lowers paperwork and manual errors, promotes clarity among all stakeholders, improves data security, complements the IoT, aids in the detection of counterfeit items, and allows for efficient traceability and management.

9.6 Challenges in the workplace

While the benefits are numerous, full-scale implementation of SC-integrated blockchain is hampered by a number of inter and intra-organizational, technological, and extrinsic constraints. Elements that influence the adoption are listed as follows:

1. Organizational preparedness
2. Technical skills
3. Digitized infrastructure
4. Scalability challenges
5. Monetary resources
6. Regulatory and legal compliance
7. Performance expectation
8. Organizational opposition
9. Standardization
10. Model security
11. Country of business

Local and international laws are becoming a hurdle for blockchain developments; thus, there is a growing need to include government agencies in the development of innovative solutions based on blockchain technology for compliance with rules and regulatory requirements.

Consumer awareness and empowerment are also important factors for adapting to the new technologies because both consumers and developers should adjust their purchasing/operations habits. Organizational preparedness and regulation need to be radically altered in order for the blockchain to have wider visibility and coverage.

Keep in mind that the legal and regulatory sides of the company are unavoidable. A lot of research needs to be done on SC technologies for developing a production-ready logistics system. As a result, while building blockchain systems for SC performance, this should be borne in mind. Organizational reluctance to exchange information on a global platform remains a key roadblock. Scaling up is hampered by difficulties such as a lack of a consistent standard across multiple enterprises, which can lead to interoperability issues. The government and industry must collaborate to produce a uniform standard to resolve the problem.

9.7 Technical difficulties

Although the blockchain is known for being a tamper-proof data storage system and one of the most secure transaction platforms, the size of each block in the blockchain could limit the performance and efficiency in the long run. The following are some of the technical problems that are frequently encountered when executing blockchain-enabled SC activities.

9.7.1 *Issues in increasing number of transactions*

Chain limits with an ever-growing number of the transaction have shown up in the need for large block sizes, even as lengthy response times and hefty development costs are some of the primary scalability difficulties in blockchain implementation. As the number of users on a blockchain grows every day, the issues of scalability grow as well.

9.7.2 *Data privacy and confidentiality*

Data confidentiality and privacy are still issues, despite the fact that data storage, security, and administration are a few of the major elements of data management in the blockchain. This is particularly true in the case of the public blockchain, which stores numerous personally identifiable information data as a public ledger.

9.7.3 *Interoperability*

The important interoperable features in blockchain architecture are:

1. Exchange of information for different block transactions.
2. Use of the same information by multiple blockchain systems.

Interoperability issues are becoming more critical as blockchain applications are growing at an exponential rate.

9.7.4 Product provenance

Product provenance refers to the history, origin, and traceability of products on blockchain, and it is an important aspect of the technology. It's crucial to know how to convey product provenance information appropriately and precisely.

9.7.5 Latency

The transaction rate, or the time it takes to process each block, as well as the security validation, consumes a significant amount of time, indicating that blockchain systems will suffer major latency issues.

Other significant elements of blockchain technology are auditing, visibility, and disintermediation, which ensure quick and trusted transactions.

9.7.6 Operational difficulties

Existing difficulties such as damage to raw materials and end-products, incorrect input data, order management issues, and so on, still exist in SC operations and logistics management. The right participation and involvement of various stakeholders, including operational efficiency, global logistics partners, maintenance costs, IT support, and big data management, would be required for the effective installation and smooth running of blockchain in SC systems. Apart from the aforementioned factors, government rules on cryptocurrencies, scalability, data warehousing, and the need for devices with high computing capacity connected to the internet are also significant obstacles.

Every transaction on the blockchain must be validated and processed by each node. The underdeveloped countries with insufficient infrastructure and the aforementioned constraints can find it difficult to achieve the aims of blockchain implementation. Standardization, adaptability, synergy, and expandability between diverse technology platforms are the other issues that must be further investigated.

9.8 Plastic recycling and circular economy—recent challenges and developments

A circular economy reimagines current production and consumption patterns in such a way that business and growth promote positive economic, social, and environmental outcomes across supply chains, business models, and life cycles, from raw material selection to product/service design to recycling and end-of-life management.

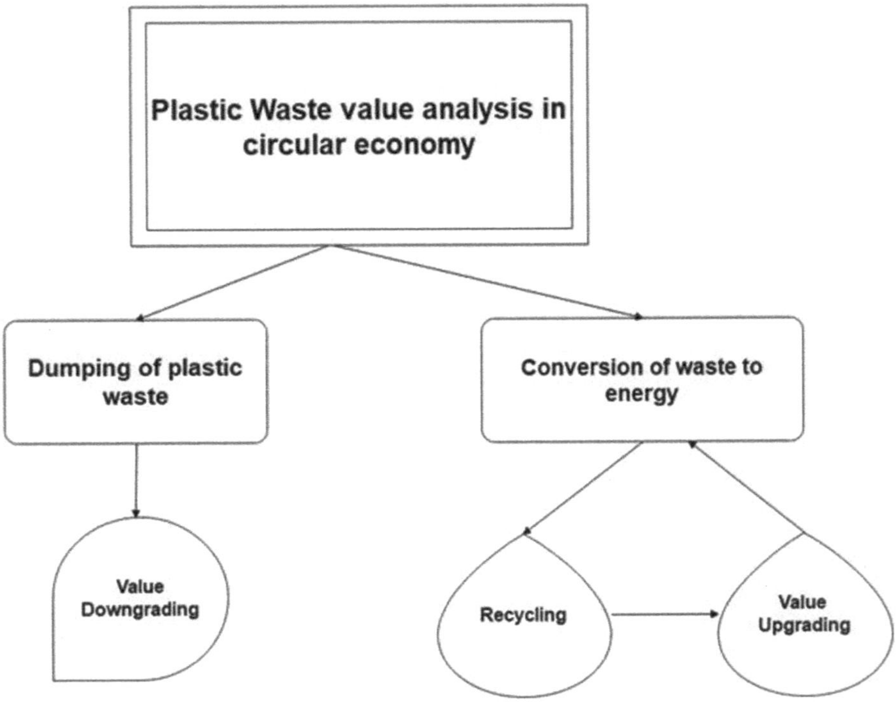

Figure 9.9 Plastic waste value analysis in the circular economy.

The coronavirus illness (COVID-19 pandemic) has not only put the world's economies and daily lives on hold, but it has also disturbed the waste value chain, resulting in a huge increase in medical and plastic trash (Sandhiya & Ramakrishna, 2020). This gives an opportunity to rebuild using more sustainable approaches, develop environmentally friendly and resilient livelihoods, and enhance technological innovation. Businesses, governments, and communities may benefit from technological innovation to find assistance in them to move to a more inclusive circular economy in a more quickly and efficiently possible way. To invest more in infrastructure and good integrated solid waste management systems is also a need satisfied by blockchain technology.

One solution is to digitize waste value chain procedures, which is shown in Fig. 9.9. The researches show that mobile application, when blended with high-level GIS data, community participation, transparency, and women's empowerment, results in higher earnings for the most vulnerable people in the supply chain.

9.9 Waste value analysis in the circular economy

In a circular economy, the value of waste is significant. Understanding the actual potential for creating value is the first step toward circular economy plastic. Total costing method takes into account the costs incurred at each stage of the circular economy, from collecting the waste to the destination. That is, the final destination of waste material may be either a recycling facility or an incinerator or dumped in the landfills. Logistics and processing are the two processes that incur a cost.

The cost of the material will gradually rise from the garbage trash to the final destination. Even if they are only going to be dumped, logistics and environmental costs must be considered. Plastics take almost 400 years to decompose, and the majority of them still persist in some form or another after being deposited on land or in the water. Between the period of 2000 and 2015, 15% to 25% of plastic garbage was burnt, while 10% to 20% was recycled, according to figures.

As a result, the bulk of waste is still dumped in landfills, and rehabilitation costs are unreasonably high. The total cost of the transaction (Total_{TC}) for the material to be processed after their individual procedures is the addition of logistics (L_{Cost}) and the processing costs (P_{Cost}).

$$\text{Total}_{\text{TC}} = L_{\text{Cost}} + P_{\text{Cost}}$$

For any recycling process to be successful, the marketing price of the material (PS) must be greater than the Total_{TC}. Further research is needed to examine the shorter, medium, and longer-term effects of the various processes and thus to account for environmental costs.

Plastic bottles can be distinguished based on certain shapes and colors. Laser diodes (near-infrared) can also be used. Plastics light absorption spectroscopy, especially in the wavelength of 300–3000 nm, revealed new possibilities for optical sensing. NIR is being utilized more and more to investigate the resonant frequencies of various polymers of plastics. These types of sensors will distinguish the various plastic bottles based on their resonance frequencies.

The absorption spectra of the plastic reach a peak at certain frequencies, which are called the resonance frequency. The resonance frequencies of different plastic molecules are different. We can determine the resonance frequency of plastic wastes by scanning them with these laser detectors over a variety of frequencies. The type of plastic can be detected based on its resonance frequency. Researchers have recently experimented with improved terahertz technology for plastic identification.

These technologies can also help with precise plastic garbage identification and segregation. For industrial-scale processes, manual separation based on color is not a viable choice. These technologies can achieve accuracy of around 99% for color-based segregation and 95%–98% accuracy for type-based plastic segregation, taking into account both the speed and accuracy required for industrial-scale set-ups.

9.10 Integration of sensors and AI with blockchain

In these multiple sensor platforms, AI's function is to reduce uncertainty and enable effective and smart separation. For instance, AI can design a system to distinguish two bottles of the same sort, which are deformed or discolored by varying degrees from one to another. In these cases, AI can be used to educate the system to recognize and separate them correctly. The current multi-sensor-driven AI system for sorting plastic garbage is put to the test in a variety of settings. Discussions of the findings of these investigations are beyond the scope of this publication and will be published elsewhere.

9.11 Segregation of plastic waste using multiple sensors integrated with AI and blockchain

It is difficult to separate commingled plastic garbage. The efficiency and accuracy with which plastics are classified are blockchain-based, taking cognizance of the following:

1. Color pigments
2. Chemical composition of the plastic-type, and
3. The origin of plastic trash directly affects the amount of plastic recycled.

Fig. 9.10 shows the integrated solution of blockchain with the supply chain for recycling plastic waste converted to a monetary benefit (Singh, Sharma et al., 2020).

9.12 Smart contracts on the blockchain

Blockchain technology functions as a trust-based platform among plastic trash collectors, buyers, and recyclers. Digital information has been distributed via the

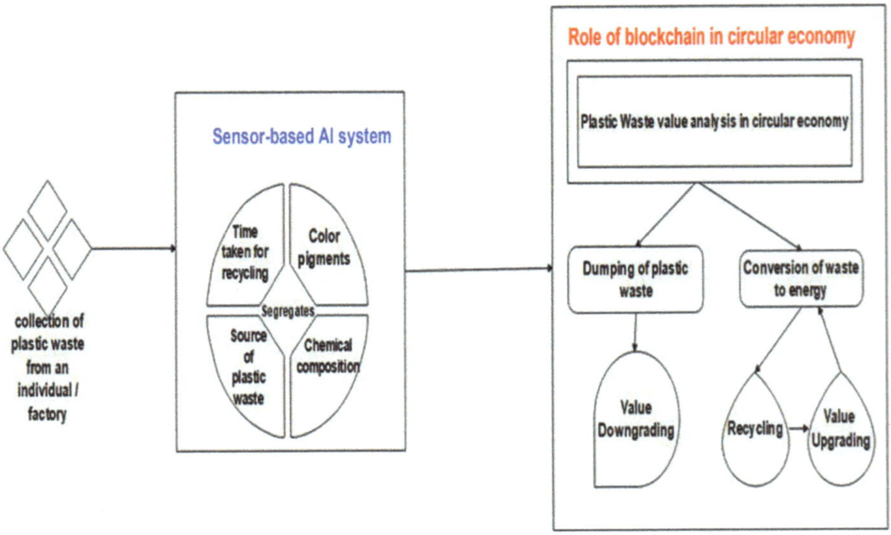

Figure 9.10 Segregation of plastic waste using multiple sensors integrated with AI and blockchain.

blockchain network, but it has not been copied. Various parties validate this digital information during each transaction. A transaction occurs when segregators, recyclers, and manufacturers exchange information. The information trade includes data on supply, bidding, demand, item specifications on quality aspects, and offer price.

Researchers are currently investigating a few use-case ideas using multiple sensor-driven Artificial Intelligence techniques (Chidepatil et al., 2020) for plastic waste separation, which in turn has been incorporated with the blockchain platform for the circular economy of recycling plastic waste. In the initial concept of a blockchain, those who participate in the process of data validation along with digital currency are rewarded. The security and transparency are traced from the fact that all data is kept on a shared blockchain ledger.

Participants can communicate and validate information regarding demand, supply, bidding specifications, and offer prices using the blockchain platform's information access. In summary, blockchain platform members are rewarded with a variety of actionable knowledge that allows for more efficient operations, cash-flow planning, and supply chain. This will fundamentally result in increased resource efficiency and a successful circular economy model for plastic trash. Two different categories of smart contracts exist in the blockchain system for modeling the plastic circular economy based on the stages of the raw material transaction.

9.12.1 Types of smart contracts

1. Type 1: Existing between a provider of separated plastic trash and a potential customer.
2. Type 2: Existing between a recycler of plastic raw material and a potential maker of plastic items.

In this example of the smart contract, there are two sorts of data: public and private. Validators can examine and approve the publicly available data. The private data, on the other hand, is restricted information that only the seller and buyer participating in this contract have access to. Only the buyer has access to the buyer's personal information. Private data often include information about price, segregation methods, waste material origin, and order type. This information is kept hidden in order to protect each seller's competitive advantage when selling identical products. Most essential, purchasers have the ability to verify and accept information about quality compliance and difficulties with purchased products. Depending on the settings provided in the smart contract, this information on quality compliance and faults can be made public or private. A validation protocol will also be used to digitally sign and timestamp each smart contract. Fig. 9.11 shows the various essential fields available in the smart contract.

9.13 The internet of things with blockchain support

The IoT is a collection of objects, sensors, and everyday items which are connected with the network with high computing power to consume (Dhiviya et al., 2018),

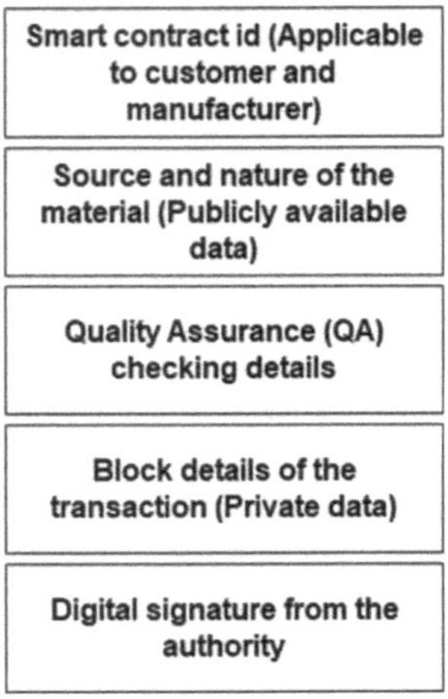

Figure 9.11 Fields available in the smart contract.

generate, and exchange data with minimal or no human intervention (Singh, Rathore et al., 2020). Cars, household appliances, televisions, drones, and even intra-body sensors, all of which are connected to the internet, may thus interact with one another to build an independent ecosystem, exchange data, and make intelligent judgments.

IHS estimates that by 2025, the overall installed base of IoT-linked devices will be close to 75 billion, a fivefold increase from 2015. The rise of DLSs is expected to serve as a significant pillar of the Machine Economy. It permits the communication of multiple nontrusting individuals over a P2P network without the need for mediators, thanks to its decentralized trust, validation, security, and privacy qualities.

By theoretically conceivable availability of 2160 distinct Ethereum addresses, the blockchain technology gives addressability to the continually expanding number of IoT-linked devices. This throws up the feasibility of a direct link between smart IoT devices, thereby clearing the way for Machine Economy rather than relying on a costly and inefficient centralized database to exchange data. It will enable governance and compliance for autonomous systems, which will be based on the distributed read-only ledger's immutability and the use of smart contracts.

9.14 Machine-to-machine economy based on blockchain

The introduction of decentralized ledger systems in blockchain technology as a potential backbone for enabling a future machine-to-machine economy is seen by experts as a near-term possibility. As a result, the machine-to-machine economy and machine-to-machine communication will be discussed more below. In recent years, the economy has shifted from a human economy to a Machine Economy, defined by fundamental interaction and ability prevailing among humans and machines. The number of devices and connections that are used on a regular basis per capita will climb from two in 2015 to more than three by 2020 on a worldwide scale. M2M connections, for example, will account for roughly 46% of all connected devices.

As a result, M2M communication is thought to be growing quicker than any other technology (Lee et al., n.d.). A count greater than 75 billion IoT-enabled devices will be in use by 2025, demonstrating the M2M device's enormous impact on society and the economy. Everything will be autonomous, smart, economically independent, networked, and most likely handled by a new technical backbone that is different from older mainframes. The M2M economy is defined as a communication network that connects objects, sensors, and machines.

The Machine Economy concept arose from solutions for basic fleet management with the goal of improving certain operations, such as the observation of valuable cargo, and where data is unlikely to be shared beyond its silo. The process of integrating stand-alone solutions with a broader corporate environment was not taken into consideration at this time. M2M economy has two markets:

1. B2B market comprises of industrial internet and constitutes the primary market for M2M
2. B2C market comprises of IoT.

Device-centric solutions have previously powered the M2M economy, but thanks to recent technological advancements, these solutions are incorporated with worldly-wise process-centric solutions that combine data streams. In addition, advanced technologies such as the IoT will eventually infect all parts of our life, including the settings we interact with and the economic activities that support our daily existence.

References

Azzi, R., Chamoun, R. K., & Sokhn, M. (2019). The power of a blockchain-based supply chain. *Computers & Industrial Engineering*, *135*, 582–592. Available from https://doi.org/10.1016/j.cie.2019.06.042.

Casado-Vara, R., Prieto, J., la Prieta, F.D., & Corchado, J.M. (2018). How blockchain improves the supply chain: case study alimentary supply chain. In *Proceedings of the fifteenth international conference on mobile systems and pervasive computing (MobiSPC 2018)/Proceedings of the thirteenth international conference on future networks and communications (FNC-2018)/affiliated workshops*, *134*, 393–398. Available from https://doi.org/10.1016/j.procs.2018.07.193.

S. Chen, R. Shi, Z. Ren, J. Yan, Y. Shi, & J. Zhang. (n.d.). A blockchain-based supply chain quality management framework. In Proceedings of the IEEE fourteenth international conference on e-Business engineering (ICEBE) (pp. 172–176). Available from https://doi.org/10.1109/ICEBE.2017.34.

Chidepatil, A., Bindra, P., Kulkarni, D., Qazi, M., Kshirsagar, M., & Sankaran, K. (2020). From trash to cash: How blockchain and multi-sensor-driven artificial intelligence can transform circular economy of plastic waste? *Administrative Sciences*, *10*(2). Available from https://doi.org/10.3390/admsci10020023.

Dhiviya, S., Malathy, S., & Kumar, D. R. (2018). Internet of things (IoT) elements, trends and applications. *Journal of Computational and Theoretical Nanoscience*, *15*(5), 1639–1643. Available from https://doi.org/10.1166/jctn.2018.7354.

X. Gong, E. Liu, & R. Wang. (n.d.). Blockchain-based IoT application using smart contracts: Case study of M2M autonomous trading. In *Proceedings of the fifth international conference on computer and communication systems (ICCCS)* (pp. 781–785). Available from https://doi.org/10.1109/ICCCS49078.2020.9118549.

C. Lee, L. Nkenyereye, N. Sung, & J. Song. (n.d.). Towards a blockchain-enabled IoT Platform using one M2M Standards. In *Proceedings of the international conference on information and communication technology convergence (ICTC)* (pp. 97–102). Available from https://doi.org/10.1109/ICTC.2018.8539724.

Dutta, P., Choi, T. M., Somani, S., & Butala, R. (2020). Blockchain technology in supply chain operations: Applications, challenges and research opportunities. *Transportation Research Part E: Logistics and Transportation Review*, *142*, 102067.

Longo, F., Nicoletti, L., Padovano, A., d'Atri, G., & Forte, M. (2019). Blockchain-enabled supply chain: An experimental study. *Computers & Industrial Engineering*, *136*, 57–69. Available from https://doi.org/10.1016/j.cie.2019.07.026.

S. Madumidha, P.S. Ranjani, S.S. Varsinee, & P.S. Sundari. (n.d.). Transparency and traceability: In food supply chain system using blockchain technology with internet of things. In *Proceedings of the third international conference on trends in electronics and informatics (ICOEI)* (pp. 983–987). Available from https://doi.org/10.1109/ICOEI.2019.8862726.

Min, H. (2019). Blockchain technology for enhancing supply chain resilience. *Business Horizons*, *62*(1), 35–45. Available from https://doi.org/10.1016/j.bushor.2018.08.012.

Sandhiya, R., & Ramakrishna, S. (2020). Investigating the applicability of blockchain technology and ontology in plastics recycling by the adoption of ZERO plastic model. *Materials Circular Economy*, *2*(1), 13. Available from https://doi.org/10.1007/s42824-020-00013-z.

Sandner, P., Lange, A., & Schulden, P. (2020). The role of the CFO of an industrial company: An analysis of the impact of blockchain technology. *Future Internet*, *12*(8). Available from https://doi.org/10.3390/fi12080128.

Shala, B., Trick, U., Lehmann, A., Ghita, B., & Shiaeles, S. (2019). Novel trust consensus protocol and blockchain-based trust evaluation system for M2M application services. *Internet of Things*, *7*100058. Available from https://doi.org/10.1016/j.iot.2019.100058.

Singh, S., Sharma, P. K., Yoon, B., Shojafar, M., Cho, G. H., & Ra, I.-H. (2020). Convergence of blockchain and artificial intelligence in IoT network for the sustainable smart city. *Sustainable Cities and Society*, *63*102364. Available from https://doi.org/10.1016/j.scs.2020.102364.

Singh, S. K., Rathore, S., & Park, J. H. (2020). BlockIoTIntelligence: A blockchain-enabled intelligent IoT architecture with artificial intelligence. *Future Generation Computer Systems*, *110*, 721–743. Available from https://doi.org/10.1016/j.future.2019.09.002.

Vanitha, C. N., & Malathy, S. (2019). Secure integration of cyber security and internet of things in addressing its challenges. *International Journal of Recent Technology and Engineering*, *8*(4), 2277–3878.

Decentralized platform for energy exchange: a case study

10

Saranya S.S.[1], *Sivaraj R.*[2], *Vijayakumar M.*[3] *and Santhosh C.*[1]
[1]Department of Computer Science and Engineering, Kongu Engineering College, Erode, Tamil Nadu, India, [2]Department of Computer Science and Engineering, Nandha Engineering College, Erode, Tamil Nadu, India, [3]Department of Computer Science and Engineering, Sri Shanmugha College of Engineering and Technology, Dharmapuri, Tamil Nadu, India

10.1 Introduction

In blockchain-based decentralized energy trading, prosumers and consumers share a common platform for storing energy exchange information without the intervention of a third-party. When there is a surplus of energy production, for example, solar power, individuals may act as prosumers while demand exceeds supply and also be active buyers to charge their vehicles. In a peer-to-peer network that promotes an incentive-based energy trading platform, prosumers can invest their surplus renewable energy in a decentralized power market. In addition, it's very difficult to integrate a P2P network with blockchain technology to support privacy. The proposed energy trading platform is proficient in trading energy without fraud, resilient, and actively promotes secure energy transactions by eliminating intermediate costs. P2P energy trading encourages consumers and prosumers to do business directly with each other. In the former centralized energy system, the operators were affected by network congestion and loss of data.

A decentralized peer-to-peer energy-centric platform promotes grid-level energy exchange. Distributed ledger technology (DLT) is premeditated to endorse distributed transactions with no third-party management. Blockchain network facilitates smart contracts for performing automated energy transactions. Instead of storing all the transactions in a single database, blockchain provides a decentralized environment where users of peer-to-peer networks share and store their transactions as a chain. A system connected to peer-to-peer networks has a similar copy of transactions, where users are capable of modifying the ledger simultaneously. A Blockchain or DLT portrays a database, and a user who successfully adds a block to the chain to be provided with an incentive as a reward for every valid state of record added by performing diverse calculations based on consensus mechanisms depicts a miner (a node of the network) in action. The transactions made on a blockchain are immutable, transparent, and tamper-proof. Abundant business opportunities and use cases have been analyzed by various researchers in the energy

Blockchain-Based Systems for the Modern Energy Grid. DOI: https://doi.org/10.1016/B978-0-323-91850-3.00002-0

sector. The energy trading platform is stipulated as a market where prosumers can trade energy based on price, and also prosumers may act as sellers and buyers based on the availability of energy. Many research works have already been carried out in peer-to-peer energy trading that has followed in the footsteps of the improvement in information and communication technologies (ICT). Energy trading using distributed ledger technologies promotes security in peer-to-peer transactions. The transactions can be made through smart contracts in blockchain-based smart grids.

10.2 Smart grid

The bidirectional flow of electricity and information is enabled by a network to discover, perform, and make predetermined changes in energy exchange platforms and to deal with multiple issues. Smart grid technology has developed at a rapid pace on the wings of Wireless Sensor Networks. The improvement in energy dispersion should be possible by embracing different smart, intelligent grids that assist with observing and pass-through of data along the nodes of a P2P network. Smart grids are empowered with a smart sensor-based monitoring system to transfer the energy between prosumers and consumers in the most promising way in the midst of manpower deficiencies and with a high precision rate. They supervise the energy transactions trapped between users and grids without any loss of information. Conventional electricity grids don't have storage capabilities being demand-driven and with a hierarchical structure. The "produce and distribute" model had many difficulties, the least many of which were not energy disruption because of climate conditions, harm from high voltage energy transmission lines, and human deaths. In smart grids, with their self-amendment capabilities, people are empowered to identify and react to framework issues and to guarantee speedy recuperation after unsettling influences, which is a "collaborate, manage and use" model. Due to the ever-growing sources of renewable energy and its demand, grid management should result in efficiency among employees, necessitating third-party support for the supply and distribution of energy. Integration of blockchain technology and grid network endorses a decentralized platform intended for energy exchanges along with its interrelated maneuvers that are an immutable history of transactions in the blockchain. So, every transaction that has been made has a timestamped block, need not be channelized through a centralized system for dynamic decision-making, and can be processed automatically with smart contracts provided preconditions are satisfied, which makes outside arbiters like a bank redundant.

10.3 Recapitulation of blockchain technology

The term blockchain is defined as a DLT, that is where a distributed ledger separated by distance and time is maintained by a number of users connected by a peer-to-peer (P2P) network. In the energy domain, transactions that are made by the prosumers or consumers are added to a chain, making it an endless chain of blocks. Each block has gone through a confirmation process by validators termed miners (nodes of the network). The miners are paid an incentive for every validation of a block on the blockchain. The transactions are cryptographically bundled together and secured, and the history of transactions is immutable by nature. The size of the block differs depending on the type of blockchain that is being used, for example, for Ethereum block sizes, there is a limit on the block size of 30 million gas units of storage capacity. The initial block in the chain is termed the Genesis block, which does not contain the previous block hash value. Every transaction in a block is related to each other through the hash value of the previous block. Technically, blocks stored in a distributed ledger are impossible to be modified but to modify a block is an extremely computationally intensive task.

The key characteristics of a blockchain are,

1. A decentralized DLT of interconnected nodes and storage mechanism where all the participants are capable of accessing information in a network to read-only.
2. The consensus mechanism ensures fault-tolerance to accomplish a concurrence among the distributed nodes of the P2P network configuration that promises all nodes are synchronized with each other to validate the transactions that have been initiated by a user. The genuine transactions which have been endorsed by the miners are then added to the blockchain as a legitimate exchange.
3. Smart contracts are supposed to be a computer program or a PC code put away in a blockchain that can be executed when certain foreordained conditions are met. They automate the execution of an agreement and also automate the outcomes of smart contracts that participants can reach rapidly with no mediation from outsiders.
4. The energy trade data is put away in a blockchain through a cryptographic method that advances security and integrity. The hash functions that are performed in each block facilitate a single view of a blockchain to all the participants of a network. Due to immutability, the contents of the blockchain remain unchanged as every block of data is associated with a unique hash value of the previous block. The hash value generated for each block is unique, and small changes in a block may lead to incorrect output cascaded across the network and putting everyone on alert.
5. Hashing function applied to each transaction that is verified by the blockchain network is guaranteed to produce a fixed size output recognized as a checksum. The blocks are embedded with the timestamp and afterward secured cryptographically with the assistance of a computed hash value. The header of each block always incorporates the previous block's data's hash value, in addition to the previous block's address in a hash pointer, which gives a chain.

10.3.1 Harmony of blockchain

The adjacent blocks are linked together by cryptographic hash values, whereas any small change that has been made by an attacker in one block, in turn, affects all preceding blocks until a new block attached to the blockchain gets added with the current block holding a hash value of its preceding block. Due to high processing computation capabilities, the blockchain is immutable.

The components of the block structure are as follows: Fig. 10.1

1. **Block Header**—Identifies a specific block on a blockchain.
2. **Timestamp**—A unique serial number that says at which time the block has been mined and validated by the miner nodes of a network.

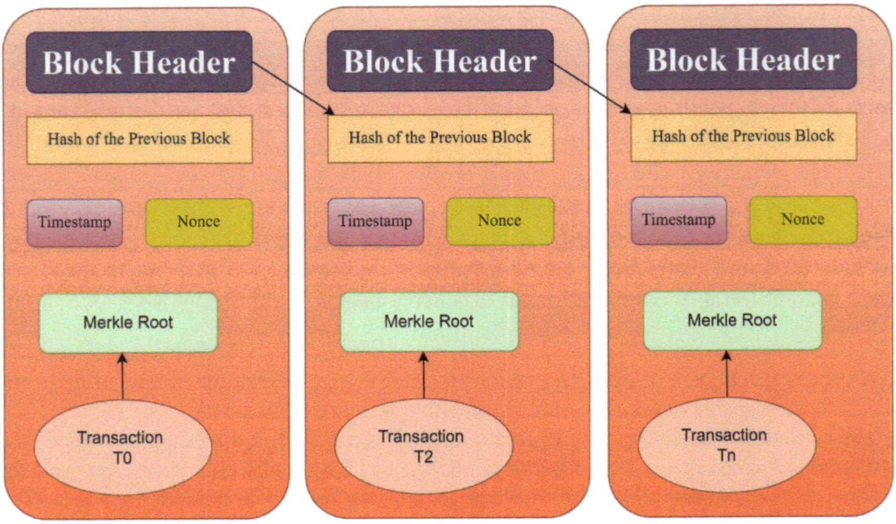

Figure 10.1 Block structure. The structure of a blockchain system.

3. **Version Number**—Type and format of data stored in a block, in addition, describes a set of validation rules to follow.
4. **Nonce**—Number Only Used Once, the number appended to the hash value of a block to elude replay attacks.
5. **Previous Block hash**—The pointer holds the address of the antecedent block in the structure along with the current hash of the data stored in the predecessor block.
6. **Merkle Root Hash**—Cumulative hash of all the hashes of all the transactions of the blockchain network.
7. **Block Body**—Stores the root of the hash of the Merkle Tree. A Merkle Tree structure is depicted in Fig. 10.2.

10.3.2 Consensus algorithms

Due to the distributed and decentralized nature of the blockchain, every node in the network must acknowledge all the transactions prior to the inclusion of a transaction in a network. The decentralized network promotes immutability, security, and integrity of the network. In the blockchain, the transactions cannot be verified and approved by a centralized authority, but verification has to be done through consensus protocols that are a part of the blockchain network. The consensus protocols are rules that all the peers of the blockchain network make a common agreement to acknowledge and use to validate each transaction being made inside a network. It promotes trust among unknown users in a decentralized environment. If a block is appended to a blockchain, only a single copy of the ledger is distributed over all peers that are connected to it. Thus, these are the working principles of a diverse and unique consensus algorithm.

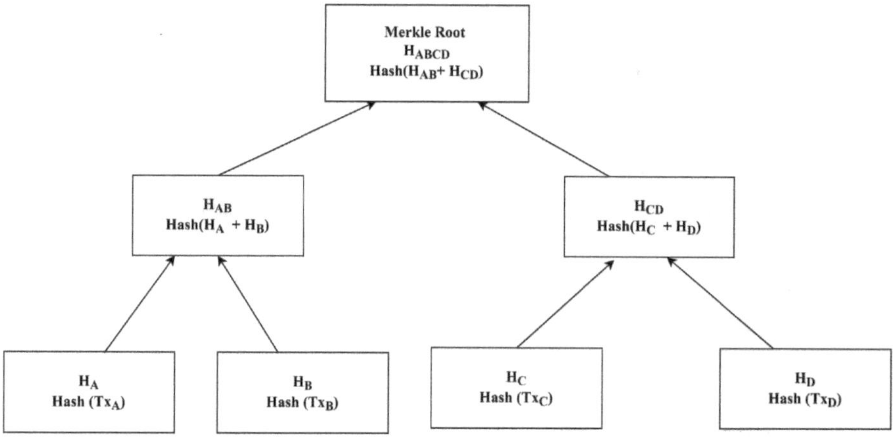

Figure 10.2 Merkle Tree. The structure of the Merkle Root Tree.

10.3.2.1 Proof-of-work

The proof-of-work (PoW) consensus mechanism deals with a complex mathematical computational problem that has to be solved by the node of a blockchain network. The node must possess high computation power and complex system capabilities similar to energy consumption nodes to solve the mathematical puzzle. The nodes who participated in the calculation are called miners. The miners of a network compete to solve the puzzle first. After computing the nonce value, the hashing necessarily must be done. Finally, the resultant hash is cross-checked with the difficulty level that has been set by the blockchain network. If the miner guessed a nonce value that perfectly matches the original difficulty level, the node acquires an opportunity to affix a new block onto the network. The level of difficulty of a blockchain network must be varied either upwards or downwards based on the cumulative computation hash power of all the peers connected to it. If the miner is not capable of predicting the accurate nonce value at initial guesses, iterations are made until the correct difficulty value is achieved. Finally, the node distributes the nonce and hash of the new block to all the peers in a blockchain network to append it to a ledger. Although the PoW consensus mechanism requires more computation power to solve a problem by appending that block to the blockchain network, it's highly secure.

10.3.2.2 Proof-of-stake

The proof- of-stake (PoS) is a revolutionary consensus protocol distinct from PoW, as it facilitates supplementary computational power to unravel the mathematical puzzles in order to envisage the difficulty level of the precise block intended to be added to the blockchain network. In proof of the staking mechanism, the distributed users of the blockchain network invest a number of coins as a stake to prove their ownership to update the present status of the network. If the user has a reasonable share of a stake in the network, then he or she is termed a validator of the blockchain. The major weight of the validators is in verifying the transactions and validating the hash value prior to addition to the network. The validators get a reward for validating the blocks, which increases the level of their stake in the blockchain network. Due to a property-based consensus mechanism, the user who has a major stake contribution has higher validation rights over a new allotted block of transactions that could lead to a vulnerability attack on a network. If the users of a blockchain network attain 51% of coins, they have the ability to attack the entire blockchain by holding a major stake. The validators are chosen randomly in a network to validate and confirm the block of transactions that have not been created. Due to impracticability, the node must possess 51% of a stake in a network to attain privileges over all the transaction validation.

10.3.2.3 Proof-of-authority

The proof-of-authority (PoA) is a similar consensus methodology similar to PoS. Rather than investing coins for a stake in the blockchain network, the validators who achieve the mining job facilitate their identity and role in the network. The miners who do not have the computational resources to solve a mathematical puzzle cannot create a block. The block creation duration can be predicted and approved by the authorized nodes in a time interval that stimulates the block validation rate to a higher range. The authorized node inside the network must prove its authorization by exhibiting stability of sustaining against DoS attacks. The node that desires to be validators of the network is excluded from the validators' list' if it fails to prove its ability to withstand DoS attacks.

10.3.2.4 Proof-of-elapsed time

In the Proof-of-elapsed time (PoET) consensus technique, every node in the blockchain network goes to sleep after generating a certain random waiting time. The node must wait until that chosen time period gets over before creating and adding the new block to a network. A node that completes the waiting time gets the chance to add the latest and the current block to an already existing network. There are two main discrepancies in the PoET that need to be addressed. The PoET consensus mechanism must ensure that every node has decided on a random waiting time rather than predicted a shorter time period intentionally.

10.4 Permissionless vs. permissioned blockchain

10.4.1 Permissionless blockchain

A Permissionless blockchain is defined as an open network where participants can join the network and interact with other participants through consolidation validation. It is a public and trustless environment where the transactions are transparent with a foundation of the open-source protocol. The network does not have a central authority. The anonymity of participants is preserved in a network. Every participant must have a common distributed ledger that is valid across the network. The validation has to be done by selective miners chosen from the list of participants. The miners get incentives as a reward for each transaction that they have validated. Once the process of validating a block has been completed and accepted successfully by all participants, it is then added to the network and updated in all participants' ledger, which is distributed in nature. The benefits of the Permissionless blockchain are transparency, decentralization, security, and resilience. In a Permissionless blockchain network, the attackers cannot possibly attack a single database as it does not have a central authority. In order to hack the network, the attackers keep track of 51% of the participants to be in their control which is almost impossible and very expensive.

10.4.2 Permissioned blockchain

The permissioned blockchain constantly guarantees the authenticity of the participants of the peer-to-peer network in order to persuade the privacy and security of the transactions. The public blockchain promotes computational-based consensus mechanisms, whereas the permissioned blockchain facilitates the collaborative consensus-based mechanism for block verification and the process of adding up blocks to the blockchain. The authentic structure has only to be visualized by the members who are legitimate before having joined the blockchain. The contents only are exposed to the legitimate users, in turn reducing the decentralization reimbursement. The permissioned blockchain has only limited participants in the network, which probably diminishes the computational process needed to reach participant consensus of the network. The performance of the permissioned blockchain is limited given the set of predetermined nodes that validate the transactions of the network. Due to the togetherness of permission less networks, not all the nodes which are connected in a network have to take the necessary action to update the new status of the blockchain. But the permissioned blockchain doesn't have to suffer the waiting time due to a limited number of participants, all working together to update the block status, which in turn speeds up the updating. The difference between Permissioned blockchian and Permissionless blockchain is decribed in Table 10.1.

10.5 Impediment of arbitrage-based system

Despite the current arbitrage-based energy exchange and the trading system, a blockchain suffers due to pricey and heritage software for maintaining the energy transaction information without providing additional functionalities that compete with an enormous consumer-centric energy market. Moreover, data processing and storage are unproductive and necessitate additional handing out time as well as the hindrance in handing in payment after transactions are over. Furthermore, it obliges

Table 10.1 Permissionless blockchain versus permissioned blockchain.

Features	Permissionless Blockchain	Permissioned Blockchain
Rapidity	Slow	Fast
Confidentiality and privacy	Public & transparent	Private connection
Validation	Authorized	Authorized
Possessor	Publicly accessible by all nodes of the network	Supervised by limited authorized nodes of the network
Decentralization	Completely decentralized	Moderately decentralized
Cost	Not cost-effective	Cost-efficient
Protection	Less secure	More Secure
Read access protection	Anyone in the network	Anyone in the network
Write access protection	Anyone in the network	Only specific internal nodes

reconciliation practices to complete the energy trading successfully. The conventional energy exchange system deliberately works in a unidirectional way wherein the consumer is capable of either consuming or producing energy no longer at the same time. Moreover, the legacy system does not promote consumers who have multiple locations for the production and consumption of energy. It fails to endorse dynamic functionalities as well as traffic monitoring based on the preferences of the consumer. The billing and payment of the traditional system influence the lack of transparency to recognize billing information and estimation cost, facilitating third-party energy retailers. The central database provisions all transactional particulars as a single point of communication leading to its being the victim of assorted attacks.

10.6 Frameworks of blockchain-based decentralized power trading platform

The framework of blockchain-based decentralized trading platform architecture consists of various layers as follows,

1. **Transportation layer**

   This layer focuses on the components like Prosumers, Consumers, and Distributed System Operators (DSOs), the last being responsible for collecting the energy resources from various energy production markets and distributing the energy to end-users and electrical components of the energy system. The components of the transportation layer connect to the blockchain network. The electrical components are as follows: Distributed Energy Resources (DERs), which is a small unit of energy generation that is associated with the smart grid and smart meters for collecting the statistics of the energy consumption/production.

2. **Contact layer**
   The contact layer depicts the communication software/applications and smart devices to gather data from numerous energy devices that may be wired or wireless. The contact layer helps to communicate and pass information from the front-end blockchain to the back-end blockchain. The smart meters act as mediators of devices that consume energy to transfer the energy utilization in terms of script into the blockchain.
3. **Blockchain layer**

It consists of all participating nodes that form a decentralized network for energy transactions. In this layer, energy bids and prices are distributed and published publicly with no hindrances. The node which is responsible for validating energy transactions is termed a miner. The miners play a vital role in the blockchain environment to append a recently created block in the network. The types of blockchain differ based on the application as the underlying technology.

10.7 Operational design of decentralized energy trading system

Based on various research works that have been published in selecting the type of blockchain for different application requirements, the Ethereum-based blockchain facilitates the peer-to-peer energy trading depicted as smart contracts. The smart contracts on blockchains are considered to be a protocol that enables an executable trading mechanism. The prosumers and consumers possibly select their own roles in the energy market, for example, buyers or sellers, and have the flexibility to trigger or initiate the transactions in a blockchain network based on their own requirements and responsibility. The entity which is the specialized authority that is solely responsible for maintaining the records of transactions is called the DSO, as depicted in Fig. 10.3.

Smart contracts were designed strictly to be executed when the conditions have been met according to the underlying algorithm (Alladi et al., 2019; Andoni et al., 2019; Esmat et al., 2021; Faizan et al., 2019; Han et al., 2020; Islam et al., 2020; Kongu Engineering College, 2021; Narayanaswamy et al., 2022; Roopa et al., 2020; Shanthi et al., 2020; Song et al., 2021; Suthar & Pindoriya, 2020; van Leeuwen et al., 2020; Yahaya et al., 2020). Ethereum technology facilitates the development of a decentralized application supporting smart contracts, where each transaction must be billed with a quantity of gas for the consumption of energy resources. While energy trading represents the integrity of the transaction, digital and financial assets must be handled with the utmost care. In a decentralized network, a system can operate with a number of nodes where any user can access the entire network. The decentralized network promotes a consensual mechanism for checking the triggered transactions and where the decision and validation of each transaction are democratic. The members of the energy network settle the payment for the energy distributed to them. The member with the highest computing power is authorized to validate and mine the transactions. DERs have grown significantly

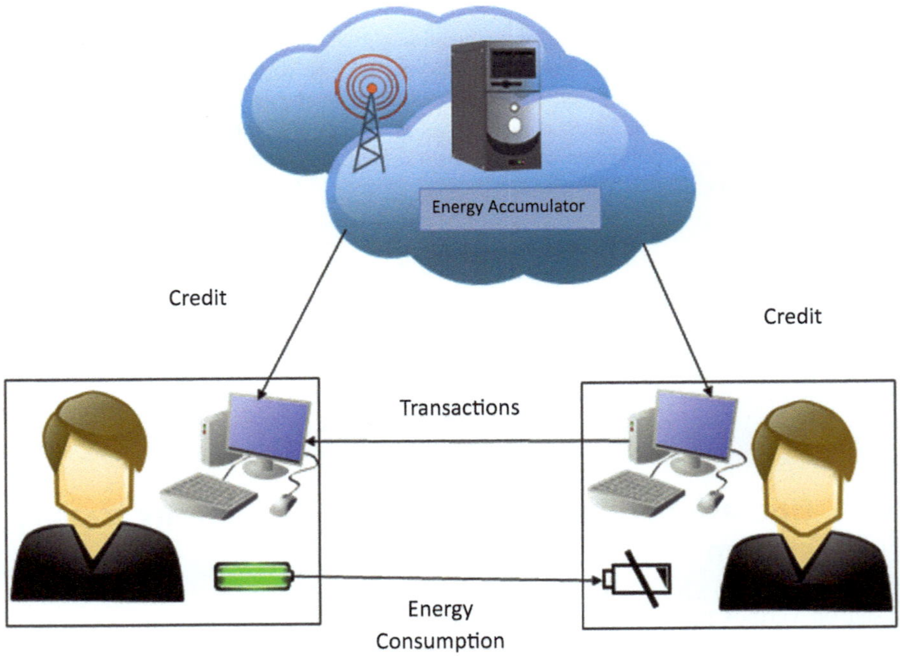

Figure 10.3 Structural design of decentralized energy trading platform.

over the past decade. The transformation made the mesh more interactive and changed a lot of small things. The traditional energy market has been balanced in a way where control is delegated only in the prosumer direction, but nowadays, energy is generated by various energy generators like solar, wind, tidal wave, etc. Technically, it's now no longer possible to leave control in the hands of the prosumer as they're dependent on the climate to generate energy. The revenue of these small-scale producers and consumers was not enough to possibly earn profits for them, especially when revenue is restricted, given that they're supposed to sell their energy produced to the intermediary suppliers at a fixed tariff. These decentralized energy trading platforms powered by Ethereum-based smart contract technology eliminate the need for an intermediary of energy trading, which, in turn, increases the profitability of the prosumers and consumers. This approach enables the prosumers to sell their energy at their own desired price range with no intervention from intermediate local energy suppliers.

The Operational Model of Ethereum-based Smart Contracts is described as follows:

1. The energy trading nodes, which are connected in a decentralized platform, generate energy, and this accumulates in the energy storage system. The smart contract acts as a smart meter to record and measure changes in energy demand and energy supply with corresponding changes in discharge and flow.

2. The prosumers and consumers of the energy trading system bid for the first time if the demand is higher than the supply in the energy market, as the price of a unit of energy has to be increased to balance the supply and demand. This approach leads the prosumers and consumers to bid for different amounts at each period of time. In order to avoid this annoyance, a single price is regulated for every stimulated period on the basis of overall demand and overall offer presented. Price may increase or decrease for a period which depends on the ratio of total demand to supply as a whole.
3. A consumer of the energy trading platform can acquire the required amount of energy from the prosumers based on the market's reaction to an increase or decrease in price.
4. The problem of double sales must be avoided as the trading of energy has to be done online. The prosumers and consumers have to ensure that a unit of energy must not be traded for more than once in an energy trading market. The Distributed System Operator is solely responsible for updating and maintaining the status of the energy unit over a period of time. The status of energy sold must be updated in a blockchain only with the private key of Distributed System Operator of the energy trading system.
5. Initially, the prosumers and consumers of the trading system trigger the request to buy or sell energy through smart contracts. It consists of contractual declarations that are automatically executed once predefined rules and requirements have been fulfilled. Transactions and states are stored in the blockchain.
6. The consumers have to make a down payment that is sufficient money to close their purchase which is already defined per the smart contract. The deployed smart contract cumulates requests from all members and computes the price of energy according to a preset and pre-decided formula.
7. Both the participants pass the penalty for deviating from the declared amount of energy if any deviation from the pre-schedule is detected when the programmed smart contract instinctively dispatches a penalty to the opponent's financial account. Finally, the status is stored as a transaction on the blockchain network.
8. When the entire demand exceeds the supply in a trade round, the final price of each unit of energy increases to demoralize demand. Dynamic pricing allows energy trading to achieve equilibrium where the whole energy supply equals the requested demand inside a smart grid.
9. The following variables depict a supply and demand scenario via the energy trading mechanism:

- **E_{sup} (*t*)—Energy Supply at time period *t***
- **E_{dem} (*t*)—Energy Demand at time period "*t*"**
- **Sup_i (*t*)—Supply of the prosumer "*i*"**
- **Dem_j (t)- Demand of the consumer "j"**
- **$Total_P$—Total number of prosumers**
- **$Total_C$—Total number of consumers**

The proposition of energy demand and supply can be depicted with the following formula:

- **Prop (*t*) = E_{dem} (*t*)/E_{sup} (*t*)**

The divergence of demand and supply of total energy is denoted as follows:

- **Diver(*t*) = E_{dem} (*t*) - E_{sup} (*t*)**

If the absolute supply of the prosumers [$E_{sup}(t)$] is appreciably overreaching total demand [E_{dem}(I)], the payment is diminished to Pay_{min}. The payment is hiked moderately slowly till the demand considerably exceeds the supply.

10.8 Life span of smart contract in Ethereum-based blockchain

Once the predefined conditions, rules, and restrictions of the energy trading system have been finalized, the prosumers and consumers must be registered in the peer-to-peer network. After successful registration, smart contracts must be tested before the implementation transaction is initiated by both the buyer and the seller; next, it must be verified by the miners of the network, while the copy of each transaction is updated in a decentralized distributed ledger. All members/participants of the decentralized communication network must have a unique copy of the same ledger. Each participant enters the network to register in the blockchain and creates a wallet for monetary transactions in the trading system. When a prosumer initiates the will to sell the excess amount of energy along with the payment price, the information is stockpiled in blockchain through smart contract computer programs. The consumer who wants to buy energy must select the application submitted by the prosumers for the sale of energy. If the amount offered by the prosumer meets the consumer's expectations, the chosen amount will be deducted from the consumer's account. The transaction needs to be updated on a blockchain, which, in turn, updates the sales request already submitted by the same prosumer to avoid the double sale of the traded energy. After successful verification of energy delivery at the consumer side, the price amount is transferred to the prosumers account after simply eliminating network charges.

10.9 Conclusion

In this paper, an attempt has been made to create an insight into a smart contract-based decentralized energy trading technique, where the buyer and the seller trade the surplus energy with the seller at an expected price, where a transaction is carried out locally without intermediary suppliers. The transactions which have been made via the blockchain are transparent, trustworthy, and immutable by nature. The recorded transactions need never be updated or modified in the blockchain, which bolsters data integrity, builds trust, and gives data security to the local energy trading system. The status of each energy request that is submitted by the seller and buyer is updated immediately on every trade. So, the double-spending problem is eliminated. The DSO is provided with the authority to verify the participants in trading to promote stability in blockchain-based decentralized energy trading. The proposed system culls the thorns in traditional intermediator-based energy trading systems and also balances the energy supply and demand. In future, the penalty-based energy trading system can be incorporated in the discussion to avoid the deviation in energy trades.

References

Alladi, T., Chamola, V., Rodrigues, J. J. P. C., & Kozlov, S. A. (2019). Blockchain in smart grids: A review on different use cases. *Sensors (Switzerland), 19*(22). Available from https://doi.org/10.3390/s19224862.

Andoni, M., Robu, V., Flynn, D., Abram, S., Geach, D., Jenkins, D., McCallum, P., & Peacock, A. (2019). Blockchain technology in the energy sector: A systematic review of challenges and opportunities. *Renewable and Sustainable Energy Reviews, 100*, 143–174. Available from https://doi.org/10.1016/j.rser.2018.10.014.

Esmat, A., de Vos, M., Ghiassi-Farrokhfal, Y., Palensky, P., & Epema, D. (2021). A novel decentralized platform for peer-to-peer energy trading market with blockchain technology. *Applied Energy*, 282. Available from https://doi.org/10.1016/j.apenergy.2020.116123.

Faizan, M., Brenner, T., Foerster, F., Wittwer, C., & Koch, B. (2019). Decentralized bottom-up energy trading using ethereum as a platform. *Journal of Energy Markets, 12*(2), 19–48. Available from https://doi.org/10.21314/JEM.2019.193.

Han, D., Zhang, C., Ping, J., & Yan, Z. (2020). Smart contract architecture for decentralized energy trading and management based on blockchains. *Energy*, 199. Available from https://doi.org/10.1016/j.energy.2020.117417.

Islam, M. M., Jahan, I., Hasan, M. K., & Jang, Y. M. (2020). A decentralized energy trading system based on public blockchain. In International Conference on ICT Convergence (Vols. 2020, pp. 658–661). IEEE Computer Society. Available from https://doi.org/10.1109/ICTC49870. 2020.9289434.

Kongu Engineering College, S. S. S. (2021). Blockchain endorsement technology—A review of future smart paradigms. *International Journal of Aquatic Science, 12*(03), 975–981. Available from http://www.journal-aquaticscience.com/article_135423_f7ba958d09cbaa-fe5a780021aae46b28.pdf.

Narayanaswamy, T., Karthika, P., & Balasubramanian, K. (2022). *Blockchain enterprise: Use cases on multiple industries. EAI/Springer innovations in communication and computing* (pp. 125–137). Springer Science and Business Media Deutschland GmbH Issues 978-3-030-76215-5. Available from https://doi.org/10.1007/978-3-030-76216-2_8.

Roopa, C., Suganthe, R. C., & Shanthi, N. (2020). Blockchain based certificate verification using ethereum and smart contract. *Journal of Critical Reviews, 7*(9), 330–336. Available from https://doi.org/10.31838/jcr.07.09.71.

Shanthi, N., Suvitha, R., & Suganthe, R. C. (2020). Blockchain based e-voting approach in P2P network. *Journal of Critical Reviews, 7*(9), 337–342. Available from https://doi.org/10.31838/jcr.07.09.72.

Song, J. G., Kang, E. S., Shin, H. W., & Jang, J. W. (2021). A smart contract-based p2p energy trading system with dynamic pricing on ethereum blockchain. *Sensors, 21*(6), 1–27. Available from https://doi.org/10.3390/s21061985.

Suthar, S., & Pindoriya, N. M. (2020). Blockchain and smart contract based decentralized energy trading platform. In 2020 21st National Power Systems Conference, NPSC 2020. Institute of Electrical and Electronics Engineers Inc. Available from https://doi.org/10.1109/NPSC49263.2020.9331883

van Leeuwen, G., AlSkaif, T., Gibescu, M., & van Sark, W. (2020). An integrated blockchain-based energy management platform with bilateral trading for microgrid communities. *Applied Energy*, 263. Available from https://doi.org/10.1016/j.apenergy.2020.114613.

Yahaya, A. S., Javaid, N., Alzahrani, F. A., Rehman, A., Ullah, I., Shahid, A., & Shafiq, M. (2020). Blockchain based sustainable local energy trading considering home energy management and demurrage mechanism. *Sustainability*, 3385. Available from https://doi.org/10.3390/su12083385.

Peer-to-peer energy trading with blockchain: a case study

11

Whig Pawan
Vivekananda Institute of Professional Studies, New Delhi, New Delhi, India

11.1 Introduction

As global greenhouse gas emissions fueled by fossil fuel burning have put a question on energy generation using traditional non-renewable sources and threaten climate by the burning issue of global warming, modern renewable energy is a dominant solution. Clean energy is known as green energy that can be used to offset conventional energy because it releases fewer pollutants and is renewable. As a result, green energy infrastructure and platforms are increasingly being built, and there is an increasing need for renewable energy expansion (Andoni et al., 2019).

In this chapter, we introduce a new energy exchange mechanism to unlock new energy resources, which could be used in the future using the blockchain. Blockchain is an open-source technology where all transactions online are registered, and all transactions can be read, submitted, or checked. Blockchain is a digitalized accounting records scheme whereby all transactions are recorded in detail according to a series of cryptographic principles to avoid irregularities like malicious interference from hackers.

Each node in a blockchain in this energy trading platform transforms into a prosumer, who is able to generate, sell, and buy electricity, as well as exchange it across a peer-to-peer network without the need for a central organization. Trading is executed remotely without the involvement of a third party because it is built on smart contracts (Alam et al., 2019).

Different exchange criteria can be set, but the simplest method of trade is when the sale and purchase amounts match depending on a price set according to the overall energy level. This allows us to use energy more effectively by establishing a non-traditional active energy trading network. At this time, even as the number of power networks and transaction records stored in the blockchain network is still unknown, we have started to create private networks that participate only in the allowed nodes without fear of node failure and malicious attacks to avoid the attacking nodes from buying and selling untraceable resources.

11.2 Blockchain

The blockchain is a distributed data network infrastructure based on P2P technology. As the blocks are clustered into chains, the blockchain gets its name. In the blockchain

Blockchain-Based Systems for the Modern Energy Grid. DOI: https://doi.org/10.1016/B978-0-323-91850-3.00004-4

per network, the user distributes and stores data from the history of transactions in block format. Since any node has a single private key and public key, the secret key and the hash function can be used to carry out a cryptographic signature on the transaction. Each node uses the public key to check that the transaction has indeed been signed by the digital signature subject (Baroche et al., 2019). The block that includes this transaction has a structure of "chains" that are constantly linked to the previous block and with a timestamp, as shown in Fig. 11.1. Each user can review their transactions by reading their own logs since every user has a transaction history. Transactions not authenticated cannot then be stored in the block. Blockchain thus has three features: data confidentiality, safety, and decentralization (Boyd, 2010).

Corporate databases are the life behind applications running at a fast pace. The more quickly and accurately the applications execute, the better, and thus data processed into information is vital to keep the infrastructure in place. Blockchain is suitable to provide the information because the storehouse of this enormity of data is an immutable ledger timestamped for an extra layer of security, shared across the network, fully open, and can be accessed by approved network members only. A network of blockchains will monitor orders, transfers, accounts, development, and much more. And because participants have a common understanding of the facts, all the transaction information is end-to-end and gives a node increased security and a high level of efficiency (Brenzikofer & Melchior, 2019).

11.3 Main elements of blockchain

11.3.1 Distributed ledger

Distributed ledger technology (DLT) is a digital-asset-transaction-based distributed ledger (DL) structure in which transactions and their data are registered simultaneously in several locations; hence, the ledger is distributed and not centralized. DLT is the only way to register the transaction. In contrast to conventional databases, DLs don't have a central store or management function.

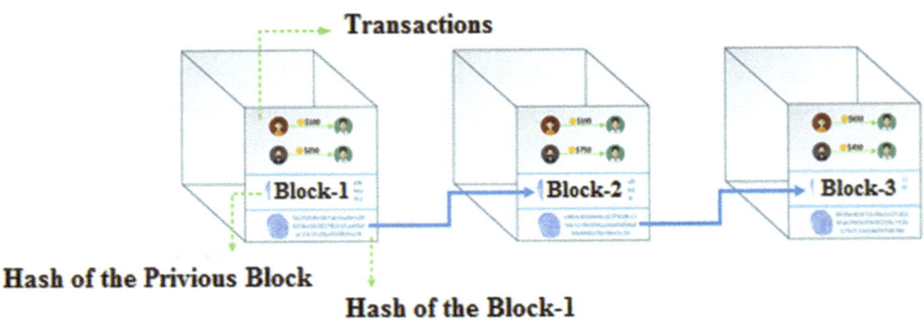

Figure 11.1 Basic of blockchain.

Each node processes and checks each item on a DL, generating a record of each item and agreement about its veracity. You may use a distributed directory for recording static records, such as a register and dynamic transactions. This computing architecture reflects a major change in record-keeping by changing the collection and communication of records.

11.3.1.1 Source of the ledger books

Ledgers—mostly transaction records and related figures—have remained on paper for thousands of years. They became digitized at the end of the 20th century with the development of computers, while computerized bookkeeping software usually reflected what used to appear on paper. However, the validity of transactions reported in the records must be checked by a central authority during the time period of the transaction. For for example, financial transfers need to be checked by banks (Buth et al., 2019).

The distributed directory of transactions that is a blockchain and its unchangeable transaction log are simultaneously accessible to all network members for reading, but to write, only the first miner who has the private key of the public key for the records will create the new timestamped block visible to all miners and thus avoiding double recording. This mutually agreed on directory of records scripts transactions only once, avoiding the repetition of effort characteristic of conventional corporate networks. The workflow in the DL is shown in Fig. 11.2.

11.3.1.2 Importance of distributed ledger (DL)

The DLT will speed up transfers because it eliminates the need for a central agency or intermediary. Likewise, distributed booklets will reduce transaction costs. Experts are also of the opinion that DLT processed information is much safer when each network node has data and creates a mechanism that is harder to hack or target effectively. Many often see a distributed block as a much more open method of managing records, as the content is being exchanged and, thereby, witnessed in a network (Cali & Cakir, 2019).

The use of DLT in financial transactions has been significant early. This is comprehensible because of the worldwide usage of bitcoin, the digital asset of the first

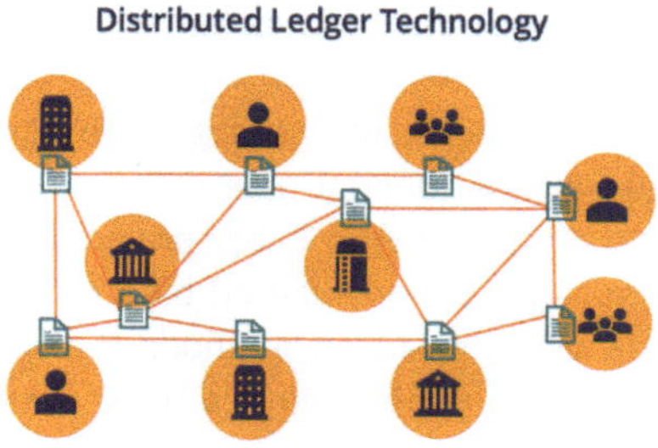

Figure 11.2 Distributed ledger technology.

blockchain, although it also shows how DLT may really function. Early innovation of blockchains took shape in banks and other financial institutions. However, in addition to financial purchases, DLT advocates claim that digital books can also be used in other fields, including government and industry. Experts think that automated blockchains will be used to collect taxes, pass land deeds, distribute social services, and also vote. They also state that DLT can be used to process and implement legal documents and the like.

Some claim that people can use this technology to collect, monitor, and selectively distribute personal data when necessary. Use cases include medical records and company supply chains. Proponents also believe digital booklets will help to monitor the rights and possession of literature, commodities, songs, movies, and more intellectual property (Cheng et al., 2019).

11.4 Future of DL systems

It is an ongoing question whether DLT innovations, such as blockchain, can revolutionize the way economies, organizations and businesses operate. Academic and financial press articles asked whether DL systems as they currently exist are secure enough to be used widely. This modern model of trade and security issues is not subject to any legislation, and this is where experts suggest practicing caution.

11.4.1 Immutable records

Once the transaction has been done and recorded in the public ledger, no user can manipulate it. If an error occurs in a transaction log, the error must be reversed by adding a new transaction, and all transactions are then available. Immutable record is represented as shown in Fig. 11.3.

The word "immutable" can almost always be found across hundreds of posts and discussions surrounding blockchain. Immutability—the ability to remain a constant, unchanging, and unalterable transaction background for a blockchain ledger—is a definite attribute that blockchain evangelists emphasize as an important advantage. Immutability will turn the audit process into a quick, reliable, and cost-effective process to increase confidence and integrity in the use and sharing of data companies on a daily basis.

Trillions of dollars would be spent on strategies for cybersecurity that can be invaded, and hackers steal our confidential data out of our sight. However, we seldom take precautions to stem the domestic cyber protection fight: to ensure that our data is not compromised, substituted, or falsified by a corporation or its employees. In certain instances, we clearly depend on methods such as private keys and user permissions to correct data. Yet, methodologically or mathematically, the knowledge is clearly flawless in a typical program database. The next (and costly) line of security is auditing (Laszka et al., 2019).

Figure 11.3 Representation of immutable record.

The introduction of blockchain will provide a regular degree of confidence for data companies—immutability gives credibility (both in its technical and primary definition). Blockchain allows customers to show that the information we used to have is not manipulated, and at the same time, the audit process can be turned into an effective, sensible, and economical operation.

11.5 What is immutability?

A brief introduction to hazardous cryptography follows: we need to consider cryptographical hashing before diving into blockchain immutability. The fundamentals are here: A hash function takes the current data and outputs a "Checksum"—a number and letter string that serves as digital signatures.

The check amount is guaranteed to indicate your exact data entry—if even one byte differs between the two files, the outputs are absolutely two strings after hashing. This can be associated with an avalanche effect—you can dramatically modify the performance in a limited way. SHA-2, Secure Hash algorithms (and its variants: SHA-256 is the most common in the blockchain world) is probably the most popular hashing algorithm. It was developed by the NSA.

11.6 Cryptography + blockchain hashing process = immutability

When a transaction is validated by the blockchain, the network is time-marked and inserted in a "block" of records, encrypted by a hashing mechanism that links the current block to the previous block that also keeps the hash of the previous block's data, thus integrating the block to the chain as a sequential update. The new block includes metadata from the previous block's hash output in the hash method for a new block. This connection during the hashing process renders the chain "unbreakable"—after validation, data cannot be manipulated or deleted, so the subsequent blocks in the chain cannot deny the change was attempted. In other words, as data is manipulated, the blockchain breaks down, and the cause can be found easily.

This feature is not present in conventional databases where details can be easily changed or erased. At a certain time, the blockchain is simply a reference to the truth. For bitcoin, these details include bitcoin transmission detail. The picture below in Fig. 11.4, illustrates how, as part of the header, the checksum of the transaction data has been applied to and becomes a checksum for the whole block (Brenzikofer & Melchior, 2019).

11.7 SC (smart contract) through blockchain

A set of rules—a smart contract—is saved and performed automatically on the blockchain for fast transactions. An intelligent contract can establish criteria for transfers of company bonds, including requirements for travel insurance and many more. A Block diagram to represent the smart contract is given in Fig. 11.4.

Intelligent contracts are self-executing computer programs bearing the terms of an arrangement between peers. The terms of an arrangement to take place, or the provisions of the arrangement, shall be included in the code of the smart contract. The smart contract runs on the open network of the Ethereum blockchain. The arrangements make capital, shares, property, or other commodities easier to trade. There are two commonly used smart contracts in Ethereum—Solidity and Serpent—programming languages. Solidity is an advanced programming language used on the Ethereum blockchain platform for integrating smart contracts. It allows developers of blockchain to review the programmed smart contract instead of reviewing it while compiling at runtime.

Two parties, traditionally, use a reputable third party's expertise to carry out the arrangement when the two parties enter into a deal. For millennia, it has been handled this way. However, the implementation of smart contracts and their associated applications automate a difficult manual operation. This chapter includes a run over

Figure 11.4 Smart contact block diagram to represent smart contract.

the technologies behind smart contracts and how they can be used. Let us first understand some main benefits of smart contracts over conventional contracts:

11.7.1 Savings for intermediaries, automation, and time

Taking days and even weeks to perform a conventional contract, the sheer number of lower and middle-level procedures in finalizing the deal delays the process.

Smart contracts can be carried out on a computer under predefined environments in just minutes since they are automated and programmed. There is no involvement of third parties.

11.7.2 Safety

Individuals with standard contracts are worried about privacy and stability. Safety can be jeopardized at any time with too many intermediary parties involved. When using smart contracts, security is retained by encryption, the public key, and private keys. The data is almost difficult to change when held in a decentralized environment. Smart contracts are entered into remotely using private keys. The public key exchanged by the participating parties cannot be decoded.

11.7.3 Precision and openness

Conditions and terms in a smart contract are predefined and pre-integrated. Once a condition has been reached, the smart contract is sent over the network and is registered automatically. If every transfer is part of a conventional contract, it is a manual procedure with workflows for consent at every level of the hierarchy. The groups concerned, peripheral bodies, and intermediaries traditionally dictate accountability. It's a device that's flawed. However, smart contracts are 100% open and available online $24 \times 7 \times 365$ days. Anyone will check and verify the transactions archived. Traditional contracts are impossible to archive since they are offline and dependent on paper and pen.

11.7.4 Fair

Compared to smart contracts, traditional contracts are costly mainly because all the intermediaries must be paid for their services. There are no intermediaries for smart contracts, and only the exchange costs derive from the underlying smart contract networks.

11.8 Benefits of blockchain

What has to be changed: activities also need expending time to maintain redundant records and to validate third parties. The record-keeping mechanism can be prone to cyberthreats from hackers and data thefts. Weak data verification can slow down computers by limiting clarity. Trade rates soared with the advent of IoT. All this is slowing down the industry, draining the end-users' expectations from the computers—which suggests that we need to move in a better direction. Here are some benefits of blockchain (Tushar et al., 2018).

11.8.1 Trust

One should be sure that one has received reliable and timely data from the blockchain as a part of Network-Only access and that private information can only be exchanged with network users to which you have specifically accessed.

11.8.2 Consensus on security

Both network participants need a security agreement on data validity and accuracy, and all authenticated transactions cannot be changed since they are forever registered. Nobody can uninstall a transaction, not even a system administrator.

11.8.3 Sustainability

The time-consuming record reconciliation is avoided with a public ledger exchanged by users of a network. And a series of laws, known as an intelligent contract, may be saved and exceeded instantly in a blockchain for pace transactions.

11.8.4 Energy storage systems requirement

Energy storage systems (ESSs) can help to improve the distributed generation (DG) operation and electricity delivery and help to eliminate system uncertainties. Conventional energy systems depend only on spinning reserves and auxiliary services rotational generators. Most DG micro-sources, especially renewable generation units, are not, however, able to meet these requirements and, therefore, depend on outside storage. The following are some requirements of the ESS in DG systems. The block diagram of ESS is shown in Fig. 11.5.

Fuel l-powered power plants are usually kept standing to ensure the spinning reserve, but they still take the appropriate time to react. ESS can help to ramp up the power supply in times of need without spinning reserve in the case of renewable energy systems. Recent advanced storage systems can increase in seconds to minutes. This means that a well-managed ESS can replace a considerably bigger reserve spinning. A consumer may also store energy for power backups for brief periods.

11.8.5 Batching

In general, the many entities that operate fastest need the highest energy supply resources and generators. It generally includes combustion engines with less efficiency and more emissions. It involves gas-powered systems. Efforts are underway

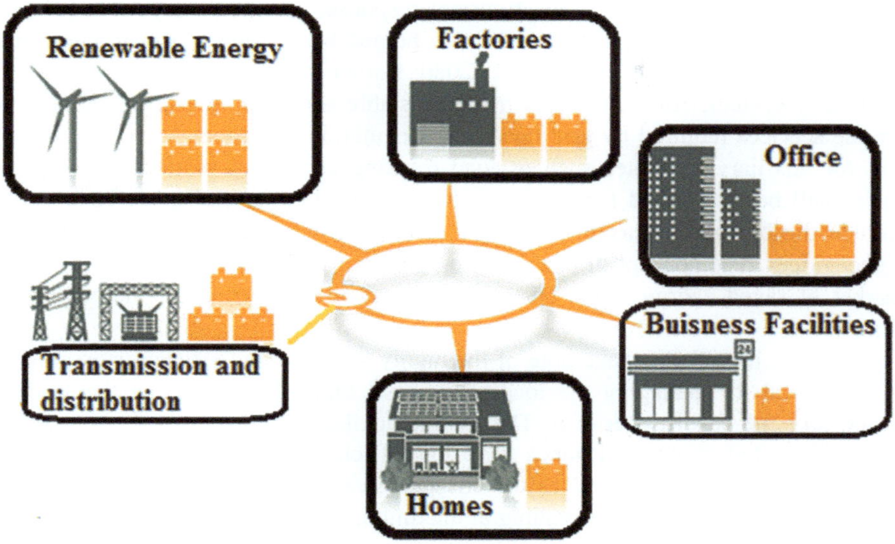

Figure 11.5 Energy storage system (ESS) block diagram to represent different ESSrequirements.

to decrease the demand curve from these entities through the improvement of end-users ' energy efficiencies, information on demand response, and highest price tactics. To store power is an attractive approach to controlling the peak demand curve. During off-peak periods, they are able to store and release energy at peak times. Therefore, the peak reserve is very sensitive and adaptable. Allowing for the direct management of local power producers, ESS—the program for demand response—will also lead to a reduction of emissions in peak shaving.

Inclusion of DG renewable sources: When the demand is low, wind power is produced mostly at night. The storage and delivery of this energy form at times of demand, therefore, increasing the DG system's efficiency. When the same storage systems are supplied on the side of the user, then, in low congestion time, wind energy can be transferred at night and stored near the supply point. T&D lines will also reduce congestion at peak periods, producing fewer failures and electricity outages. In the same way, just so far as the sun is brilliant is solar energy accessible, so that the generation is saved at peak time and at night time.

Increased solar energy penetration led to a critical situation known as the cane curve. The system providers had enormous problems ramping up supply to meet demand shifts. One of the advantages of ESS is that it can help in boosting supply at high times in such scenarios. It also improves penetration by eliminating variability and intermittent use of renewable generators.

Support for power quality: Renewable energy integration poses a wide range of power quality problems, from swells, slugs to fluctuations and spikes. Storage systems can be implemented, which can quickly respond to fluctuations in the system to maintain the standard of power quality according to the grid. The ESS can help to remove harmonics, LVRT, and transient response. The frequency, voltage control, and also quality of energy in DG are helped by flywheels, superconductors, and fast-response batteries (Jiang et al., 2020; Kathuria & Whig, n.d.).

Annex services: Annex services maintain stable and reliable grid operation and are the services required by grid operators. Regulation of frequency is very important for auxiliary attributes such as load tracking energy. A platform for energy trades will be established for the future market based on several other factors that include the charge at peak times at low prices and the delivery at peak times at higher prices. The concept of peak shaving goes hand in hand, but the focus on the market of saved energy is the best possible way to accommodate fluctuations in demands.

However, it is vital to take note of the highly sensitive and efficient storage systems utilized for managing the loading process. Otherwise, losses in the storage would take the benefits out of it. The routine auxiliary service creators are usually run as the rated power and hence have low efficiency and significant emissions. Thus, ESS may be considered as the cheaper green alternative for auxiliary tasks that liberates generators from running at maximum efficiency.

11.9 Energy storage systems type

For many decades, humanity has been practicing energy storage and conservation. Traditional electricity storage faces issues that are based on hydrogen and electrochemical storage. The various ESS can be classified as shown in Fig. 11.6 based on the technique involved in it. Fig. 11.7 demonstrates the global share of different installed capacity-based energy storage technologies.

11.10 Pumped hydro-storage

This energy storage technology puts storage on the grid and is commercially installed and operated most extensively. Pumped hydro-storage utilizes two tanks at varying heights that use inexpensive power from an off-peak grid that pumps water from below to above. If required, the stored water will be released into the bottom tank, and the revolving turbines create power. They last around 50–60 years and can have huge energy capacity for several months. Their efficiency varies between 70%–80% in terms of plant capacity, height difference, and turbine type. A considerable period of investment and planning, and gestation is required. New elements, such as speed pumping, are included in the new advances in hydro pumping.

11.10.1 Storage of compressed air power

In vast subterranean tunnels and mines, compressed air is stored as a sort of potential energy. When necessary, they are combined with natural gas and burned to

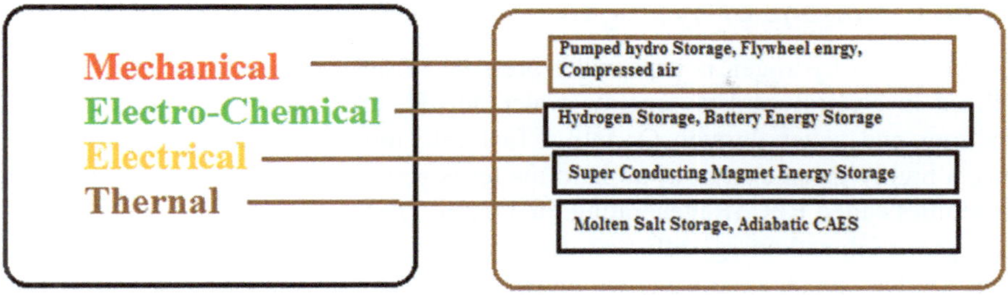

Figure 11.6 Energy storage systems (ESS) classification of ESS.

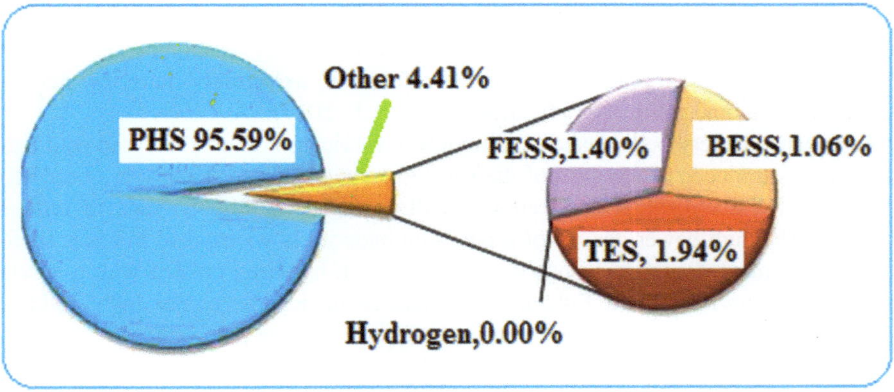

Figure 11.7 Energy storage systems (ESS) global share of different ESS.

generate electricity using expansion turbines. Diabatic compressed air power with a low-efficiency value of approximately 50% is commonly dubbed Compressed Air Energy Storage (CAES). When heat is withheld during compression and employed as a type of thermal energy during the discharge process to heat air, it is called an Adiabatic CAES (Khorasany et al., 2019; Laszka et al., 2019; Li et al., 2019). These technologies continue to be investigated and, if realized, will attain a very high efficiency of around 75%. It's utilized in numerous places, such as North America and Australia. With energy capacity and long-term storage, CAES provides tremendous advantages. But they also need enormous expenditure and efficient for the round journey.

11.10.2 Various energy storage system (flywheel)

FESS has excellent effectiveness, long life, high power, and other desirable features, at least without the need for maintenance. High self-discharge generated via deceleration and robust spinal losses are the major drawbacks.

11.10.3 Storage of hydrogen

The generated hydrogen is kept under pressure in separate tanks, and electricity is utilized in electrolyzers to separate water into H_2 and O_2. For the production of electrons and water through O_2 into a fuel cell, this hydrogen is useful. This is, thus, a highly green energy as no hard matter is generated as an output. This is also sometimes called RFC (Regenerating Fuel Cell). These cells are very modular, scalable, powerful, and energy-efficient. However, 50 to medium efficiencies are available, and they are automatically deployed.

Battery Energy Storage Systems (BESSs) are the most frequently installed and used storage systems for commercial purposes. The basic notion is that power is converted into some electrochemical shape and stored as an electrolyte in it. The electrolytes react during the discharge, and the reverse effect produces electrical power. Over the years, a number of different battery types have been created, making the battery storage technique, as illustrated, extremely adaptable and diverse (Fig. 11.8). Non-rechargeable batteries are referred to as primary batteries, mainly used in medical and military applications (Liu et al., 2017).

To enhance battery performance and minimize servicing in the 1890s, lead-acid batteries were introduced in the form of the latest innovations. Ni-Cd battery has been utilized for wind stabilization application but discouraged because it leads to serious pollution. The high-power density of lithium-ion batteries is widespread in smart small mobile equipment. Recently Li-based polymer system has been created with a strong renewable integration feature. Enhanced NaS battery applications can be found to have applications in high energy situations and are efficient in being renewable.

Recent research into metal-air batteries has had many applications. With high specific energy, lithium-air batteries are the most enticing, followed by zinc-air batteries. Metal-air batteries, on the other hand, are particularly vulnerable to fire due

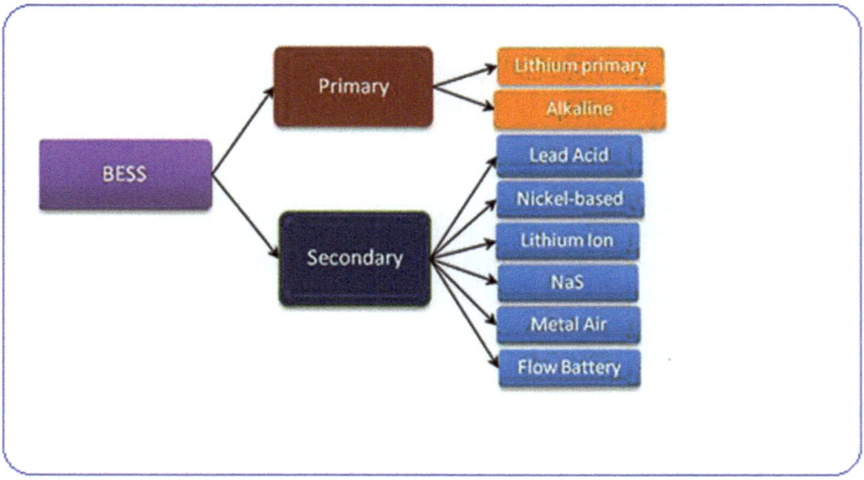

Figure 11.8 Types variety of battery types.

to their high metal humidity and air reactivity. Rechargeable batteries are those that store electrical fluids in separate tanks outside of the battery. This contributes to an increase in the power capacity of many plugs. To increase their capacity, include highly modular and scalable electrode cells. The electrolytes in the tank may be simply replenished with instant recharging.

Thanks to hundreds of cycles of recyclable electrolytes, the system is very efficient. They are also termed redox batteries since, during charging or discharge, the battery interacts with reduced oxidation. Vanadium redox (bromide polysulfide, zinc bromide, and iron-chromium) batteries are the specific varieties of flow batteries. Complete detail and characteristics of battery storage devices are provided in [7].

11.11 Magnetic energy store that are superconducting

Every super conducting magnetic energy storage (SMES) unit features a low superconductivity belt, a power converter unit, and a cooling unit. When supercooled metals demonstrated superconductivity, electromagnetic energy was discovered many years ago. This characteristic for the stockpiling of electric energy began at the end of the 1960s when powerful superconducting coils such as four Kelvin magnetic fields were studied. Materials that can also work at the 100 Kelvin were recently researched (Pop et al., 2018; Sorin et al., 2019). The system is highly efficient, but it is becoming less and less powerful. Furthermore, cooling systems raise system expenses and maintenance and are still in the early phases of development.

11.11.1 Dual-layer electric condensers or super condensers

Electrostatic charges are used to store electricity in these devices. It is usually constructed of a parallel dielectric construction that retains the charges. This technique bridges the gap between typical electrical capacitors and batteries due to its massive rechargeable batteries' power capacity. They are extremely quick and can do rapid load and discharge cycles since they are not transformed. These are incredibly efficient (90%), environmentally safe, easy to recycle, and can handle a wide range of applications. It has a good self-rate and is rapidly being used in electric vehicles and propulsion devices.

11.12 Thermal energy storage

Thermal storage has gained popularity as solar energy penetration has risen. Excess heat could be kept in buildings or outside of buildings as alternative power sources for future use. The snow or liquid salt techniques are now being developed, and it is expected that once operational, they will provide a significant amount of the electricity required for heating. Thermal storage on ice, which includes latent heat storage, is said to reach exceptional efficiency when utilized in solar production.

11.12.1 Energy management energy storage systems for distributed generations

There are various issues with the integration of renewable systems and the consequent ESS needs. It provided an overview of the many applications that can benefit from ESSs. Some applications, such as power smoothing and frequency regulation, necessitate the use of short-term storage devices capable of charging and unloading large amounts of power. Arbitration and maximum saving systems require an increase in energy storage to ensure minimal self-discharge energy from stored energy.

Depending on the applications, the suitable storage type and size must be selected and scaled for integration. Because storage solutions are costly, they must be carefully managed to guarantee longevity and performance; otherwise, they can cause facilities and equipment (F&E) difficulties, decreased performance, and premature failure. To ensure that distributed energy resources are used efficiently, proper management of both storage and producing devices is essential. As a result, power consumption has become a hot issue among renewables scientists and producers (Pop et al., 2018; Sorin et al., 2019). The various components of the energy management system are shown in Fig. 11.9.

Power control strategies are a vital component in the design of a self-contained and self-sustaining renewable energy system. The storage solution may be run to handle various applications stated in a microgrid, and a management technique will be offered to achieve the preset goals while minimizing the effect on both the storage system and the renewable energy system. According to Nykamp et al. (Nykamp, Bosman, & Molderink, 2017), the use of a storage device is primarily focused on peak shaving, and renewables integrated uses from the standpoint of a Distribution Network Operator (DNO).

A private energy trader, on the other hand, aims to maximize income through arbitration and supplementary services by utilizing the storage system. The simulation results from a direct comparison of several uses of the same storage in a distribution grid were offered. Power consumption is often dependent on the state of energy storage in order to reduce charging complications. Reihani et al. created a

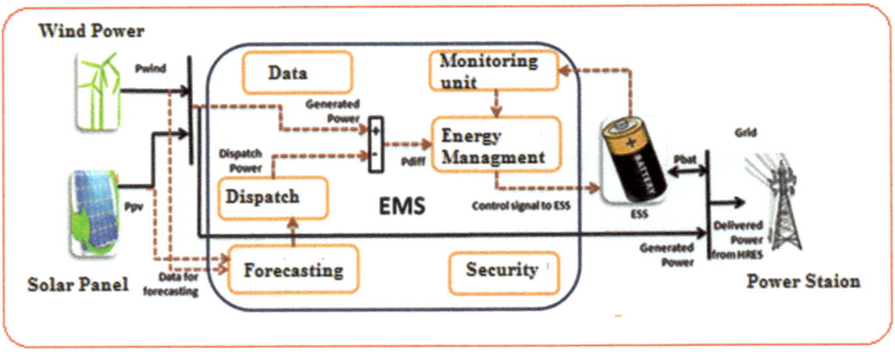

Figure 11.9 Components of an energy management system.

simulation of battery ride-scale storage for peak load shaving, power smoothing, or voltage management in distribution systems.

Tewari and Mohan looked at the possibility of using the battery sector while assessing and controlling a sodium sulfide battery to transmit wind energy between peak-to-peak times. To achieve wind energy smoothing, Daz-González employed a vector-controlled flywheel-based storage system for grid integration, whereas Chandra et al. used a frequency oscillation damping battery system. In order to enhance hydrogen system performance, Zhou et al. evaluated the size of an independent PV hydrogen system utilizing an optimum power management technique (Sorin et al., 2019; Whig, 2018, 2019).

Garcia et al. created a smart energy management method where the proposed hybrid system's operating costs to establish power exchange among hydrogen and energy storage outperformed standard state-based energy management techniques. Cheng et al. used particle swarm optimization (PSO) in conjunction with a roulette wheel re-distribution mechanism to incorporate a balance of power equality constraints into the energy management technique. The power management technique compensates for the effects of battery storage depth depletion and provides prolonged service life by penalizing state of charge (SOC)-based charge and discharge.

Gamarra and Guerrero provide an in-depth examination of the different optimization approaches for energy management and microgrid design. Wang et al. introduced and evaluated a hierarchical energy management method to reduce operating costs by minimizing wind and load uncertainty on an RT Lab practical platform. Mohammadi et al. improved the handling of a microgrid to minimize costs and control vulnerabilities using Hong estimations and stochastic-based frameworks . The Commission proposed a microgrid commitment formulation rely on the cuckoo search method.

Chaouachi et al. have tried an online strategy for energy management with smart, multi-objective optimizer technology to minimize cost and emissions for hybrid microgrids. In a DC microgrid integrated with a new source, Chen et al. used the intelligent fuzzy management of the SOC battery, which improved battery life. Feroldi et al. designed control based on a horizon retreat strategy which places the highest priority on wind energy and the lowest priority on hydrogen energy. The strategy improved the power supply by approximately 88%. In the wind-p V-diesel hybrid storage system, R. Palma-Behnke et al. used the rolling horizon strategy to evaluate set points of production and storage for optimized microgrid operation based on demand-side management. The results of this strategy have been developed. Dynamic programming based on stochastic models has been employed to adapt to wind energy uncertainties and changes in the market price (Pop et al., 2018; Whig & SN, 2018; Zaidi & Hong, 2018).

Due to transmission system constraints, ESSs can also find uses for the control of renewable energy reduction in dispersed systems. Fu modified and ran a distribution feeder as a microgrid to assess and monitor power quality concerns using renewable resources and ESS. While integrating energy storage technology for grid market management applications, Silva Monroy and Watson faced some critical challenges. Abdeltawab et al. have presented a market-oriented wind-battery-storage management approach aimed at maximizing profit within a deregulated market framework.

11.13 Benefits of peer-to-peer energy

The buying and selling of energy between two or more grid-connected parties is known as peer-to-peer energy trading (P2P). Any extra energy, which is frequently in the form of solar energy, maybe exchanged and leased to the other customers through a stable platform. Consumers may choose who they buy power from or sell it to through peer-to-peer energy markets.

Excess energy from the sun is being sent to the network for a minor feed-in tariff. However, as more individuals seek flexibility and control over how their resources are dispersed, this technique is becoming outmoded.

Energy is traded on a secure network, which frequently employs blockchain-based techniques. Blockchain is a network platform for processing and storing data, such as asset transactions. Renewable credits, which may be exchanged through the database, are one example of these assets. Consumers will be able to share their extra energy amongst themselves and decide how it is delivered through microgrids using peer-to-peer power trading platforms like blockchain.

Anyone without solar panels may still buy renewables from their neighbor at a fair price, and those who sell their extra energy can sell it for a premium cost than they would earn as a given tax through their supplier.

Electricity transportation expenses are reduced since energy does not have to be delivered from centrally situated power facilities. According to Aurora Energy, 41.1% of the consumer's electricity bill is spent on monitoring the poles and cables that carry power from plants to consumers' homes.

Renewable energy may be used to generate electricity, which offers a number of advantages. Electricity may be purchased from a reliable source. This offers a variety of options for dealing with other customers and eliminates the middleman.

All activities on a blockchain are transparent, and then once recorded, they cannot be changed in any way without validation by the network, ensuring complete transparency.

11.14 Conclusion

Although peer-to-peer energy trading isn't quite ready for widespread adoption now, it's gradually gaining traction as an option for the future. Consumers will no longer depend on energy retailers for their power consumption and will be able to make wise, sustainable decisions about when they use and transmit energy as a result of this technological advancement.

This chapter broadly covers the use of blockchain in peer-to-peer energy trading, and its discussion can prove very helpful for researchers working in the same field.

References

Alam, M. R., St-Hilaire, M., & Kunz, T. (2019). Peer-to-peer energy trading among smart homes. *Applied Energy*, *238*, 1434–1443. Available from https://doi.org/10.1016/j.apenergy.2019.01.091.

Andoni, M., Robu, V., Flynn, D., Abram, S., Geach, D., Jenkins, D., McCallum, P., & Peacock, A. (2019). Blockchain technology in the energy sector: A systematic review of challenges and opportunities. *Renewable and Sustainable Energy Reviews*, *100*, 143–174. Available from https://doi.org/10.1016/j.rser.2018.10.014.

Baroche, T., Pinson, P., Latimier, R. L. G., & Ben Ahmed, H. (2019). Exogenous cost allocation in peer-to-peer electricity markets. *IEEE Transactions on Power Systems*, *34*(4), 2553–2564. Available from https://doi.org/10.1109/TPWRS.2019.2896654.

Boyd, S. (2010). Distributed optimization and statistical learning via the alternating direction method of multipliers. *Foundations and Trends® in Machine Learning*, 1–122. Available from https://doi.org/10.1561/2200000016.

Brenzikofer, A., & Melchior, N. (2019). Privacy-preserving P2P energy market on the blockchain. arXiv. https://arxiv.org

Buth, M. C., Wieczorek, A. J., & Verbong, G. P. J. (2019). The promise of peer-to-peer trading? The potential impact of blockchain on the actor configuration in the Dutch electricity system. *Energy Research and Social Science*, *53*, 194–205. Available from https://doi.org/10.1016/j.erss.2019.02.021.

Cali, U., & Cakir, O. (2019). Energy policy instruments for distributed ledger technology empowered peer-to-peer local energy markets. *IEEE Access*, *7*, 82888–82900. Available from https://doi.org/10.1109/ACCESS.2019.2923906.

Cheng, S. F., De Franco, G., Jiang, H., & Lin, P. (2019). Riding the blockchain mania: Public firms' speculative 8-k disclosures. *Management Science*, *65*(12), 5901–5913. Available from https://doi.org/10.1287/mnsc.2019.3357.

Jiang, Y., Zhou, K., Lu, X., & Yang, S. (2020). Electricity trading pricing among prosumers with game theory-based model in energy blockchain environment. *Applied Energy*, *271*115239. Available from https://doi.org/10.1016/j.apenergy.2020.115239.

Kathuria, K., & Whig, P. (n.d.). Macroeconomic Implications of the Monetary Policy Committee Recommendations: An IS-LM Framework. ACTA SCIENTIFIC AGRICULTURE, 2020 2, pp. 1–7.

Khorasany, M., Mishra, Y., Babaki, B., & Ledwich, G. (2019). Enhancing scalability of peer-to-peer energy markets using adaptive segmentation method. *Journal of Modern Power Systems and Clean Energy*, *7*(4), 791–801. Available from https://doi.org/10.1007/s40565-019-0510-0.

Laszka, A., Eisele, S., Dubey, A., Karsai, G., & Kvaternik, K. (2019). TRANSAX: A blockchain-based decentralized forward-trading energy exchanged for transactive microgrids. In Proceedings of the International Conference on Parallel and Distributed Systems—ICPADS (Vols. 2018, pp. 918–927). IEEE Computer Society. Available from https://doi.org/10.1109/PADSW.2018.8645001

Li, Y., Yang, W., He, P., Chen, C., & Wang, X. (2019). Design and management of a distributed hybrid energy system through smart contract and blockchain. *Applied Energy*, *248*, 390–405. Available from https://doi.org/10.1016/j.apenergy.2019.04.132.

Liu, N., Yu, X., Wang, C., Li, C., Ma, L., & Lei, J. (2017). Energy-sharing model with price-based demand response for microgrids of peer-to-peer prosumers. *IEEE Transactions on Power Systems*, *32*(5), 3569–3583. Available from https://doi.org/10.1109/TPWRS.2017.2649558.

Nykamp, S., Bosman, M. G. C., Molderink, A., et al. (2017). Value of storage in distribution grids-competition or cooperation of stakeholders? *IEEE Transactions on Smart Grid*, *4* (3), 1361–1370.

Pop, C., Cioara, T., Antal, M., Anghel, I., Salomie, I., & Bertoncini, M. (2018). Blockchain based decentralized management of demand response programs in smart energy grids. *Sensors*, *18*(1), 162. Available from https://doi.org/10.3390/s18010162.

Sorin, E., Bobo, L., & Pinson, P. (2019). Consensus-based approach to peer-to-peer electricity markets with product differentiation. *IEEE Transactions on Power Systems*, 994–1004. Available from https://doi.org/10.1109/tpwrs.2018.2872880.

Tushar, W., Saha, T. K., Yuen, C., Liddell, P., Bean, R., & Poor, H. V. (2018). Peer-to-peer energy trading with sustainable user participation: A game theoretic approach. *IEEE Access*, *6*, 62932–62943. Available from https://doi.org/10.1109/ACCESS.2018.2875405.

Whig, P. (2019). Artificial intelligence and machine learning in business. *Engineering Reports*, *2*, 8–13.

Whig, P. (2018). Simulation & performance analysis of various R2R D/A converter using various topologies. *International Robotics & Automation Journal*. Available from https://doi.org/10.15406/iratj.2018.04.00108.

Whig, P., & Ahmad, S. N. (2018). Novel pseudo PMOS ultraviolet photo catalytic oxidation (PP-UVPCO) sensor for air purification. *International Robotics & Automation Journal*, 393–398. Available from https://doi.org/10.15406/iratj.2018.04.00155.

Zaidi, B. H., & Hong, S. H. (2018). Combinatorial double auctions for multiple microgrid trading. *Electrical Engineering*, *100*(2), 1069–1083. Available from https://doi.org/10.1007/s00202-017-0570-y.

Blockchain-enabled electric vehicle charging

12

M. Jajini[1], *N. Kamaraj*[2], *M. Santhiya*[3] *and S. Chellam*[1]
[1]Velammal College of Engineering and Technology, Madurai, Tamil Nadu, India, [2]Thiagarajar College of Engineering, Madurai, Tamil Nadu, India, [3]Department of Electrical and Instrumentation Engineering, Kongu Engineering College, Erode, Tamil Nadu, India

12.1 Introduction

Blockchain is the emerging technology in the fields of secured sharing of medical data, cross-border payments, logistics and food selling, etc. Simply put, it is a decentralized, shared network computing that demonstrates the overall origin of the digital item. Blockchain is trying to extend its features in numerous new industries, given its high potential. With its expertise proven in the supply chain, banking, digital authentication, investment management, and the arts and music sector, it is now expanding into the fast-growing market of EV. The foremost barrier to EV adoption is range anxiety. The majority concern is about power capacity running out on long road journeys. This is due to the fact that electric chargers are not readily available everywhere and, as a result, of the absence of charging infrastructure, potential consumers remain hesitant to switch to EVs. As already said before, blockchain is a peer-to-peer (P2P) technology that enables users to share their chargers when not in use. Moreover, in turn, charger owners will be benefited by getting paid for using the chargers. It also allows people and organizations to exchange, purchase, and transfer extra energy between their selves without the intrusion of a central authority or general public having control over how and when transactions can happen.

Experts think that by allowing P2P charging, this new blockchain enterprise will minimize drivers' range anxiety and boost the desirability of EVs in the future. This chapter shows the relationship between the broad field of machine learning (ML) and blockchain. Since ML is a venture of artificial intelligence (AI) where systems can analyze information processed from data, identify patterns, and make decisions with little or no human intervention, ML with Blockchain Fig. 12.1 can create a deadly combination if utilized properly. Both are more or less like the two sides of a coin if used together. While blockchain helps to store correct data that is unaltered and permanent, ML can utilize this data to analyze data patterns and give accurate predictions.

In blockchain's P2P network, mobile apps are used by users to access participating homeowners' locations to charge their EV. It always keeps track of the time taken by each vehicle to fully charged and the amount of energy consumed for the same. Further, a ledger account helps in the implementation of payment in digital

Blockchain-Based Systems for the Modern Energy Grid. DOI: https://doi.org/10.1016/B978-0-323-91850-3.00008-1

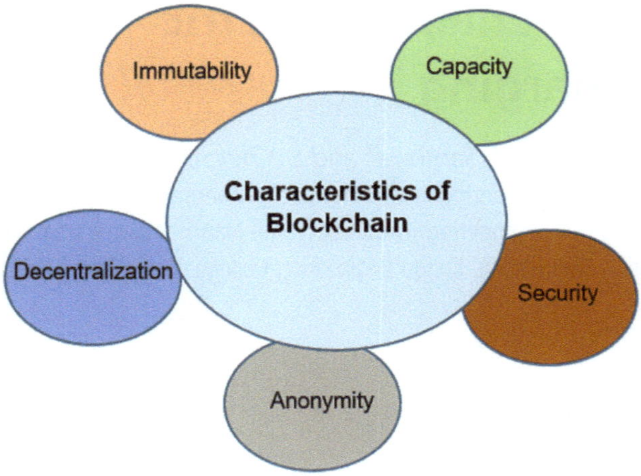

Figure 12.1 Characteristics of blockchain.

mode from the driver to the charger owner linked in the network. With the data collected from the blockchain, the data can be used by the ML algorithm to predict the availability of fast charging ports nearby. This will reduce the waiting time and charging time of the vehicle and will also lead to hassle-free transportation even as drivers can get rid of anxiety rising from driving range issues.

Features of Blockchain:

12.2 Types of blockchain technologies

12.2.1 Public blockchain network

Anyone can join public blockchains, and all the information is accessible by them. This makes blockchains suitable for firms that value transparency and trust concerns in data storage. As a result, NGOs and social support organizations can benefit from the public blockchain. The public blockchain is depicted in Fig. 12.2.

Because of its public nature, the data stored by a public blockchain cannot be used for private-sector operations since the information is not secured for the use of just one entity or organization but can be accessed by anyone. Furthermore, public blockchains can be costly to maintain since they require nodes to function as miners and run either proof-of-work or proof-of-stack (PoS) algorithms (PoS).

12.2.2 Private blockchain network

Because it provides a private network, the private blockchain is the right opposite of the public blockchain. It's ideal for companies who desire a private network while still reaping the advantages of blockchain. They're also centralized, which

Figure 12.2 Types of blockchain.

means a firm may govern the network without having to keep it accessible to the public.

They provide all of the key benefits of a blockchain, as well as a method for corporate members to develop trust through immutability and security.

The company can develop its own set of rules and operate the network according to its own needs, secure in private networks.

12.2.3 Consortium blockchain network

Next, rather than a centralized node (private) or a decentralized (public) network, a consortium blockchain network is functioned by a group of organizations or nodes. A consortium blockchain is advantageous since it includes nodes that have been pre-selected.

It's great for a solution that necessitates cross-departmental cooperation. Supply chain, food, and medicine, for example, would all necessitate cross-brand collaboration.

12.2.4 Hybrid blockchain network

Finally, we have the hybrid blockchain, which is the most recent blockchain technology type. This blockchain combines all of the advantages of both private and public blockchains with the least number of drawbacks.

In actuality, the hybrid blockchain, like other blockchain technology types, has drawbacks though fairly minor ones.

Hybrid blockchains appear to be ideal for a variety of prospective business models.

12.3 Blockchain applications

12.3.1 Peer-to-peer based energy exchange between renewable microgrids

Grid-connected microgrids, as demonstrated by TERI, will be used for access to energy, P2P energy markets, and utility network management. The blockchain verifies the transactions and ensures security, safety, and while maintaining efficiency in each step of transaction processing with the help of digital ledger basics. The blockchain is a chain of several digital units that collects and shares replicas of a transaction such that each node recognizes what transaction actually happened at a specific point in time.

A "transaction" is a method of exchanging data between two physically located nodes, the digital version of which is saved as a record on the blockchain, as shown in Fig. 12.3. The notion of cryptography, which falls under the domain of information security, is one of the bases on which blockchain technology is developed. Each physical node's information is encrypted before being disseminated on a blockchain network using a private and public address key (Ping et al., 2020).

12.3.2 Peer-to-peer based vehicle charging

With this strategy, the blockchain has been proposed as a supporting tool for accelerating the adoption of EV in India. Several vehicles commuting through their itineraries might be charged at privately created EV charging stations, and also the transaction of charging site owners' bids, energy flow, accounting, and legitimate transaction closure might all be handled by the blockchain.

The blockchain-enabled application could be used to increase the utilization of privately owned EV chargers through the proper linkages between EV customers and EV private owners in an attempt to compensate for a shortage of charging stations and to make greater use of existing facilities. EV owners can use the P2P mode of communication for accessing nearby charging stations to charge their vehicles' batteries.

Owners of electric car charging (EC) stations can enroll themselves on blockchain-based mobile apps and can have the charging stations connected to the

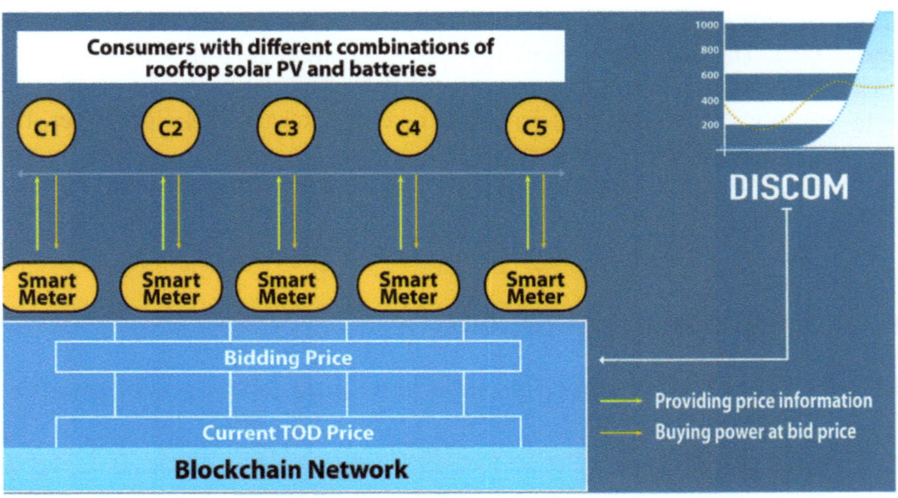

Figure 12.3 Energy exchange topology.

public directories and render such charging stations accessible to the public if they're not in operation. In return, they might be compensated for offering charging stations to other EV owners who meet certain criteria. The EV owner can schedule a charging session using his or her own encrypted passwords, a card, or online payment methods, even as the blockchain will handle the invoicing and accounting on a real-time basis Fig. 12.4.

12.3.3 *Aggregating distributed energy storage: a blockchain-based topology*

Electric vehicle, like power plants, will be used as decentralized energy storage systems capable of producing electricity in response to demand analysis. Many of these EV charging stations might help the localized distribution network by operating as a virtual power plant (VPP) in the region, and the blockchain could help secure and simplify the transactions. The EC aggregator may pool the available stored energy from several EV charging stations, letting them compete for power supply and alleviating stress on the local DT. (Dinh et al., 2017) and (Huang et al., 2020) refer Fig. 12.5.

As a result of the swapping of electricity units and monetary value, a huge number of transactions can take place. When the local transformer level is balancing the grid, the efficiency and availability of transactions in a secure way will enable optimal use of electricity stored in EVs through blockchain technology. This would also create a market that is both competitive and transparent. Each and every transaction will be informed to each player aggregator and grid operator, making participation simple and easy.

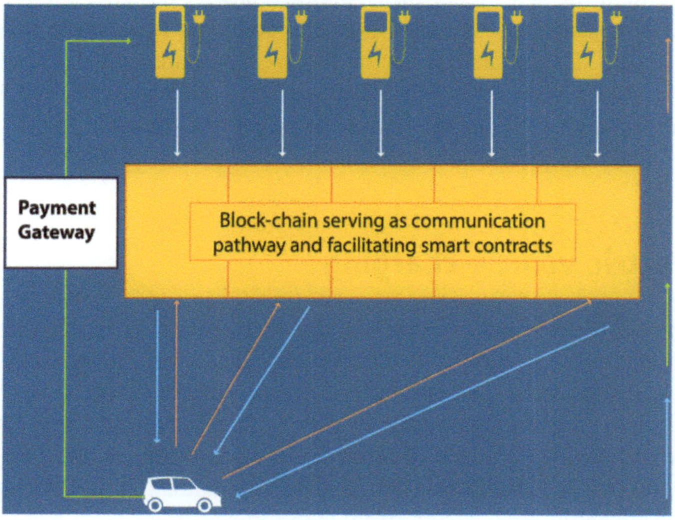

Figure 12.4 Vehicle charging topology.

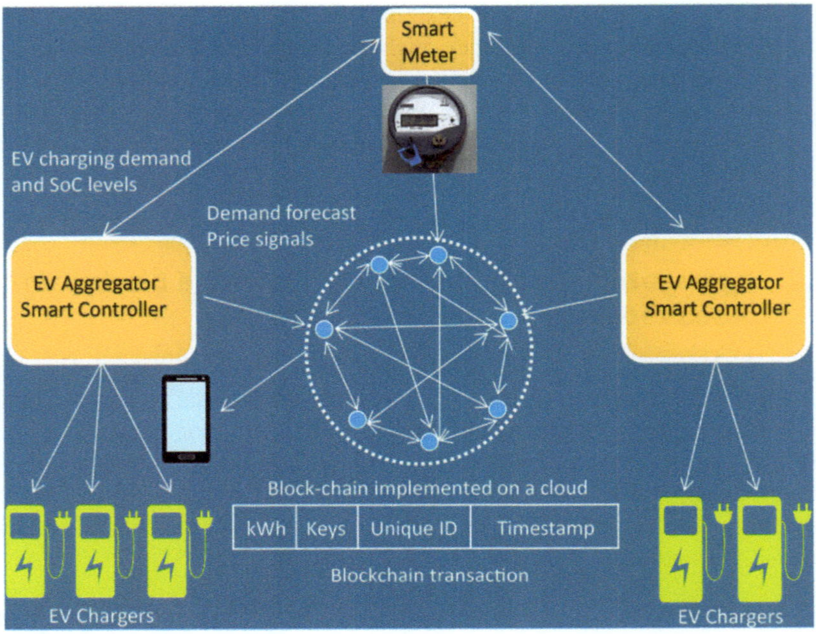

Figure 12.5 Distributed energy storage.

12.3.4 Blockchain—financing for rooftop solar PV

(Wang et al., 2021) A blockchain may be used to develop a rooftop solar setup incorporating transactions in a better way. Blockchains may be used to provide a level playing field for investment firms, entities, project designers, and prospective buyers to strike a settlement between financial companies and project designers as well as customers. This is also projected to give transparency throughout the loan process, from loan award through EMI payment to loan payback and energy bill reduction, all using smart contracts on the blockchain as shown in Fig. 12.6.

12.4 Electric vehicle charging

Electric vehicles are expected to represent the next generation of mechanical power for transportation and progress. These vehicles have the ability to alter energy usage, age, and redirection. Electric vehicles can overcome the negative environmental impact of conventional vehicles (EV). Traditional vehicles emit a large amount of fossil fuel by-products that are released into the air and inhaled by the general public, making individuals immune to contamination and ozone-depleting substances. An electric car is a significant step forward in terms of making a significant contribution to the globe (Li et al., 2014; Tang & Zhang, 2016).

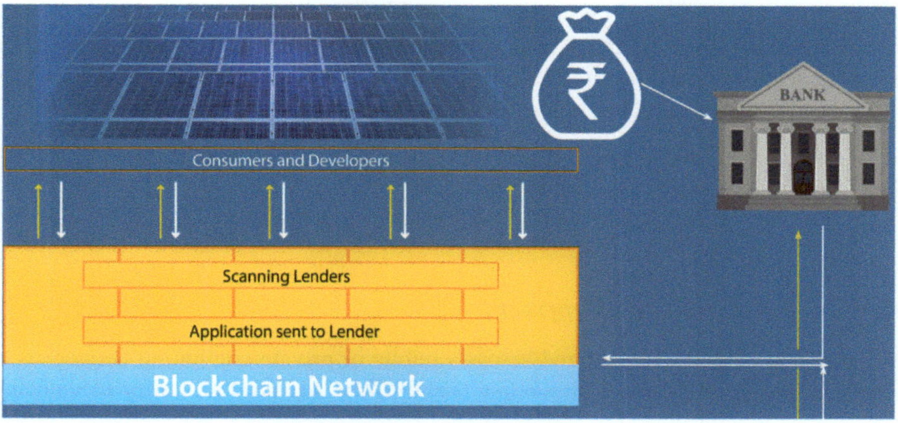

Figure 12.6 RoofTop solar PV.

The introduction of blockchain technology for cryptographic money has given the world yet another venture. It provides a reliable record that keeps track of transactions in a safe manner. Blockchain is a technology that underpins bitcoin, Ether, and other common digital currencies. The blockchain has applications that are beyond the use of digital currency. Blockchain is a type of decentralized record, and not all decentralized records use block or chain swapping by default. This article proposes a blockchain-enabled P2P energy exchange and payment system for EVs (Mengelkamp et al., 2018). As the process outlined below shows, the outcomes of the system can provide a lot of time and resources to enter into these projects now, and they are less difficult and risky comparatively:

- Activate P2P EV energy trade and local energy balancing with DLT.
- Make it easier to pay at charging stations.
- EV owners get access to real-time pricing data.
- A mobile application is needed.
- Both EV and charging station owners must register in the mobile app.
- As a P2P exchange layer, the architecture uses an e-wallet and brilliant agreements on the public Ethereum Blockchain, including the Euro-based Mobility Token.
- P2P services allow EV and charging direct owners to lease their charging framework

We've proposed a blockchain-based platform for EV charging installment plans. The proposed framework is practical because it addresses the issue with EVs, which are deemed more environmentally friendly than vehicles based on petroleum derivatives. Due to the safe and reliable decentralized dataset supported by a blockchain, the suggested shared energy exchanging and charging installation framework for blockchain-based EVs can gain validity and broad applications in real-world transportation settings (Zhu et al., 2020).

Clients can manage their own data, EV statistics, and charging usage under this framework. We used IBM's Hyperledger texture, which is an open-source stage. This is a simple-to-use portal for a private blockchain. Clients can use the

Hyperledger texture's smart contract feature to describe the set of points they need to execute organically. When a smart contract is started, the standards embedded in it will be measured as a result.

This article's core commitments shall be summarized (Fig. 12.7) as follows:

- Suggesting a framework for distributed energy swapping and charging for EVs.
- To build trust between clients using blockchain technology.
- Making use of smart contracts for the electronic payment of charge bills.

12.4.1 Relevance and impacts

A distributed ledger that is shared across a public or private computing network is a catalog of data. The system eliminates the need for a central authority to validate transactions among different parties to avoid double-spending. Before being added to the ledger, each piece of data is mathematically encrypted and attached as a new "block" to the historical records chain, which is then validated by different consensus methods. Smart contracts are a club of conditions stored on the blockchain that activate transactions when the requirements are met and can also be encoded into the blockchain (Xiang et al., 2018).

12.4.2 Related works

Several works use smart contracts to implement exchanged charging. Huang et al. (2020) suggested one decentralized security protocol based on lightning organization and brilliant agreement 90 innovation to work on the security of trades between EVs and CPs. According to Xiang et al., 2018, every charging point has the brilliant meter that will monitor the blockchain network and identify the charging station's strength, and also the implicit correspondence module collaborates information with

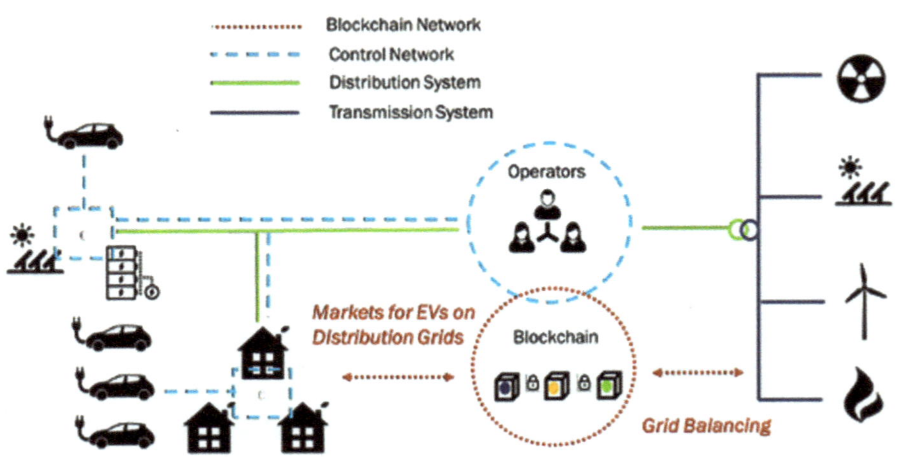

Figure 12.7 Electric vehicle charging mechanism.

the keen agreement on the blockchain. Finally, the miners (nodes of the network) combine the information and add it to the blockchain. The article of Su et al. includes 95 blockchain and clever agreements for executing safe charging administrations for EVs.

In various studies, they classified EV charging into three varieties: uncontrolled charging, delayed charging, and smart charging. Uncontrolled charging is the process in which charging stops only when the battery reaches full battery capacity. Oli Systems GmbH, Stuttgart, is assisting with this endeavor. detaches it at some point in the future. This is a kind of manual that is occasionally exchanged. Delayed charging is where service providers use the responsibility to encourage EV owners to shift their EV load from peak to off-peak hours as per their convenience. Lastly, smart charging is a process of charging in which the EV charging station is limited by a mathematical formula. The conveyance framework can initiate smart charging.

Blockchain-enabled simply means that the end-user has a simple and secure payment system that uses digital currencies like bitcoin or Ether and can be paid with either a conventional wallet application or an RFID/NFC card (Aste et al., 2017). The following are the primary advantages of paying for a charging cycle using one of these ways:

- There is no need to pre-register with a local charging provider.
- There is no outside installment or charge card provider; thus, there are almost no costs.
- EV charging could become the standard, both locally and globally.
- The charging station may be operated by governments, organizations, and private individuals (P2P) on the same homogenous blockchain framework, thanks to the incorporated blockchain innovation.
- On a global scale, data collection (for example, energy charge per charging cycle) becomes a reality.
- The charging station is a claimed IoT Oracle for catching ecological information, such as air pollution, and is foreordained to be the choice of smart urban populations of the future (Huang et al., 2020).

Instructions for making a payment for a charging cycle.

12.4.3 Payment can be made in two ways

To begin with, the EV owner can use a wallet program such as Jaxx, Coinbase, or Free wallet. Clients scan a QR code shown on the charging station's touchscreen and enter their required amount. When the transaction is confirmed, the touchscreen instructs the car owner to attach the charging cable to his or her vehicle. The charging process begins automatically.

The EV owner can also use an NFC card to pay. This card is known as the Ethan Card. The NFC card functions as a wallet with a wallet address. The customer can purchase a pre-loaded card with various amounts, such as $10, $25, $50, $100 USD, EUR, or other fiat currency. The amount purchased by the client is saved as a digital note such as bitcoin or Ether. The client can buy a blank card in any way and send any amount of Ether or bitcoin to the card's wallet address. The Ethan Card can be re-energized at any time with any amount. The card features a 4-digit PIN code to prevent unauthorized use. Ethan's NFC card for simple and secure installments is described in this (Nakamoto, n.d.).

12.4.4 Difficulties we need to confront while growing such a framework

Charging an EV is a delightfully perplexing procedure. Different mechanical principles are used, as well as numerous types of vehicles. The vehicle communicates with the charging station while charging, and each manufacturer must follow this convention. The ecological requirements are extremely stringent. The charging station should be used in the temperature range from $-20°C$ to $+60°C$ (-4 up to 140 Fahrenheit). Every component of an EV charging station must meet these requirements and work properly in these punishing conditions. We eventually found a solution to each problem we encountered. We were able to allocate various improvement projects to isolated engineers due to the seclusion of our EV charging station. Our charging station concepts will not receive configuration funds, but we expect to be able to work out the station's suitability to serve current scenarios with architects. However, as a proof-of-concept, the charging station should work as best as can be wanted.

12.4.5 System model

Our suggested charging framework for EVs based on blockchain has three elements: an administrator, an energy aggregator (EAG), and a client/EV, and it consolidates four stages: introduction, enlistment, confirmation, and charging as shown in Fig. 12.8.

1. To obtain charging administration access, EV enlists their personality with the administrator.
2. Electric Vehicle and EAG cross-check each other.
3. Energy Aggregator facilitates trade.
4. EAG verifies the legitimacy of the transaction and logs it on blockchain network circles.
5. Methodology.

Blockchain is a complex budding technology that is still in development, not well understood, and, in most cases, not ready for critical applications. As a result, we provide a robust method for identifying if an application use case is fit for a blockchain execution. Furthermore, many requests that could benefit from a blockchain backend rely on legacy programming, making essential engineering modifications impossible. We propose the following system for embedding blockchain innovation into applications, and we explain its implementation in the following case study using Fig. 12.9:

Stage 1: Find the elaborate gatherings and the trust relationships that exist between them. If the level of trust in a connection is not enough to achieve the application's goal, or if it blocks an activity necessary to achieve that goal, name it as a trust issue.

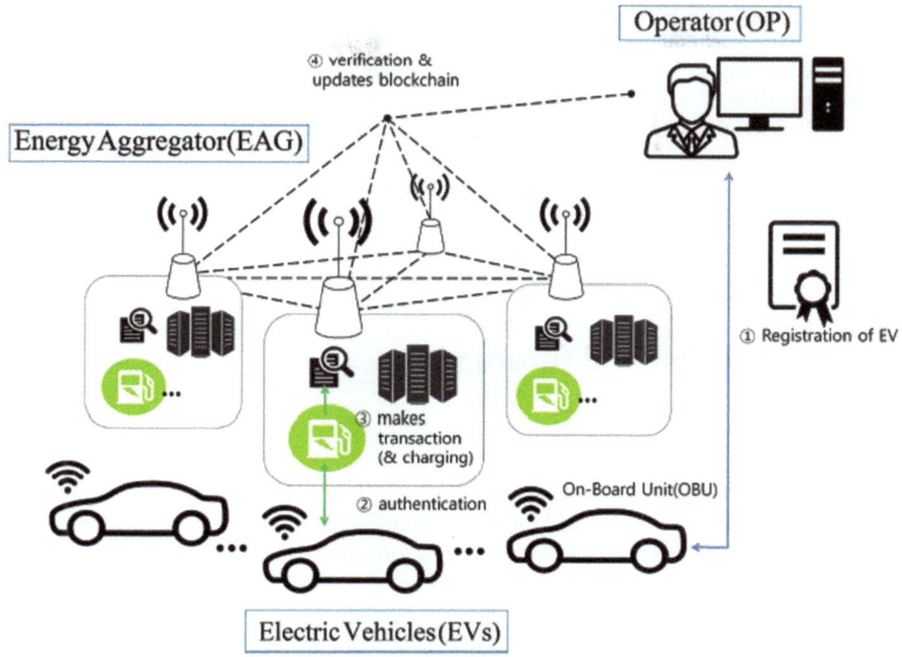

Figure 12.8 Elements of blockchain incorporating the proposed charging framework for electric vehicles.

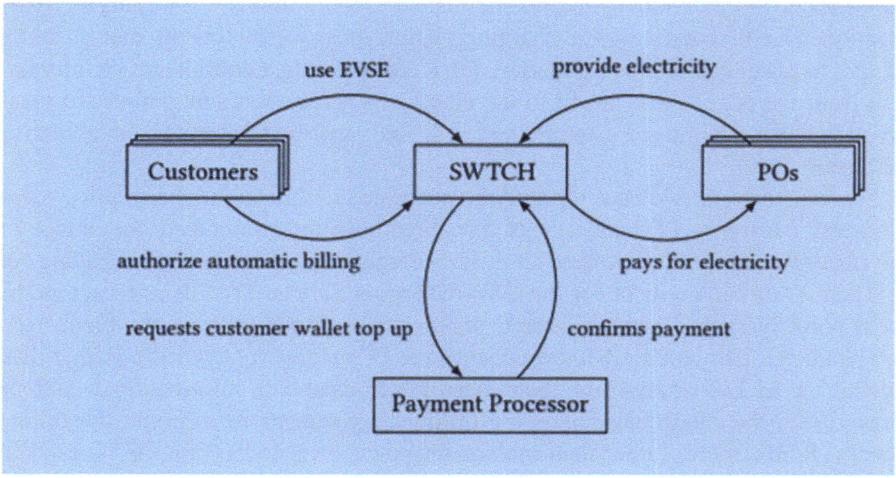

Figure 12.9 Stages in blockchain execution embedded into applications.

Stage 2: Develop a minimal blockchain framework, complete with smart contracts, that addresses the trust issues identified in the first phase. If parts of a legacy topology need to be replaced, the new topology should closely resemble existing interfaces so that conditions may continue to function with reduced adjustments.

Stage 3: Once the trust-relieving blockchain is in place, the rest of the structure can be shifted in stages over time. This allows the technique to evolve from a heritage/blockchain hybrid to a truly decentralized system in the long run. This technique shall be used in our contextual investigation for the remainder of this work.

12.5 Conclusion and future vision

We provide a technique for evaluating the availability of blockchains in decentralized energy systems. Our strategy is based on the assumption that resolving trust concerns is a unique feature of blockchains, as opposed to various distributed information topologies. We next offer a study identifying trust difficulties, how they might be addressed with blockchains, and a proof-of-concept development using Hyperledger Fabric (Musavi & Wilson Young, 2014). The proposed execution is consistent with the current inheritance backend to put it another way.

Embracing DLT is putting a lot of energy in front of a lot of different ventures. Outside of the financial sector, the energy sector is one of the industries where DLT could have the most unusual and beneficial, disruptive effects. DLT-based meter-to-cash robotization might completely eliminate the need for discount to-retail go-betweens, lowering exchange and operational costs. DLT has the potential to accelerate the development of the P2P energy sector, through which energy prosumers trade on a local scale. DLT may be able to assist the matrix executives by organizing smart contracts. DLT enables real-world trades to operate naturally in accordance with individually defined norms. DLT could improve EV charging coordination by collaborating with charging station energy installments and enabling drivers to reach continuous estimation. DLT can assist the coordination of information from the edge of the lattice to the cloud in a secure and straightforward manner, allowing data to be captured, stored, and analyzed later for collaborative purposes.

Moving forward, we can improve the suggested architecture by including more blockchain features. POs and clients link directly in this 'smoothed out' image of the plan of action, with one providing EV charging and the counterpart paying for it. There is no requirement for the CSP (Charging Service Providers) to act as the focal point in such a genuine shared organization. Chain code on the blockchain can handle all partnerships between clients and POs (Property Owners). Rather than relying on ad hoc analysis of their mass meter data, this information would be entered into the blockchain, allowing brilliant agreements to examine the billing system. Furthermore, rather than maintaining their own node(s) on the blockchain network near the CSP's hubs, all the hubs shall have oversight responsibilities by POs, with the CSP's server topology facilitating this.

References

Aste, T., Tasca, P., & Di Matteo, T. (2017). Blockchain technologies: The foreseeable impact on society and industry. *Computer*, *50*(9), 18–28. Available from https://doi.org/10.1109/MC.2017.3571064.

Dinh, T. T. A., Wang, J., Chen, G., Liu, R., Ooi, B. C., & Tan, K. L. (2017). BLOCKBENCH: A framework for analyzing private blockchains. In *Proceedings of the ACM SIGMOD International Conference on Management of Data* (Vol. 127746, pp. 1085–1100). Association for Computing Machinery. Available from https://doi.org/10.1145/3035918.3064033

Huang, Z., Chen, D., Lai, C. S., Zhao, Z., Lai, L. L., & Wang, M. (2020). A distributed transaction mechanism for electricity market with electric vehicles and blockchain. In 2020 8th International Conference on Power Electronics Systems and Applications: Future Mobility and Future Power Transfer, PESA 2020. Institute of Electrical and Electronics Engineers Inc. Available from https://doi.org/10.1109/PESA50370.2020.9344037.

Li, Z., Kang, J., Yu, R., Ye, D., Deng, Q., & Zhang, Y. (2014). Consortium blockchain for secure energy trading in industrial internet of things. *IEEE Transactions on Industrial Informatics*, *14*.

Mengelkamp, E., Notheisen, B., Beer, C., Dauer, D., & Weinhardt, C. (2018). *A blockchain-based smart grid: towards sustainable local energy markets, Computer Science - Research and Development* (Vol. 33, pp. 207–214). Springer Verlag Issues 1–2. Available from https://doi.org/10.1007/s00450-017-0360-9.

Musavi, F., & Wilson Young, E. (2014). Overview of wireless power transfer technologies for electric vehicle battery charging. IET Power Electronics,.

Nakamoto, S. (n.d.). Bitcoin: A peer-to-peer electronic cash system. Available from: https://bitcoin.org/bitcoin.pdf (accessed 27.05.19).

Ping, J., Yan, Z., Chen, S., Yao, L., & Qian, M. (2020). Coordinating EV charging via blockchain. *Journal of Modern Power Systems and Clean Energy*, *8*(3), 573–581. Available from https://doi.org/10.35833/MPCE.2019.000393.

Tang, W., & Zhang, Y. J. A. (2016). A model predictive control approach for lowcomplexity electric vehicle charging scheduling: Optimality and scalability. *IEEE Trans. Power Syst.*

Wang, L., Jiao, S., Xie, Y., Mubaarak, S., Zhang, D., Liu, J., Jiang, S., Zhang, Y., & Li, M. (2021). A permissioned blockchain-based energy management system for renewable energy microgrids. *Sustainability*, *13*(3), 1317. Available from https://doi.org/10.3390/su13031317.

Xiang, K., Chen, B., Lin, H., Shen, Y., Du, Y., & Yan, T. (2018). Automatic demand response strategy of local pure electric vehicle with battery energy storage system based on blockchain technology. In *2nd IEEE Conference on Energy Internet and Energy System Integration (EI2)*, Beijing, China.

Zhu, S., Cai, Z., Hu, H., Li, Y., Li, W., & Zkcrowd. (2020). A hybrid blockchainbased crowdsourcing platform. In *IEEE Transactions on Industrial Informatics*.

Blockchain-based systems for modern energy grid: a detailed view on significant applications of blockchain for the smart grid

13

Kaushik Sekaran[1], *J. Kalaivani*[2] *and M. Nikhil raghava rao*[3]
[1]Department of Computer Science and Engineering, ICFAI Deemed University, IcfaiTech, IFHE CAMPUS, Hyderabad, Telangana, India, [2]Department of Computer Science and Engineering, SRM Institute of Science and Technology, Chennai, Tamilnadu, India, [3]Vignan Institute of Technology and Science, Deshmukhi, Telangana, India

13.1 Introduction

A smart grid is a new generation of energy distribution technologies that include grid operators, smart meters, smart appliances, and renewable energy resources. The smart grid provides greater controllability and observability than traditional energy systems, as well as a solution for low energy consumption rates. Hence, the energy distribution business can make needed distribution and pricing adjustments. Internet technology is integrated with the smart grid, which is formally known as Internet of Energy (IoE). Such novel approaches strive to ensure that energy is available at all times and in all places. Because connectivity in Internet-enabled smart grids, aka IoE, is becoming more abundant, some issues have arisen. The IoE's centralization makes it vulnerable to single point failure and data tampering at a central node. As a result, in smart grid technologies, decentralization (Dinh et al., 2018) is a new and increasing trend, though numerous issues remain as a result of the Internet's involvement in the smart grid, such as user privacy and the security of power grid data saved and shared with a central node. As a result, ensuring the security and privacy of user information and power grid data is a top responsibility.

We propose an approach, which is a blockchain-based security protocol, to address all of these difficulties with smart grid technologies. Informally, blockchain is the digital currency's underlying technology (e.g., Bitcoin, Ether, etc.). It's a decentralized, tamper-proof, immutable, and transparent system that allows for peer-to-peer (P2P) transaction security without the need for a third party. The user identity and transactional data are permanently recorded in the form of blocks in the blockchain. Merkle root in the block header confirms whether anyone has

Blockchain-Based Systems for the Modern Energy Grid. DOI: https://doi.org/10.1016/B978-0-323-91850-3.00011-1

tampered with the transaction data or not. As a result, the key qualities of blockchain technology, where blocks are cryptographically structured in a particular data structure, are a tamper-proof and distributed shared ledger. By recognizing the blockchain features, the problems of traditional centralized smart grid technologies can be solved.

13.2 A blockchain framework for smart grid

The usage of blockchain in smart grids could enhance our current and future electric power systems in a variety of ways. The characteristics and operational principles of the blockchain in relation to the electrical power system provide more security, improved resilience, better transparency, greater scalability, less bureaucracy, and increased processing capacity.

13.2.1 Benefits of blockchain technology in the power/electricity industry

The financial services industry's impact on the blockchain has been thoroughly studied. Blockchain is gaining traction in the energy sector, where blockchain's effects on the transformation of the energy sector appear to be enormous. Innovations in the electrical industry, such as rooftop solar and electric vehicles, have constantly improved the industry. Blockchain is the next emerging technology to drive growth in the industry through its smart contracts and systems interoperability. The primary advantages of blockchain in the energy sector are cost savings (Sekaran & Krishna, 2017), environmental sustainability, privacy for stakeholders, etc.

The following are examples of blockchain products that can be adapted to address various energy or environmental applications:

- Electricity distribution at wholesale,
- Energy trading on a peer-to-peer basis,
- Data management for electricity,
- Trading in commodities,
- Utility company's exploration of oil and gas resources,
- Storage and transportation of oil and gas resources,
- Resource management and sale that is precise.

The usage of blockchain in smart grids could enhance our current and future electric power systems in a variety of ways. The characteristics and operational principles of blockchain, as represented in Fig. 13.1, in relation to the electrical power system (Tushar et al., 2015), provide more security, improved resilience, better transparency, greater scalability, less bureaucracy, and increased processing capacity.

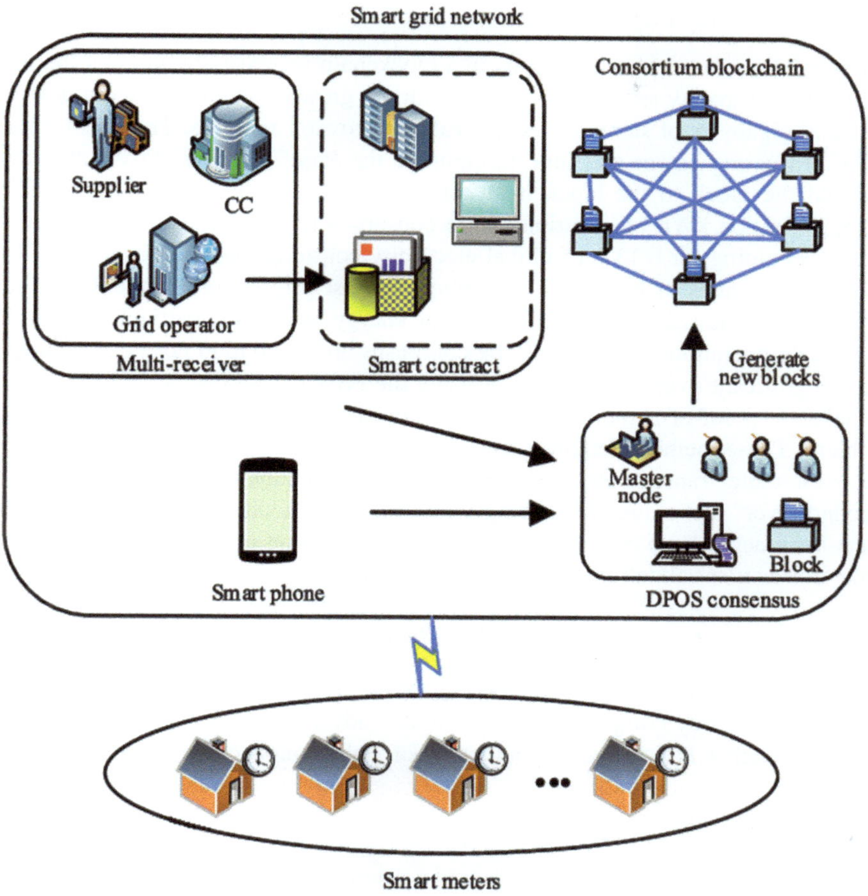

Figure 13.1 Architecture of blockchain framework for smart grid. Blockchain framework for smart grid through internet.

13.2.2 Challenges in implementing blockchain for power generation and storage

Both the demand for electricity and the need for electrical energy as an energy carrier are rising in our modern society. Due to socio-economic changes, renewable energy sources will play a significant part in future energy generation. Our current electrical power grid will face new challenges as a result of these distributed energy supplies. The arrival of a new grid user, the prosumer, who consumes and creates electrical energy (Tushar et al., 2015), is one of the difficulties posed to our current electrical power system. Another issue is the inconsistency of the supply of renewable energy sources, like solar and wind power. Blockchain applications in the electrical power system could provide solutions to some of the difficulties that the future electrical power system will encounter.

13.2.2.1 *Scalability issues*

Every day, the more the transactions in a blockchain increase, the more is the need felt for high-capacity storage to keep up with the growing volume. Bitcoin now has over 200 gigabytes of storage, whereas Ether has over 1 terabyte. The nodes need a lot of processing power and time to compute the Proof-of-Work (PoW) algorithm and add the block to the blockchain network. According to a study published in 2015, 30 billion kWh of electricity were consumed to conduct 30 million transactions, accounting for 0.13% of global electricity consumption (Tushar et al., 2015). Because thousands of consumers are buying and selling energy at the same time, the number of transactions per second in the energy business is particularly high for large-scale operations. As a result, the consensus and validation nodes are susceptible to significant overhead. Instead of using the PoW consensus technique, a proof-of-stake or proof-of-authority consensus procedure might be used to tackle this problem. These methods require a lot less computer processing power and can handle a lot more transactions. The community is also looking into other so-called "second-layer" techniques for coping with scaling difficulties. Off-chain and side-chain approaches have been created to reduce the amount of transactions and parallelize transaction validation, as depicted in Fig. 13.2.

13.2.2.2 *Centralization's probability*

Blockchain applications in the energy sector are a new technology that is open to attacks from energy companies looking to take advantage of it. The issue of power waste and delay has fueled the adoption of public blockchain designs. Because

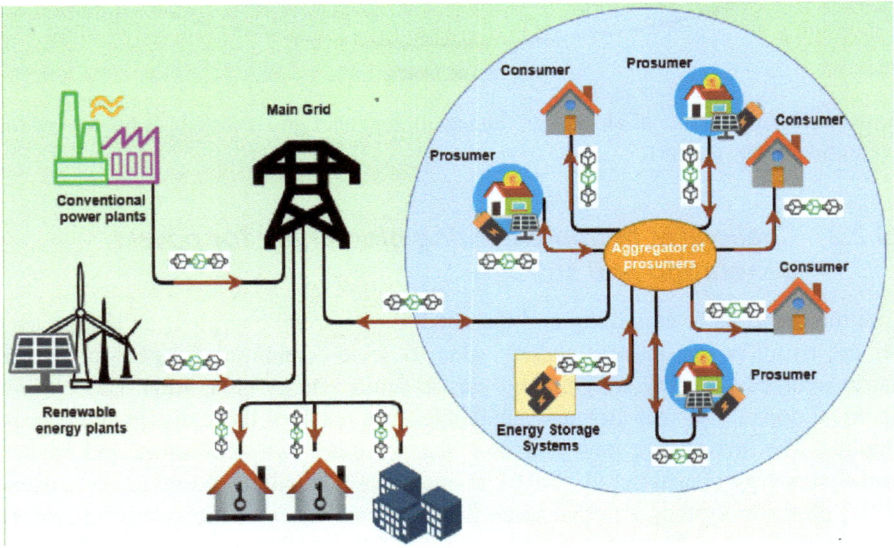

Figure 13.2 Challenges in implementing blockchain for power generation and storage. Challenges to implementing blockchain for power generation and storage.

public blockchains rely on a preset selection of nodes for validation and consensus, there is a risk of fraud. As a result, to assure security, strong government supervision should be adopted, particularly in the early phases.

13.2.2.3 Development and infrastructure costs

The grid's current infrastructure was developed with huge research, and it produces optimal outcomes at a far lower overall cost. In comparison to the blockchain, the grid communication system (Li et al., 2019) today employs more advanced and less expensive telemetry technology. As generation evolves, the technology and development regarding wireless networking and mobile networking will increase day by day. Portable networking has turned into the request of the day. The portable system is a correspondence system where each connection is remote. As the era advances, the innovation and improvement in regards to remote systems administration and mobile systems administration will build up every day. Significant media communications suppliers have given the component of voice and data cell frameworks over most of the region domain in the Earth. A key characterizing normal for a Mobile Networking Operator (MNO) is that an MNO must claim or access a radio range permit from an administrative or government element. A second key normal for an MNO is that an MNO must possess or pick up the components of the system base, which are important to give administrations to the endorsers over the authorized range of networks.

Universal or Pervasive Computing (UC) has risen to a great extent as an energizing new worldview. That incorporates framework for Pervasive computing and distributed registering which give figuring and correspondence administrations at whatever time and at any place. It is vital that smart frameworks work reasonably for UC. A standard convenient arrangement of servers and clients with a gigantic gathering of adaptable clients that move all through the land is the scope of the framework. In a distinction Mobile Ad hoc Network (MANET) is an immense social occasion of compact servers and clients. Each one of the center points are remote convenient and battery controlled. The topology can change more a significant part of the time. Traditional convenient frameworks incorporates the servers in all data correspondence. MANET gadgets are allowed to move freely in any heading, and can change its connections to numerous different gadgets in the system all the more every now and again. The versatile processing and specialized gadgets are driving an upheaval of change that is crucial for the information society. While framework-based systems give an incredible approach to cell phones in getting system administrations, it requires some investment and really high cost to set up the vital base. As the exploration group hunts down and preparing the stage after the Pervasive computing era, now the systems of remote sensors have turned out to be very intriguing in another environment where to look for examination challenges. These have been empowered by the substantial union of three key advances like computerized hardware, remote interchanges, and Micro Electro Mechanical Systems.

In every territory, progression in equipment innovation and building plan has prompted diminishing in size. Remote portable impromptu systems are a remarkable kind of system that have various specialized difficulties. Web portal hub viably stretches out web administrations to the non-framework zone. This broad writing devotes to the extension in the sensor systems. Now, we found a great and complete scope of sensor systems in this Internet era. The significance of these systems are shown by the two particular applications. The first is bury vehicular correspondence, where self-ruling vehicles try to speak with each other keeping in mind the end goal is to give data in regards to area, speed, and course to goal. In military and defense applications, numerous directing calculations have been proposed. The essential goal of a versatile transporter is to keep clients from changing to other portable bearers. Because of serious focused conditions in the business sector, there are couple of contrasts in connection to charges, with the telephones companies and discretionary administrations industries given by three portable bearers such as Docomo, au company and SoftBank. Along these lines, versatile bearers require valuable approaches to comprehend the portable transporter decision conduct with a specific end goal to build the quantity of clients. Most regular utilizations of remote sensor systems (WSNs) include the accumulation of tangible information from a Base Station (BS) frequently under the condition that certain information gathering delay limitations.

A major test for those systems is to bolster opportune information gathering with least vitality utilization at the sensor hubs (sensor networks (SNs)). In the primary approach single jump correspondence is required, in this way vitality utilization ought to be minimized, in any case, to the detriment of long information conveyance delay.

Portable Networking has turned into the request of the day. Versatile system is a correspondence system where each connection is remote. As era advances, the innovation and improvement with respect to remote systems administration and mobile systems administration will build up every day. Convenient Networking has transformed into the demand of the day. Convenient framework is a correspondence framework where every association is remote. As period advances, the development and changes with respect to remote frameworks organization and Mobile frameworks organization will be enhanced. Huge media represents that the providers have given the part of voice and information cell systems over the majority area in the Earth. A key portraying for a Mobile Networking Operator (MNO) is that the MNO must claim or get to a radio range and it must be allowed from a managerial or government component. Also the MNO must get the parts of the framework based on organizations that are in the endorsers approved range.

Among the applications and organizations which are controlled by mobile phones, the MNO framework affiliations supplying the data through organizations that have been given by the adaptable customers. While structure based frameworks gives a mind blowing way to deal with mobile phones in getting framework organizations, it requires some venture and truly high cost to set up the key base. Overall an another choice way to deal with pass on the organizations have been risen (Paudel et al., 2019). These were locked in around having the PDAs interface with each other in transmission extended through customized course of action. Inside the off the cuff frameworks organization field, remote convenient sensor frameworks had a phenomenal impact. A sensor framework is made out of a broad number of little sensor center points, which are usually thickly included inside the district in which a ponder is being checked. As the investigation amass chase down the get ready stage past the PC, now and then frameworks of remote sensors have ended up being exceptionally interesting in another environment where searching for examination challenges. These have been engaged by the generous union of three key advances like mechanized equipment, remote exchanges and Micro Electro Mechanical Systems.

Presently, gigantic measures of data is about to represent informal communities and social collaborations that are recorded, which can permit researchers to study social connections on an expansive scale and at a level of the detail that has never been conceivable. For instance, the procedure by which the general population find, get together, and organize gatherings of the people with the same interests shared with interpersonal organizations. Informal organizations as of now turn into an extremely noteworthy medium for the far-reaching dispersion of the news in a mass union occasions. For example, presidential races, and crises like the landfall of the Hurricanes. Be that as it may, there are still huge mechanical parts and the object is to show a brief review on the dissemination of data in versatile interpersonal organization and about portable systems administration. MSNs, for example, Facebook and Twitter have likewise been outstanding for giving an extraordinary accommodation amid the late exhibits in the Middle East. In light of these striking occasions, understanding the data dispersion in MSNs has turned into a basic research objective.

13.2.2.4 Legal and regulatory assistance

To support such services, new forms of contracts, specifically for the P2P trading system, as well as energy rate adjustments, are necessary. Under the current grid structure, such concerns are highly regulated. As a result, while blockchain technology has demonstrated its utility in the creation of microgrids, bringing it into the main grid framework without updated legal procedures is extremely challenging.

13.3 Electricity data management

Data management encompasses all disciplines concerned with data management as a useful resource. Energy management software monitors and reports on facility energy consumption, allowing businesses to reduce their energy costs. Energy management software typically provides tools for changing energy usage to achieve more use efficiency and lower carbon emissions.

Energy data management systems collect real-time usage data using internet of things devices and sensors. Rather than manually compiling this data, the EDMS automates data presentation where readable data is presented to be analyzed. Manufacturers can utilize this data to make active changes to their energy consumption strategies in order to improve user efficiency. An established EDMS is the cornerstone of a completely automated production. It converts data into digital form and offers clear evidence of energy consumption. Manufacturers can use an EDMS to improve their operations in a variety of ways, from environmental effects to cost savings (Fig. 13.3).

EDMS for manufacturers

Integrating an EDMS into your production line can provide your company with a wide range of capabilities and advantages (Sekaran & Krishna, 2016). You can:

Collect data from various sources: Multiple pieces of equipment and systems with varying power consumption are used in a typical industrial activity. Energy usage statistics throughout manufacturing lines and facilities management are an important part of comprehensive data collecting. When connected to the organization's application backbone, an EDMS makes this collection possible.

Receive an overview of energy consumption: An EDMS provides a picture of an organization's energy usage (Miglani et al., 2020) by combining data from numerous sources. It gathers all of this data in one place, allowing the owner to see trends in consumption and adjust operations accordingly.

Monitor data from new and existing equipment: Old equipment may fall behind in efficiency compared to more recent models. With an EDMS, the owner can compare usage between new and existing machines to increase efficiency and ensure effective consumption monitoring.

Identify areas to improve energy efficiency and reduce consumption: An EDMS gives the organization access to trend statistics and equipment performance, allowing the owner to identify and improve inefficient aspects. The owner can arrange repairs if they discover a machine that consumes a lot of energy but produces little output.

Evaluate deviations to avoid costly errors: For maximum output, each piece of machinery requires a specific amount of energy. The owner loses productivity and money when a machine's energy flow is clogged. When these discrepancies arise, an

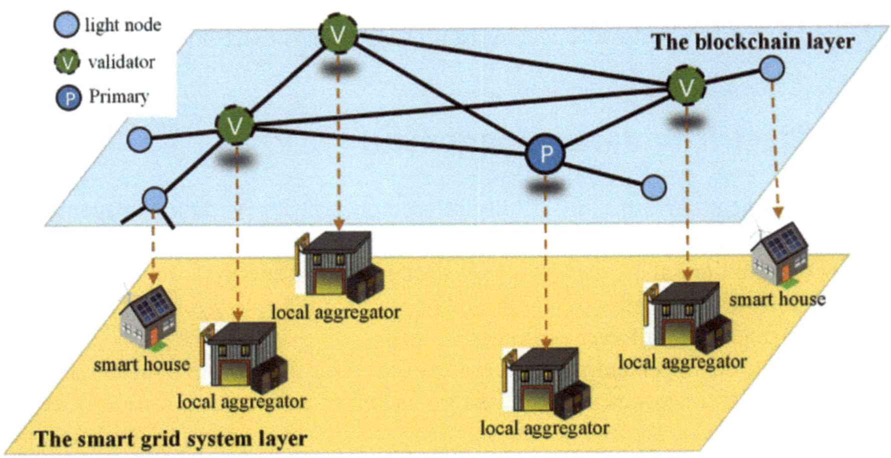

Figure 13.3 The smart grid system layer. Blockchain layer.

EDMS can notify so that the owner may correct them as quickly as feasible. It also helps to use energy management solutions to improve upon these inconsistencies.

13.4 Electricity sector

Over 37% of India's total installed capacity is made up of renewable energy installations, such as major hydropower plants. India's utilities generated 1383.5 TWh of gross power in fiscal year (FY) 2019–20. The gross power consumption per capita in FY2019 was 1208 kWh. The use of electric energy in agriculture was the highest in the world in FY2015 (17.89%). Despite its low electricity rate, India's per capita electricity consumption is modest when compared to most other countries.

13.4.1 Power generation

In the electric energy production efforts, it is the phase occurring just before the supply (transmission and power distribution, etc.) of electricity to end customers or the storage (Swan, 2015) of electricity. Because electricity cannot be found naturally, it must be "produced."

13.5 Teleporting electricity data through a smart grid

Sharing data across big networks raises a slew of challenges and risks, including security and usability concerns. While data availability, integrity, and secrecy are critical, our team believes that enabling non-repudiation of shared data utilizing blockchain technology would dramatically improve microgrid resistance to The Certificate Authority's centralized management environment that now determines the security of electronic credentials. On the other hand, the existing ecosystem's reputation is fading, and it is nearing the end of its useful life. Smart contracts, for example, give distributed ledgers increasing programmability, allowing them to perform additional functions and significantly expand their functionality. A Hyperledger Fabric network is being designed to let microgrids share information in a way that can grow to a large industry grid.

13.5.1 Network of power grids

We planned and designed our project around an already completed system for wide-area monitoring and control (WAMC). This project described and discussed the technical underpinnings of an existing smart grid testbed. Seven Phasor Measurement Units (PMUs) are located on each power system bus and provide phasor measurements. Through OpenPDC, all PMUs data is pooled into a Microsoft SQL database. A noticeable lag can also be noticed, which can last anywhere from seconds to minutes.

13.5.2 Scalability

Changes in the environment would have an impact on latency. These adverse effects can be avoided when the algorithm evolves and is fine-tuned. The likelihood of the control system being unstable grows as the amount of delay increases. Our data sharing experiment will allow us to record and analyze data sharing latencies in order to predict the stability margins of a connected set of microgrids so that different WAMC algorithms (Sekaran, Chandana, et al., 2020) can be used.

13.6 Blockchain technology's conceptual background

The fourth industrial revolution considers distributed ledger technology (DLT) to be a significant component. The purpose of creating a decentralized ecosystem is to eliminate business-related trust difficulties. Blockchain technology is employed in a range of areas due to its distributed database design and audit trail potential. DLT is essential for the growth of traditional higher education. Sony Global Education has also established a blockchain-based transparent evaluation system for data storage and administration services. In addition, MIT created a digital badge for online learning based on a distributed ledger. The Holberton School was one of the first schools to hold degrees and educational data using DLT.

Because of the unique nature of e-learning, evaluation requires not only traditional assessment of students' academic results but also assessment of their learning mechanisms through the evaluation of students' e-learning data, such as online meetings, presentations, and conferences, especially during pandemics. A self-contained e-learning method is essential for relieving the burden on online teachers while also broadening the scope of e-learning evaluation. The Certificate Authority's centralized management environment (Natoli et al., 2019) now determines the security of electronic credentials. On the other hand, the existing ecosystem's reputation is fading, and it is nearing the end of its useful life. Smart contracts, for example, give distributed ledgers increasing programmability, allowing them to perform additional functions and significantly expand their functionality.

13.6.1 Taxonomies of blockchain system architectures

Blockchain is a game-changing technology that aims to build a secure record-keeping system using cryptography and consensus procedures, that is a system members can communicate with, store, and verify transactional data, eliminating the requirement for verification of transactions from a third-party authority.

13.6.2 Distributed ledger

A distributed ledger differs from the most prevalent type of ledgers used by businesses, which is a centralized ledger. Because it is a single point of failure, a centralized ledger is more vulnerable to cyber assaults and fraud. Bitcoin is based on the blockchain, which is a distributed ledger.

Ledgers have been used to record economic transactions, such as contracts, payments, buy-sell agreements, and the transfer of assets and property, since the beginning of time. Information stored on computers is evolving into increasingly advanced formats that are cryptographically secure, quick, and decentralized as technology advances. One of the ways that businesses can benefit from this technology is through distributed ledgers.

13.7 Blockchain applications for smart grid

13.7.1 Devices and equipment's tracking

The main use of blockchain technology in the smart grid is that we can use the grid fruitfully with the integration of digital currency with the smart grid for paying bills. The company which accepted Bitcoin as a method for payment for energy bills was BASNederland. This led to a slew of other companies to build cryptocurrency-based billing and metering solutions, with some even offering incentives to consumers who pay with cryptocurrency rather than fiat money. Meanwhile, other businesses, such as the South African firm Bankymoon, is developing smart meters that accept Bitcoin payments. Jouliette is a blockchain-based coin (Wood, 2019) for energy sharing developed by Spectral and Alliander, both based in the Netherlands. This currency allows users to spend their energy tokens from their e-wallets, enabling P2P electricity transactions.

The first test of the initiative, which featured five prosumers and five clients, was the first time blockchain was used to record energy transactions. Smart meters installed in prosumers' homes convert surplus energy absorbed by rooftop photovoltaic (PV) panels into tokens that may be utilized directly for energy market trading. Depending on the user's preferences, this platform keeps track of the method of transactions or tokens. A smartphone application is being created to make it easier for consumers to interact with the platform.

Users can use their online money to monitor real prices and conduct transactions on this public network, which handles billing automatically. For a long time, eMotorwerks in California has used JuiceNet, a blockchain-based platform, to lease out charging stations to electric vehicle users. The platform maintains track of all transactions and makes the necessary payments to the owner of the charging pod. EV drivers can also use JuiceNet's mobile app to find a charging station from a list of local charging stations.

13.7.2 Smart grid data sharing

13.7.2.1 Applications concerning energy management

Because of the two-sided movement of electricity and data in the smart grid, all can participate actively in grid management and auxiliary services. Smart grids must analyze huge volumes of data efficiently and intelligently to provide real-time energy management. The reference outlined in (Sekaran et al., 2019) several steps for extracting data from large datasets in order to manage energy in smart grids.

13.7.2.2 Improvement of smart grid reliability and stability

PMU big data can also be utilized to track asset health in real-time and estimate stability margins. The use of a PMU to detect data-driven mode oscillations is proposed. A defect location technique based on PMUs is proposed in this article. The proposed big data methods help to improve the reliability and stability of the electricity system. Cloud based Operators in the Distribution Industry Operators of Transmission Systems, Data Hub (Sekaran, Rajakumar, et al., 2020) for the Generation Control Center Demand Response Load Forecasting Real-time Energy Management, Asset management, Planning, Forecasting Generation Planning, Economic Load Dispatch (Sekaran et al., 2019), Wide area control State Estimation, Stability enhancements, Market Operators Energy Forecasting, Price Forecasting, Risk Analysis Revenue Management Aggregators/other Businesses Flexibility Assessment, Forecasting Generation Planning economic Load Dispatch.

13.7.2.3 Applications

Because the smart grid is such an important piece of infrastructure, any cyber or physical flaws might have far-reaching consequences. A contingency study is often performed by a power system planner to provide resiliency in the event of a system failure or natural disaster. Due to the considerable interconnection of electricity and communication networks in the future, they will be increasingly vulnerable to hostile attacks. Close cyber-physical coupling is required to protect power systems from the threat of loss of both cyber assets and physical assets.

13.7.3 Real-time applications

Electricity Generation: Blockchain technology gives deploying firms complete real-time visibility into the whole health of an electricity grid. It aids businesses in developing shipping plans that boost earnings.

Energy Exchange: By eliminating the need for market middlemen, the implementation of blockchain in the energy sector exchange process helps to save time and effort. The frames within the chain in P2P trading schemes track the modules of electricity produced, allowing owners and buyers to make a quick sale. Rather than relying on an indirect operator, it provides users (customers and merchants) with personal freedom, expectations, and expenses. Consumers and holders alike benefit from a blockchain-based "intermediary-free" electricity exchange.

Electric Vehicles: Powered vehicles and smart grids have become a hot topic in recent years. When it comes to connecting to the electrical grid, the largest issue is electric charging because the unorganized loading of such vehicles puts strain on the grid. As a result, several solutions have been proposed to address this issue, including blockchain technology. Automobile integration is smoother when using the same strategy to reduce energy volatility and loading prices. Using blockchain technology helps to discover a local charging station where electric vehicles can be charged.

Cyber Security: The smart grid's installation has resulted in a number of flaws that can be exploited or targeted in various parts of the cyber-physical grid. The "FDI attack," in which an attacker manipulates or injects fraudulent data into measures or command messages in order to affect the energy grid dynamics, is a significant cyber-physical device danger. Many blockchain techniques have been studied in order to improve grid resilience to cyber-physical threats. A blockchain-based architecture that is used to allow consumers to regulate how energy is used without fear of abuse by either group is an example.

Protection and Security: Power distribution security is increased by blockchain. The implemented computational securitization, when combined with the consensus technique, ensures that the information on the blockchain is atomic. When an electricity transaction/information is incorporated into the blockchain, it will be difficult to edit or remove the transaction for unlawful reasons, resulting in a network that is extremely safe and stable. The electric power network will be more durable as a result of the blockchain.

References

Dinh, T. T. A., Liu, R., Zhang, M., Chen, G., Ooi, B. C., & Wang, J. (2018). Untangling blockchain: A data processing view of blockchain systems. *IEEE Transactions on Knowledge and Data Engineering*, *30*(7), 1366–1385. Available from https://doi.org/10.1109/TKDE.2017.2781227.

Li, Z., Bahramirad, S., Paaso, A., Yan, M., & Shahidehpour, M. (2019). Blockchain for decentralized transactive energy management system in networked microgrids. *Electricity Journal*, *32*(4), 58–72. Available from https://doi.org/10.1016/j.tej.2019.03.008.

Miglani, A., Kumar, N., Chamola, V., & Zeadally, S. (2020). Blockchain for internet of energy management: review, solutions, and challenges. *Computer Communications*, *151*, 395–418. Available from https://doi.org/10.1016/j.comcom.2020.01.014.

Natoli, C., Yu, J., Gramoli, V., & Esteves-Verissimo, P. (2019). Deconstructing blockchains: A comprehensive survey on consensus, membership and structure.

Paudel, A., Chaudhari, K., Long, C., & Gooi, H. B. (2019). Peer-to-peer energy trading in a prosumer-based community microgrid: A game-theoretic model. *IEEE Transactions on Industrial Electronics*, 6087–6097. Available from https://doi.org/10.1109/tie.2018.2874578.

Sekaran, K., & Krishna, P. V. (2016). Big cloud: A hybrid cloud model for secure data storage through cloud space. *International Journal of Advanced Intelligence Paradigms*, *8*(2), 229–241. Available from https://doi.org/10.1504/IJAIP.2016.075731.

Sekaran, K., & Krishna, P. V. (2017). Cross region load balancing of tasks using region-based rerouting of loads in cloud computing environment. *International Journal of*

Advanced Intelligence Paradigms, *9*(5–6), 589–603. Available from https://doi.org/10.1504/IJAIP.2017.088151.

Sekaran, K., Chandana, P., Krishna, N. M., & Kadry, S. (2020). Deep learning convolutional neural network (CNN) With Gaussian mixture model for predicting pancreatic cancer. *Multimedia Tools and Applications*, *79*(15–16), 10233–10247. Available from https://doi.org/10.1007/s11042-019-7419-5.

Sekaran, K., Khan, M. S., Patan, R., Gandomi, A. H., Krishna, P. V., & Kallam, S. (2019). Improving the response time of M-learning and cloud computing environments using a dominant firefly approach. *IEEE Access*, *7*, 30203–30212. Available from https://doi.org/10.1109/ACCESS.2019.2896253.

Sekaran, K., Rajakumar, R., Dinesh, K., Rajkumar, Y., Latchoumi, T. P., Kadry, S., & Lim, S. (2020). An energy-efficient cluster head selection in wireless sensor network using grey wolf optimization algorithm. *Telkomnika (Telecommunication Computing Electronics and Control)*, *18*(6), 2822–2833. Available from https://doi.org/10.12928/TELKOMNIKA.v18i6.15199.

Swan, M. (2015). Blockchain: Blueprint for a new economy.

Tushar, W., Chai, B., Yuen, C., Smith, D. B., Wood, K. L., Yang, Z., & Poor, H. V. (2015). Three-party energy management with distributed energy resources in smart grid. *IEEE Transactions on Industrial Electronics*, 2487–2498. Available from https://doi.org/10.1109/TIE.2014.2341556.

Wood, G. (2019). Ethereum: A secure decentralised generalised transaction ledger, Yellow Paper, Ethereum.

Privacy-preserving in smart grids using Ethereum and Hyperledger blockchain

14

R. Anusuya, D. Karthika Renuka and L. Ashok Kumar
Information Technology, PSG College of Technology, Coimbatore, Tamil Nadu, India

14.1 Introduction

The smart grid concept was developed as a modern perspective for the traditional power infrastructure to determine the most effective strategy for incorporating renewable and green energy technology. The smart grid is developing as a novel solution for ensuring energy access whenever required and from any location. The overriding goal of such developments is really to foster the development of a sustainable society (Yoldaş et al., 2017). Nevertheless, connecting and managing a rapidly rising number of users and connections seems to be a challenge in the existing centralized grid architecture. As a result, the smart grid is evolving from a centralized (Gensollen et al., 2016) to a decentralized topology. On the other hand, the blockchain offers several advantageous characteristics which can benefit the smart grid paradigm. We intend to present a complete assessment of blockchain applications in smart grids in this study. As such, we highlight security solutions blockchain can resolve in smart grid.

Throughout the last several years, customary centralized energy systems based on fossil fuels have encountered several significant obstacles. For solving the difficulties like environmental damage, CO_2 emissions, distant transfer of energy, and for sustainable development, the two main potential solutions are to enhance the usage of renewable energy from various sources and/or to improve the efficiency of power usage. Over the last few years, the smart grid concept has been used in a variety of ways to alleviate the power crisis. To start with, it incorporates communication and energy technologies, improved control systems, and smart meters. The phrase "smart grid" refers to a new vision for the traditional grid which allows bidirectional power and data interchange to identify the most efficient means of distributing, controlling, and merging sustainable and renewable energy technologies. Unfortunately, the smart grid prohibits the expansion of distributed and scalable energy resources as well as does not ensure security. To overcome its existing limits, Smart Grid 2.0, also known as the Internet of Energy (IoE), was established, which integrates the Internet with the smart grid. IoE is an Internet-based approach to power issues that incorporates the IoT, advanced communication technologies, electricity grid components, and other power generation systems. IoE can provide

Blockchain-Based Systems for the Modern Energy Grid. DOI: https://doi.org/10.1016/B978-0-323-91850-3.00016-0

energy connectivity wherever and at any time. As connectivity expands, a significant challenge is combining and managing numerous connections inside the traditional centralized grid system, like the growing number of distributed energy providers, users, electric cars, smart gadgets, and cyber-physical systems. Centrally managing such a rapidly developing network would need complex structures.

Hence, decentralization is a critical necessity for IoE, and it enables the dynamic integration of all its elements. Having said that, the smart grid system integrated with the Internet is vulnerable to security and privacy issues due to its massive number of complicated linkages, necessitating the development of new and inventive solutions. However, blockchain as a novel and upcoming technology paves the way for decentralized systems' development. This blockchain technology is decentralized, which means that no central trusted authority is necessary to operate it; instead, various entities in the network can construct and preserve blocks among themselves. Each entity shall validate the data integrity and the order of the chain. The decentralized method renders all systems reliable and resistant to failure and cyber-threats, as well as resolving other issues associated with centralized architecture. While the blockchain was initially established and designed to support cryptocurrency, its superior qualities have attracted significant interest in a variety of other nonmonetary uses. Simultaneously, blockchain is advancing the implementation of safe, privacy-preserving, and trustworthy smart grid advancements toward decentralization, in addition to digital currencies. Smart grids are the building blocks of future communities. When intelligent gadgets are stacked on top of intelligent devices, smart buildings and intelligent cities are created. To operate at full capacity, they must all be backed up by an intelligent power supply. Blockchain technology has a lot of advantages in power architecture, including optimizing the usage of power, and exchanging and monitoring power. This chapter describes employing smart contracts for the smart grid that help to implement functions like payment and compares the performance of smart grid implementations on the Ethereum and Hyperledger Sawtooth blockchain.

Today, the energy sector's need for power is greater than ever. Renewable energy sources are eventually gaining traction in power generation. Wind and solar energy look to be the largest contributors to total energy production. The conventional smart grid system is incapable of solving a slew of issues. The primary obstacle is the introduction of a new entity called a prosumer, capable of both energy production and consumption. Smart grids are becoming increasingly important as the world develops in technological innovation. Smart grids facilitate the decentralized implementation, use of intelligent meters, energy transmission in both directions, and digital platforms. Such characteristics will help prosumers and consumers in effective monitoring of energy and more efficient energy trading. Smart grids can help create a digital utopia as the globe moves toward smart cities. The smart grid integrates all power sources while assisting domestic energy generation. Traditionally, power systems have been operated in a centralized manner where the architecture was controlled centrally by a single entity. A particular void existed for decentralized architectures with dispersed control. For consumers and prosumers engaging in the smart grid, a decentralized system is plainly required (Strasser et al., 2014).

Blockchain technology has the potential to disrupt established centralized institutions. This technique enables multi-node peer-to-peer (P2P) networks to operate independently of a single entity, allowing decentralized applications to collaborate anonymously. The emergence of smart contracts represents a paradigm shift in the way systems are managed without recourse to a central authority. Smart contracts enable the deployment of completely autonomous, P2P systems, which can eventually do away with the need for any third party. After the advent of the smart contract in 2013, the real-world uses of blockchain began to concretize. A smart contract is a software code that is performed over a network and enables us to develop decentralized service applications at ease. Blockchain technology serves as a platform for the creation of bespoke cryptocurrencies, digital assets, and complicated apps that enable users to manifest any arbitrary circumstance. As a result, Decentralized Autonomous Organizations (DAOs) based on blockchain technology are being formed. Ethereum introduced smart contracts for the first time in 2015, making it the first blockchain-based platform to do so. It provides a platform for deploying smart contracts developed using a Turing-complete programming language that is built-in. This enables the development of safe, cost-effective, and intelligent applications with user-defined rules. Blockchain technology enables decentralized marketplaces and smart contracts. Its broad use enables the formulation of rules sans third part intervention, with all participants deciding on a democratic consensus procedure. Due to the fast growth of renewable energy resources, it is critical to evolve the traditional energy distribution system toward more intelligent, robust, and optimized energy grids. P2P energy trading platform is an essential aspect of smart grids. Additional benefits of P2P energy trading include users' preference for energy sources, participant empowerment instead of a centralized authority, an overall capacity increase, and a consensual market system that can help with load balancing, power failures, and other electricity grid issues.

14.2 Literature survey

Blockchain implementation in smart grid is garnering significant attention as many research works have been conducted, especially for using blockchain to solve security issues in smart grid. For example, Miglani et al. (2020) explore the applicability of blockchain to the IoE. The system of intelligent infrastructure, management, and protection are the three key systems (Gungor et al., 2012) discusses. In each system, recommendations for future directions are listed. The challenges and potential solutions of an IoT-based smart grid are examined in this study by (Bekara, 2014). The work focuses on several forms of security attacks and gives an overview of the smart grid's cybersecurity situation. By focusing on the physical layer of the cyber-physical smart grid, Zhang et al. (2021) proposes an obfuscated and integrated state-space model in which cyber-physical threat and defensive prototypes can be effectively generalized. Work by Gunduz and Das (2020) concentrates on network vulnerabilities in smart grids, attack countermeasures, and security requirements debate and examination and identifies the technical challenges that blockchain can solve in various applications and briefs the projects that are currently using blockchain. Moubarak et al. (2018) provide several potential attacks scenarios in the blockchain and evaluate different solutions for preventing network vulnerabilities. Aitzhan and Svetinovic (2016) discuss how to secure transactions in decentralized smart grid energy trading without depending on third parties. The article by Butt et al. (2021) provides an overview of "Smart Grids," including their features and implications for the power distribution business. Mengelkamp et al. (2018) propose a decentralized market platform enabling energy prosumers and consumers to trade local power production, also without a central intermediary.

14.3 Privacy concerns in smart grid implementation

When the grid is modernized, the number of personal details collected and then used will dramatically increase. Smart devices have the potential to create a data explosion of personal everyday life facts. However, it is unclear now who will have access to the customer data other than the customer's utility provider. Energy measurements may be made at significantly shorter durations in a smart grid. Even if statistics on power usage are not obtained on a regular basis, information may be gathered by continuous monitoring of energy usage undertaken at a slower pace. Consequently, sensitive data like the number of individuals living in a home, their availability at home, their gadget usage frequency, and sleep time may all be obtained or assumed. Most people dislike collecting such type of data as it violates the "sanctity of the home," and some might object that such sensitive information about an individual's everyday routine must not be exposed.

The cybersecurity working group of the United States has reinforced these previously mentioned privacy issues through a privacy impact assessment. The research identifies the following challenges and concerns in the smart grid about consumer data transfers in the United States (Siddiqui et al., 2012).

- There is a lack of clarity on the privacy implications.
- The rules of personal data used by the utility sector are vague and imprecise.
- Smart meters and distributed energy systems have the potential to expose information about residential users and household activities.
- Devices connected to the smart grid that roam (e.g., the electric vehicle being charged at another place) may generate additional private data.
- Roaming Smart Grid devices may generate additional personal data.

The IoT is gaining tremendous popularity among advocates of the smart grid because of its implementation in projects like smart cities. The data collected over the various intelligent devices should be optimally processed. The energy market application should be optimized since it is a critical component of the development of smart cities. Our chapter aims to analyze the performance of blockchain-based smart grids. It addresses the reasons for developing a smart grid based on blockchain technology and the accompanying limitations. Privacy concerns are a significant factor in smart grid adoption. This chapter in the following sections presents a brief history of blockchain technology and provides the details of important terminologies of blockchain-like definitions of distributed ledger, smart contracts, blockchain types, and consensus methods. This chapter examines the possibility of privacy enhancement in a blockchain-based smart grid deployment. DAOs that utilize blockchain technology result in more expensive platforms than those that use traditional techniques. The improvisations are examined, and the article's central subject is the possibility of optimization. It covers the major research issues associated with various components and the use cases of the smart grid. It makes a comparison of smart grid blockchain implementations in Ethereum and Hyperledger Sawtooth. It also highlights existing challenges that have been explored and the approaches that have been used in conjunction with blockchain.

14.4 Blockchain background

14.4.1 History of blockchain

In 1991 researchers Stuart Haber and W. Scott invented blockchains to save cryptographically protected time-stamped records. In 1992 the method was made more efficient by using Merkle trees where more than one document was contained in one block but was in practice not utilized. A computer scientist named Hal Finney established a Reusable Proof of Work (RPoW) in 2004, which set the template for cryptocurrencies' development. A blockchain model was built in 2008 on a white paper and was published under the identity "Satoshi Nakamoto" on a cryptography email list (Nakamoto, 2008). This may be an individual or development group that works on the undiscovered blockchains. The presented system is a decentralized, P2P bitcoin currency mechanism. The bitcoin source code was made open source in 2009. The first block was mined on January 3, 2009, by Satoshi Nakamoto (genesis block). Satoshi Nakamoto sent Hal Finney ten bitcoins as the first recipient of a bitcoin transaction on January 12, 2009. In 2013 the Russian Canadian programmer, cryptocurrency investigator and author, VitalikButerin, introduced Ethereum (Blockchain 2.0).

Recognizing the necessity for a programming language in bitcoin to enable the development of decentralized apps, Buterin founded Ethereum, a new blockchain-based computing platform that supports smart contracts that record not just cash but also other assets. Ethereum is the second biggest public blockchain platform next to Bitcoin and was established on 30 July 2015. Ether is the name of the Ethereum cryptocurrency. In December 2015, to work together across sectors, the Linux Foundation announced Hyperledger by constructing blockchains and distributing ledgers with increased reliability and performance. A lot of Hyperledger frameworks were published later in 2018 and after. Blockchain technology continues to expand and develop with enormous potential for incorporation into a variety of sectors and generating popular acceptance (Bashir, 2017).

14.4.2 Blockchain types

The following categories apply to blockchain protocols as described in Shrivas and Yeboah (2018):

- **Public blockchain**: A public blockchain, such as Bitcoin or Ethereum, is accessible to everyone in both read and write mode.
- **Private blockchain**: This is appropriate for businesses. Individuals are only authorized to read or write to the blockchain by a single authority, which can grant participants specific permissions based on their function within the company.
- **Permissionless blockchain**: In this type, no one is restricted from creating blocks.
- **Permissioned blockchain**: This type of blockchain requires prior permission from a centralized authority to publish, read, or validate a block. This sort of blockchain is ideal for banks, businesses, and organizations where the allowed members and their access rights may be set in advance to ensure seamless commercial operations.

14.4.3 Elements of blockchain

Cryptographic hashing functions, asymmetric keys, a shared ledger, transactions, blocks, and consensus all play a crucial part in the blockchain process. Below is a summary of these components.

14.4.3.1 Cryptographic hash functions

Cryptographic hash functions are distinguished from other types of hash functions by their suitability for applications involving cryptography, security, or privacy. They are used for a variety of purposes, including encryption, password management, digital signatures in e-commerce protocols, and bitcoin creation. It is a mathematical function that accepts any input, such as a file, text, or image, and returns a single fixed-size output. Thus, the collection of mathematical changes performed on the input data results in a compressed bit string referred to as a digest, hash, tag, or hash value.

This output value is generated individually for each input, and any changes to the input data will result in a completely new output digest value. Collision resistance is the feature that no two distinct inputs hash to the same output, and the digest output generated is random and unrelated to the input data. Due to collision resistance, predicting useful information about the incoming data using the digest value is extremely difficult. Hashing functions are used in blockchain to generate addresses, create unique identities, digitally sign, and authenticate block contents and block headers. Additionally, cryptographic hash algorithms should be computationally efficient and rapidly generate results.

14.4.3.2 Asymmetric key cryptography

In a blockchain, asymmetric key algorithms shall be utilized (Wang & Wu, 2017). It is asymmetric in that each user holds two distinct keys, referred to as the public key and private key, which are used in the encryption and decryption operations involved in blockchains. This, in conjunction with cryptographic hash algorithms, is used to create digital signatures for the purpose of signing transactions and operating in untrustworthy contexts. Digital signatures are a digital replacement for handwritten signatures that ensure data integrity, verification of the data's provenance, and nonrepudiation.

The Elliptical Curve Digital Signature Algorithm is applied (Abidi et al., 2014) to produce digital signatures. Digital signatures are divided into two stages: the signing stage and the verification stage. When a user initiates a blockchain transaction, the person's private key is kept hidden, but copies of the legitimate public key are available to all other users on the network.

Prior to transmitting a message, mathematical transformations and cryptographic hashing algorithms are performed to the message, and the resulting message digest is merged to generate the digital signature, which is a sequence of integers. The user receiving the message decrypts it using the public key supplied to them and confirms it by comparing the hash value.

14.4.3.3 Transactions

A transaction occurs when two distinct parties transfer data over several nodes. When data is communicated via a network, a transaction occurs. This data includes the sender's identity, digital signature, and public key. The input to the transaction is the data that will be transmitted, and the outputs are the receiver account that will receive the digital asset. Additionally, it provides information about the receiver's identification and the amount of data that are to be delivered. A transaction block is a group of related transactions received during a certain time. The transaction's and block's sizes define the number of transactions that are kept in the block. The transactions are hashed in pairs to generate a single digest value. When a user initiates a transaction, this message is communicated to all network nodes together with other transaction messages.

After validation, transactions are compared to the digest value and entered into the ledger book. The transaction will indicate the source of the data, perhaps a reference to prior events or if it is a new transaction, the source event itself. Since transactions are in plain text, consumers may view all data.

14.4.3.4 Ledgers

Blockchain technology makes use of a distributed ledger database to store a group of transactions that are exchanged and synchronized between many locations and individuals. Cryptographic algorithms securely preserve transactions and contracts that are dispersed, obviating the need for central authority approval and verification. Blockchains are resistant to cyberattacks because once records are saved, they become unalterable (Pilkington, 2016).

Additionally, owing to the distributed nature of records, copies are dispersed across several computers, making it difficult to edit a single record without affecting all other copies. Ledgers facilitate the flow of information, making it simple for accountants to follow an audit trail.

14.4.3.5 Blocks

Blocks are unchangeable recordings that contain details about transactions that collectively comprise a blockchain. They are like ledger pages in that they contain a collection of transactions. Blocks are generated by publishing nodes, which are then filled with transactions. A blockchain is conceptually a chain of unique digital information kept in the form of blocks, each of which may be joined by a chain, like in Fig. 14.1. Every single block has a header holding block information and data holding the collection of transactions and other associated data. Except for the very first block, each block header carries a link to the immediately previous block's header. Typically, a link is a pointer to the previous or parent block's hash value.

Thus, the hash value gets altered when the content of a block is changed, which in turn reflects in the following blocks. The initial block is referred to as the Genesis block. A new block is produced when a block achieves a predetermined number of authorized transactions. The exact contents of the header and data may change depending on the implementation of the blockchain. The header of the block contains:

1. **Version**—determines the set of block validation criteria to use.
2. **Timestamp**—indicates the current approval time for a certain block in seconds since 1 January 1970.
3. **Parent block** hash- the hash value of the preceding block with size of 256 bits.
4. **Merkle tree root hash**—the hash of all transactions inside a block.
5. **nBits**—goal threshold, which represents the difficulty of mining or creating a valid block hash.
6. **Nonce**—a four-byte field that is typically initialized with 0 and increases with each hash computation. It modifies the block's hash output.

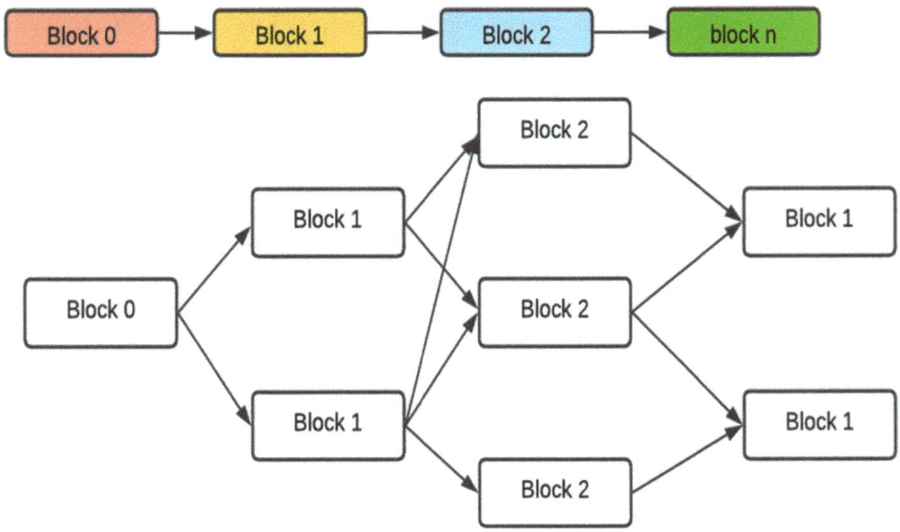

Figure 14.1 Linear chain and DAG-based logical structure. Blocks are connected in a linear fashion in a blockchain.

Data block comprises:

1. **Transaction counter**—keeps count of all the transactions that have been successfully completed.
2. **Transaction data**—this varies according to the nature of the transactions and the data recorded in the fields of business data, healthcare data, and Bitcoin data.

14.4.3.6 Merkle tree

The Merkle binary tree is used for storing the hash values of various blocks in a big data set, allowing for an efficient verification procedure of the set's whole. Each block has a digital fingerprint that summarizes all transactions. Participants can validate transactions using the Merkle tree regardless of whether they belong to a certain block. The Merkle tree's hierarchical structure is constructed by hashing pairs of child nodes, and this process is repeated until the Merkle root node is produced as in Fig. 14.2. Between pairs, hashing is conducted via a bottom-up methodology starting with each transaction. The leaf node stores the current hash value, whereas the nonleaf node stores the prior hash value. The root node aggregates all transactional data, maintaining data integrity. Because the blocks are interconnected, every change to the transactional data affects the root node. If a transaction in the hashing pairs is changed, the root node is also changed. If an attacker or hacker disrupts a block's transactions, it is much easier to detect or identify illegal access by recalculating the block's hash value.

14.4.3.7 Consensus

Consensus is an agreement between groups of nodes on their data, and consensuses are critical in determining the system's efficiency and scalability. The nodes in blockchains are decentralized, and there is no single node that verifies the ledgers' consistency; specific protocols are required to achieve this. Additionally, when there are untrustworthy nodes present in a group of dispersed nodes, getting a consensus on who publishes the next block is challenging. This predicament is analogous to the Byzantine Generals (BG) Problem from antiquity. The next section discusses various strategies for achieving consensus.

14.4.3.7.1 Proof-of-work

The primary objective of this is to select one of the several contesting distant nodes to publish a block to the chain. Each node competes by doing several computations, which establishes the node's legitimacy and requires a significant amount of user effort. Calculations are performed by frequently computing the block header's hash value while altering the nonce. The node with a hash value less than or the same as a specified target value is chosen. The chosen node is then able to publish the block and gets compensated for his efforts. The transaction block is validated by all other nodes, and once validated, it is added to the chain.

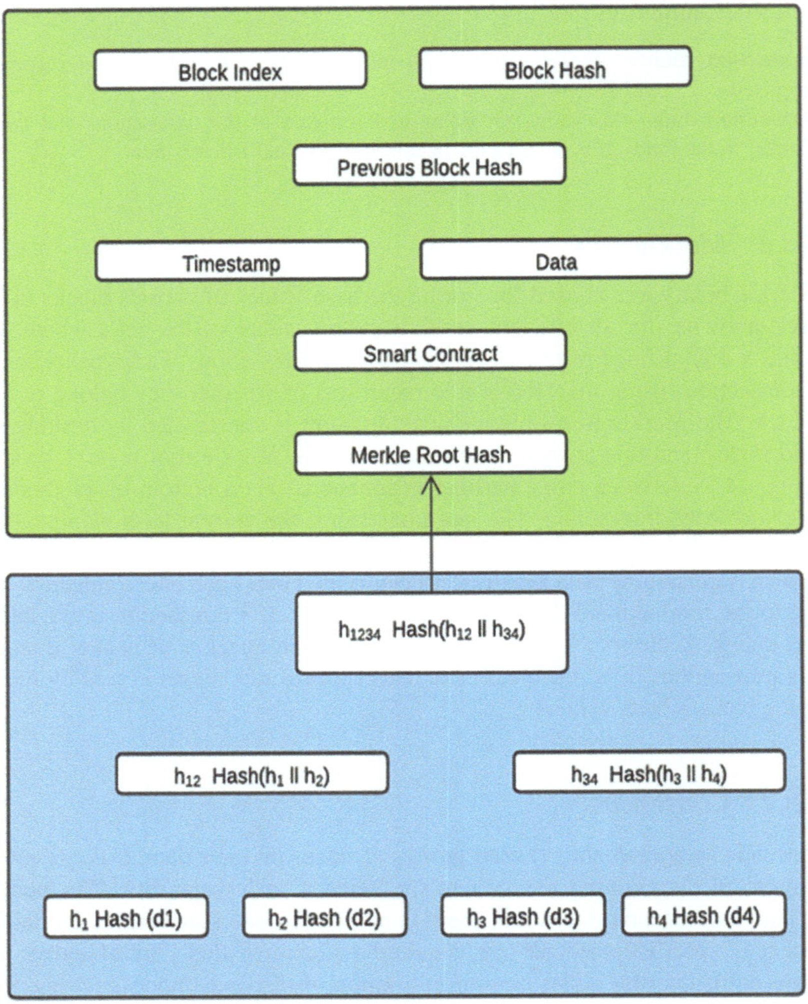

Figure 14.2 Merkle hash tree and block structure. The header of a block contains various fields, which include the previous block's hash, timestamp, and Merkle tree root hash. The Merkle tree's hierarchical structure is constructed by hashing pairs of child nodes, and this process is repeated until the Merkle root node is produced.

14.4.3.7.2 Proof of stake

When the stake or cryptocurrency holdings of an individual user node are taken into account, we have PoS. The users stake their cryptocurrency to participate in a bidding process for the right to publish the block. The individual with the most stake or bitcoin is deemed more legitimate and is granted the opportunity to publish the block. The transactions are validated by the other nodes. When the total number of attestations received from other nodes meets a predefined threshold, the block is added to the blockchain. The nodes that publish the block, as well as the nodes that validate the transactions, are rewarded in this operation. The primary disadvantage is that the individual with the largest stake has an advantage over other users; however, this may be mitigated by modifying the strategy formula, for example, by mixing stake size and user age.

14.4.3.7.3 Practical byzantine fault tolerance

Byzantine Fault Tolerance allows for establishing consensus (agreement on the same value) even when some network nodes either do not react or respond with erroneous information. The phrase originates from the Benzantine Generals Puzzle. It can deal with up to one-third of harmful user nodes in a node group. The publishing node is chosen in rounds. A primary node is selected to carry out transactions in each round, while all other nodes are secondary nodes. The procedure is divided into three stages: pre-planning, preparation, and commitment. A node can progress to the next phase if it receives two by third of the votes from all other nodes in each phase.

14.4.3.7.4 Delegated proof of stake

Given that in PoS, the block producers are chosen according to their stake, it follows that in DPoS, there is PoS distinguished from DPoS by the stakeholders electing the delegates or representatives for block formation and validation. In DPoS, the system requires only a few nodes for validation, resulting in simple block confirmation. Nodes are rewarded for their efforts and granted the position of publishing node, and they are prohibited from acting maliciously due to the possibility of being voted out.

14.4.3.7.5 Round robin consensus model

This is the consensus mechanism utilized in permissioned blockchains, and under this paradigm, nodes publish blocks only when they receive their turns. A node must wait numerous block creation cycles before getting its chance. If a node is not accessible for publishing within the time limit specified, other nodes may publish.

14.4.3.7.6 Proof of authority (identity) model

This consensus technique is usually employed where the identity of the publishing user is validated and added to the blockchain. The user's reputation can be created based on their activity and can influence their ability to post a block.

14.4.3.7.7 Proof of elapsed time consensus model

When a request for publishing a block is made in this manner, the software assigns a random wait time. The node then waits for the specified amount of time to expire before publishing a block. A signed certificate is generated for the randomly assigned time and is published alongside the user's block. This procedure is quite transparent and notifies other users when a malicious user is idle for an extended length of time and then begins broadcasting his blocks sooner.

14.4.4 Bitcoin/cryptocurrency

The term "cryptocurrency" refers to a digital asset designed for community electronic trade. This enables the creation and distribution of money units. Bitcoin, often referred to as a cryptocurrency, was the first decentralized program based on blockchain. Satoshi Nakamoto, in 2008, announced the concept of Bitcoin in his white paper. His work laid the foundation for the electronic P2P currency system. Bitcoin and similar digital payment systems have been a component of the blockchain technology ecosystem since 2009 (Mukhopadhyay et al., 2016).

14.4.5 Ethereum

Ethereum introduced smart contracts for the first time in 2015, making it the first blockchain-based platform to do so. It provides a user-friendly platform by deploying smart contracts written in an integrated Turing-complete programming language, which enables the construction of apps that are safe, cost-effective, and intelligent, with user-defined rules. Ethereum's 20-byte account address size allows checking balances, storing data, and executing contracts. The digital currency is Ether, and smart contracts are self-executing agents that facilitate inter-account transactions. The EVM (Ethereum Virtual Machine) is used to offer a runtime environment for smart grid applications. EVM creates complete nodes for the purpose of verifying new blocks. Ethereum typically takes about 15 seconds to generate a block. To safeguard against cyberattacks, Ethereum charges a transaction fee, often known as gas consumption. In other terms, it can be said, the Ethereum transaction is the gas usage times the gas price.

14.4.6 Hyperledger Sawtooth

Hyperledger Sawtooth is an enterprise blockchain platform for building distributed ledger systems. Emphasizing the importance of distributed ledgers and the security of smart contracts, particularly for corporate usage, the domain of the application developers does not need to be acquainted with the underlying architecture of the core system to express the business rules required for their application in the language of their choice.

Sawtooth enables applications to customize transaction rules, permissioning, and consensus algorithms to meet their specific business requirements.

14.5 Blockchain in smart grid

The possibility of blockchain technology in smart grids is discussed here. We begin by delving into the specifics of the future smart grid system (Agung & Handayani, 2020; Mollah et al., 2020). Following that, we will outline the blockchain capabilities and the privacy and security objectives that blockchains help to resolve in smart grids (Mylrea & Gourisetti, 2017).

14.5.1 Decentralized smart grid system

The smart grid upgrades the traditional grid into a more precise, effective, and intelligent energy consumption and distribution network. Climate changes led to the demand for renewable energy sources, and smart grids paved the way for it by integrating more renewable and distributed energy resources.

Recent developments include the introduction of the Energy Internet (EI) concept in the smart grid. The Internet technologies that enable the creation of the future smart grids combining data, energy, and business are referred to as EI. The EI is designed to provide the substantial potential for the integration of numerous clean and renewable energy sources into the grid, as well as foster enhanced interaction between different components of the electricity grid to create a complete smart energy network. EI's central concept is to enable the exchange of energy and information in a manner that answers questions like how data is shared on the Internet? Although the smart grid and EI are designed to adapt to dispersed and centralized power generation, the major drawback of its present design is that it is based on centralized or intermediate organizations for energy generation, transit and delivery, and markets. The smart grid's components communicate and interact with centralized organizations that can observe, acquire, and analyze data, as well as send appropriate control signals to all the grid's components. Additionally, energy is often sent across a lengthy network to distribute energy to end consumers via the distribution network. However, as renewable energy consumption increases and the number of elements continue to grow, the existing architecture of the smart grid system creates significant issues. Scalability, expandability, high computational and communication costs, and availability assaults are a few of the notable concerns. Future power systems with many components will be impossible to govern. As a result, decentralizing the smart grid is becoming increasingly popular as a means of delivering more dynamic, intelligent, and proactive features. A completely automated network with decentralized topologies is being developed to dynamically increase interconnections between all smart grid components. Also, EI's connection and accessibility help make the smart grid more affordable, efficient, and reliable.

For ages, ledgers have been used to record and account for transactions. The notion of an unchangeable digital ledger is a game-changer in the banking industry that happened with blockchain technology. Smart grids powered by blockchain technology are a perfect match for the characteristics necessary for an optimal energy market. By removing the central entity and dispersing authority among stakeholders, the use of blockchain enables the creation of a model that is more cost-efficient. Contemporary grids combined with cryptocurrencies like Ether (Ethereum's digital currency) may provide a more long-lasting and efficient solution to electricity concerns in particular settings, as well as during periods of severe weather conditions. Blocks are used to keep track of all transactions that may be accessed, forming a set of time-stamped and chronologically connected transactions forming a blockchain. The blockchain-based approach enables such grid designs. Both the market mechanism and the payment function are virtual operations that take place on the blockchain.

14.5.2 System architecture

Numerous strategies for evaluating the performance of smart grids using blockchain have been developed recently. Fig. 14.3 depicts a blockchain-incorporated smart grid's overall design.

Dynamic real-time power usage is mediated via blockchain in the proposed system design. Smart meters act as regulatory authorities, regulating data and providing consumers with access to monitor their power usage. Smart meters serve as entry points to the smart grid's decentralized blockchain-based infrastructure.

The objective of this section is to describe the components of the smart grid. This section also details the method to build an energy-based marketplace capable of the bidirectional flow of energy and trade. Smart contracts govern the autonomous energy market. The purpose of smart contracts is to develop and combine valuable aspects of the energy market, as well as to perform the below-mentioned functions:

1. The smart grid is facilitated by the purchase and sale of electricity using cryptocurrencies.
2. Blockchain technology ensures trustworthiness in the smart grid.
3. Smart contracts can check the balance, transfer funds, and add new customers, among other things.
4. The existing smart grid is devoid of regulatory norms. Blockchain-based smart contracts may enable decentralized energy solutions that are beneficial for tracking energy and carbon impact.

The design illustrated enables participants to engage in active energy markets without relying on central authority (energy providers or banks). The architecture

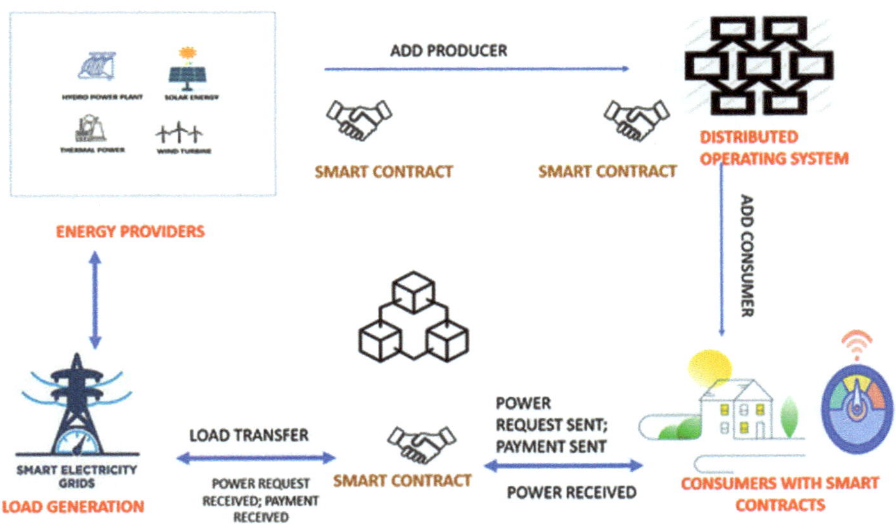

Figure 14.3 Smart grid using blockchain. A blockchain-incorporated smart grid constitutes energy providers, smart electricity grids, consumers, prosumers, and computer systems.

creates a transparent marketplace for suppliers and customers. With blockchain as the core entity, smart contracts secure the integrity of deposits, facilitate energy trading, and operate as a middleman between energy providers and customers. Along with market dynamics, blockchain enables the introduction of micropayments via cryptocurrencies. Ethereum, Blockchain's protocol, inherits the "ether" payment platform. Similarly, practically all blockchain protocols support micropayments for executing transactions across the blockchain network.

14.5.3 Energy trading

The smart grid is composed of several distinct entities and components. The distribution system operators (DSOs) and a relatively limited number of local energy providers now control most of the stake in energy markets. With the introduction of blockchains, due to the bidirectional energy flow capability of smart meters, local providers can transmit excess energy back to the grid, but instead of DSOs, the blockchain actively controls the energy transaction process between prosumers and consumers (Pee et al., 2019).

The purchasers may be prosumers or consumers. Smart meters that are powered by smart contracts help sellers and buyers enter an agreement that assures compliance with all transaction terms and conditions. The blocks created throughout the mining process contain all the transactions. All transactions and inquiries are time stamped, and the blockchain ensures that data cannot be altered. As seen in the design, smart meters mine blockchain transactions. The smart meters and blockchain connect the grid's components and operate as gatekeepers, ensuring that any illegal energy usage is prohibited. Smart contracts determine which transactions are accepted and automatically begin the necessary data functions. DSOs initially implement the complete blockchain technology in a virtual smart grid.

14.5.4 Distribution system operators

Customers and local energy providers use DSOs as middlemen, handling market processes and payment functions. DSO is a blockchain miner that ensures market consensus while ensuring the privacy of data from prosumers and consumers (Khajeh et al., 2020). Because DSOs had the role of centralized authority in the traditional energy trading system, they had played the role of the sole governing body in charge of all market operations.

14.5.5 Local energy providers

The smart grid is fed with power produced by local providers using renewable energy sources like wind and solar energy.

Additionally, local energy suppliers serve as blockchain miners, guaranteeing agreement and putting the decentralized market concept into action. The concept is based on micropayments, with energy input being compensated with preset tokens through smart contracts that enable blockchain transactions when specific conditions are fulfilled.

Users who pay for the power consumed are the consumers. They participate in the blockchain as nodes. Each home is equipped with a smart meter that reads the energy usage of consumers, which smart meter readings are recorded by smart contracts on the blockchain network. Consumers can purchase power from a local energy supplier or a DSO using defined tokens.

14.5.6 Consumers

Consumers are the users who pay for the power consumed. They participate in the blockchain as nodes. Each home is equipped with a smart meter that reads energy usage by a home, which readings are recorded by smart contracts on the blockchain network. Consumers can purchase power from a local energy supplier or a DSO using defined tokens.

14.6 Characteristics of a smart grid

The suggested design of a smart grid is predicated on the assumption that all customers are linked to the grid via smart meters. Through micropayment functionalities, smart meters enable users to sell or buy energy. DSOs ensure that the blockchain network is only accessible to genuine smart meters.

Consumers can purchase energy using smart contracts when a payment feature is activated. Local energy suppliers sell electricity via smart meters, and the payment feature ensures that the transaction is completed. DSOs retain a majority share and are accountable for all off-grid tasks like maintenance. They are not the exclusive central authority anymore, as local energy suppliers participate in the consensus process prior to data being stored on the blocks.

14.6.1 Grid architecture

The steps involved in grid architecture are as follows:

1. DSOs establish a blockchain and implement smart contracts in smart meters.
2. Users register and create an account through smart contracts on the blockchain. Each consumer is assigned a profile.
3. Data from the smart meter is encrypted and cannot be manipulated while it is transmitted across the blockchain network.
4. Local energy suppliers and DSOs function as miners to assure participant integrity and consensus, therefore producing a decentralized market process without a central authority.
5. Energy is purchased and sold via a token-based trading mechanism.
6. DSOs are also in charge of data that are delay-tolerant.

14.6.2 Peer-to-peer electricity trading

Peer-to-peer trading can be used in a case where a customer wishes to purchase 5 kW of power from a prosumer. Following the offer, the smart grid determines the validity of the transaction based on rules in the smart contract. Each user's transactions are maintained in electronic wallets recognized by a unique identifier. The blocks are permanently updated and cannot be changed when the validation procedure is completed. This ensures a safe exchange of power between two users may be implemented without the involvement of a third party.

14.6.3 Blockchain-based peer-to-peer electricity trading difficulties

There have been many studies conducted to investigate ways to overcome the obstacles encountered while implementing blockchain technology in a smart grid design. Most of the time, blockchain technology is selected on the basis of its consensus algorithm and operational characteristics. It may differ based on the nature of the case study. Several obstacles remain to be overcome:

1. Cost analysis is critical for any technology to have a substantial impact. At present, Ethereum and Hyperledger implementations cannot compete with existing power rates. The system's total cost is enhanced because of the blockchain's high transaction cost.
2. Without a doubt, the most significant problem is the smart grid's throughput capacity. The blockchain protocol used is totally dependent on the amount of data.
3. Blockchain is susceptible to cyberattacks. Even if blockchain provides a safe environment for storing data in possibly unmodified blocks, there is still an element of risk involved with P2P power transactions between buyer and seller. There is no assurance of a successful conclusion. Privacy can be violated if a third party has access to a user's daily activities and transactional records.
4. According to a market case study, incentives for local energy suppliers that simultaneously operate as miners must be standardized. These incentives may result in increased investment in the renewable energy sector.
5. As with any growing technology, regulatory norms are required, and blockchain currently lacks regulatory structures. Without standards and compatibility, it will be difficult to persuade smart grids to adopt the technology, particularly major businesses.
6. Now, blockchain protocols rely on algorithms that consume a lot of energy and have a huge carbon footprint. The physical deployment of the smart grid has a considerable influence on its design. Additional study is necessary to address this issue.

When utilized as a distributed ledger and mediator, a blockchain-based smart grid delivers the following benefits:

1. Credibility and honesty are ensured using an immutable blockchain kept in the distributed system as proof of transactions.
2. Disclosure of inconsistencies because each unauthorized effort is tracked and documented.
3. Energy data and transactions are subjected to a series of agreed-upon conditions and verification procedures before being attached to the blocks, enhancing the system's resilience.

4. The ability of blockchain-based smart grids to inherit the capability of conducting real-time settlement via micropayments is a beneficial disruptive force.
5. A blockchain-enabled smart grid ensures data integrity via consensus, authentication of users, etc., making it extremely profitable for any security breaches to be intercepted in advance.

14.7 Implementation

14.7.1 Ethereum blockchain

This section provides information on writing smart contracts and building Ethereum private chains by deploying the contract to the blockchain.

14.7.1.1 Creation and initialization of related member variables

Smart contracts exist in the blockchain permanently after deploying. The contract gets executed according to the agreement. Power management organization deploys the smart contract, and as the only administrator, it can manage the generation and consumption units and release dispatching information. Fig. 14.4 pictures the various entities involved in building a smart contract using Ethereum for a smart grid. The definition and initialization of the administrator are as follows: Here, an address type variable is defined as the administrator address, and it is initialized to be the contract creator's address through the constructor when the contract is first deployed and cannot

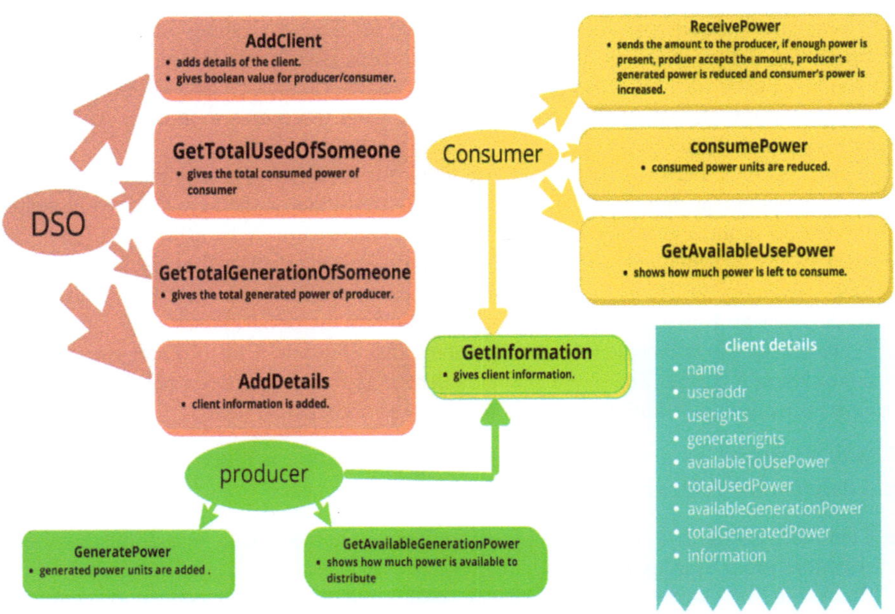

Figure 14.4 Ethereum blockchain workflow for smart grid. The structure shows various entities involved in building a smart contract using Ethereum for the smart grid.

be modified later. Power generation is the constructor, that is, the function with the same name as the contract class. The user structure is defined below, which contains the basic user details like user name, address, and other information, as well as the user's initial total power consumption, total power generation, power consumption right, generation right, etc., and also includes dispatching information. The power management organization can send the power dispatching information to the user through this variable. Finally, an array container is declared, which maps the Ethereum address to the user, and each Ethereum address corresponds to a unique user structure. Ordinary users can register as consumption users or generation users in the contract publisher. The function starts to verify the permissions to ensure that the caller of the function is the contract publisher, and then the parameters are passed to the corresponding user structure. Addr is the unique Ethereum address of the user to be registered, which is written into each meter as the unique identification of the meter or user. Here, the administrator can grant the power generation rights and power consumption rights to the user through Generate rights and User rights member variables. The initial energy consumption is initialized to 0, and the dispatch information is initialized to null. These variables are controlled by various functions.

14.7.1.2 Implementation of GeneratePower function

14.7.1.2.1 Struct user

The user structure contains the credentials such as Username, User Address, Available GenerationPower, Total GeneratedPower, Available power to use, Total used power, Generator rights, User rights, and dispatching Information. Finally, an array container is declared, which maps the Ethereum address to the user, and each Ethereum address corresponds to a unique user structure. Fig. 14.5 shows the pseudocode for creating user structure.

```
Struct User
---
name;
useraddr;
AvailableGenerationPower;
totalGeneratedPower;
AvailableToUsePower;
totalUsedPower;
generaterights;
userights;
information;
---
Array container
---
mapping (address => User) user;
administrator <- msg.sender;
```

Figure 14.5 Pseudocode for user structure. A data structure named "User" is created to store user details.

AddClient

The function AddClient starts to verify the permissions to ensure that the caller of the function is the contract publisher, and then the parameters are passed to the corresponding user structure. Addr is the unique Ethereum address of the user to be registered, which is written into each meter as the unique identification of the meter or user. Here, the administrator can grant the power generation right and power consumption right to the user through generaterights and UseRights member variables. The initial energy consumption is initialized to 0, and the dispatch information is initialized to null. These variables are controlled by various functions. The pseudocode for the function is shown in Fig. 14.6.

14.7.1.2.2 GeneratePower function

After getting the power generation right, the power generation users can generate electricity according to the relevant regulations and plans and upload the real-time generation data to the blockchain by calling the incremental generation function in the contract.

The function is implemented as follows: The function takes the generated capacity as a parameter and has no return value. The function first verifies the authority to make sure that only the authorized generation users can call the function. Then, the function adds the generated capacity of the caller to the total energy generation variable in the user structure and the current required generation variable. Since the variables in the contract are changed in the process, the call will be made in a similar ratio. The form of token transactions is permanently recorded in the blockchain. The pseudocode for generate power function is given in Fig. 14.7.

14.7.1.2.3 ReceivePower

This function is used by the consumers to receive power from the prosumers. The pseudocode for the function is shown in Fig. 14.8.

```
Function AddClient (memory_name, useraddr, generaterights, userights)
If (msg.sender == administrator) {
user[_usersddr].name_<- name;
user[_useraddr].useraddr <-useraddr;
user[_useraddr].AvailableGenerationPower <-0;
user[_useraddr]. totalGeneratedPower<- 0;
user[_useraddr].AvailableToUsePower <-0;
user[_useraddr].totalUsedPower <-0;
user[_useraddr].generaterights <- _generaterights;
user[_useraddr].userights <- _userights;
user[_useraddr].information <-" ";}
```

Figure 14.6 Function AddClient. The function is used to add a new client.

```
Function GeneratePower(_value)
If(!user[msg.sender].generaterights)
Return;
user[msg.sender.generatePower +=_value;
user[msg.sender].totalGenerationPower+=_value;
```

Figure 14.7 Function GeneratePower. The function adds the generated capacity of the caller to the total energy generation variable in the user structure.

```
Function ReceivePower (payable_from, value)
If(user[_from].AvailableGenerationPower >=_value)
user[_from].AvailableGenerationPower - = _value;
user[msg.sender].AvailableToUsePower + = _value;
```

Figure 14.8 Function ReceiverPower. The function is used by the consumers to receive power from the prosumers.

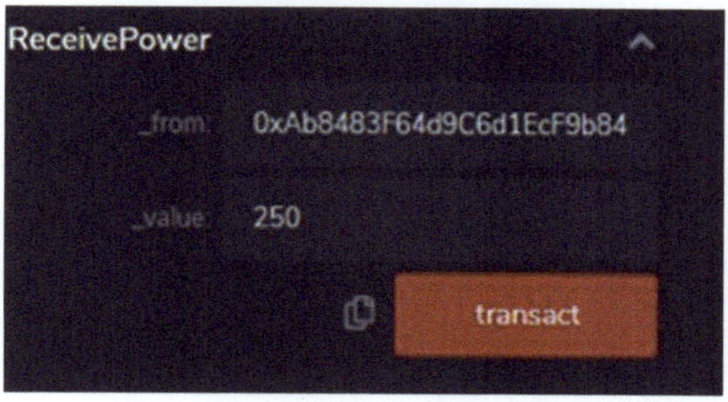

Figure 14.9 ReceivePower. Implementation screenshot of ReceivePower function.

14.7.1.2.4 Before transaction

Figs. 14.9 and 14.10 are the screenshots before the transaction of ethers from the sender's account to the producer's account.

14.7.1.2.5 After transaction

Fig. 14.11 shows the value after the transaction from the sender's to the producer's account.

14.7.1.3 Implementation of consume power function

After the power users purchase the electricity from the power generation users, they will have the available electricity at their disposal. At the same time, the users can automatically call the power consumption function in real-time to deduct the corresponding electricity from the user's account.

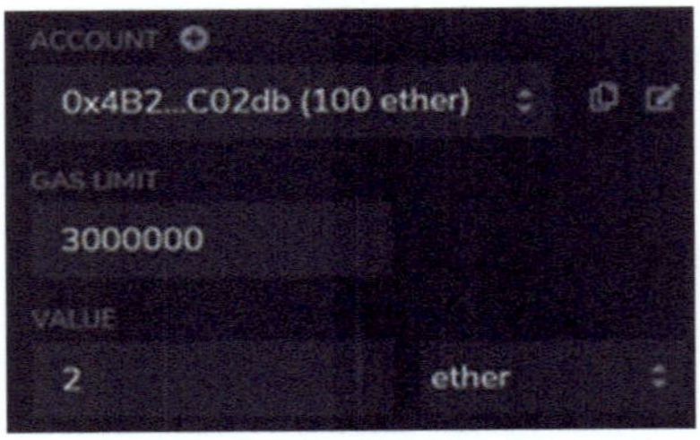

Figure 14.10 Transaction. Screenshot showing value before transaction.

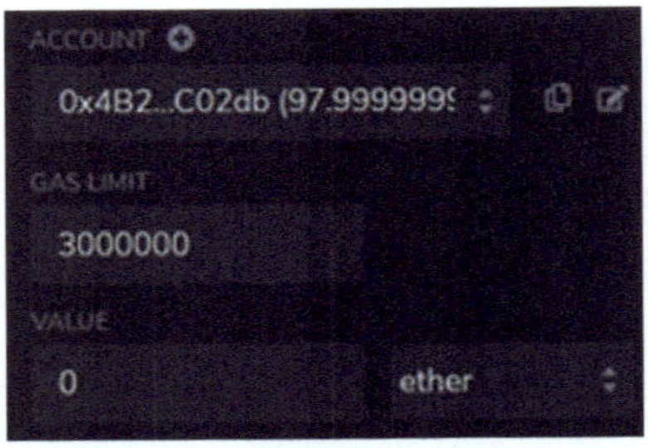

Figure 14.11 Transaction. Screenshot showing value after transaction.

Function consumePower(power)

```
If (user[msg.sender].userrights) &&
If(user[msg.sender].AvailableToUsePower >=power)
User[msg.sender].AvailableToUsePower ← AvailableToUsePower – power;
User[msg.sender].totalUsedPower ← totalUsedPower – power
```

Figure 14.12 Consume power algorithm. The algorithm verifies for valid user and checks whether the user has enough power.

The specific implementation of the function is as follows: The function starts with two verifications, one is to verify whether the user has the right to use electricity, and the other is to verify whether the user has enough power at his disposal. The input parameter of the function is the power consumed by the user in real-time. The corresponding energy is subtracted from the corresponding user's account. Because the value of the internal variable of the contract is changed, this operation will be permanently recorded on the blockchain. Fig. 14.12 depicts the pseudocode for Consume Power function.

14.7.1.3.1 AddDetails

Fig. 14.13 shows the screenshot of the implementation of AddDetails function, which is used for adding client details.

14.7.1.4 Implementation of query information function

The contract reserves the following query interfaces, through which users can perform verification of the data stored on the blockchain. Electricity users can query the remaining electricity in their current accounts to determine whether they need to buy electricity. At the same time, power users can also query how much electricity they have used in total.

The functions are GetAvailableUsePower() and GetAvailableGenerationPower(): To view their own residual power function, there are no input parameters, but only to query the function caller's residual power, which ensures the privacy of users and avoids being used by criminals. To view the user's total electric energy function, there is an address variable as the input parameter, which represents the address of the user to be checked. However, at the beginning of the function's commercial use, the function was limited to ensure that only the user or the administrator could view it, which also avoids the leakage of user information and data being used by unauthorized people. Fig. 14.14 shows the algorithm for retrieving the available use power.

Generation users can also query their current demand for power generation and the total amount of power generation. The specific implementation of the function is as follows: There are no input parameters for the function to view the current required generation, which ensures that only the caller's own current required generation can be queried. There is an address variable as the input parameter for the

Figure 14.13 AddDetails. Implementation of AddDetails.

```
Function GetAvailableUsePower()
Return (user[msg.sender].AvailableToUsePower);
```

Figure 14.14 GetAvailableGenerationPower() function. Algorithm for retrieving balanced power and consumed power.

function to view the total amount of power generation, and only users and administrators have permission to use it. The algorithm for the function implemented is given in Fig. 14.15.

Fig. 14.16 shows the screenshot of implementing a function for retrieving available generated power.

14.7.1.4.1 GetInformation

In addition, both power users and power generation users may receive dispatching information. The function of viewing their own dispatching information is as follows:

The function also has no input parameters to ensure that only its own scheduling information can be viewed. The smart meter calls the function periodically. Once the dispatching information is read, the corresponding processing is made at once according to the dispatching information. Consumption users and generation users can query data at any time by calling the above functions. Fig. 14.17 shows the algorithm for GetInformation function implementation.

```
Function GetAvailableGenerationPower()
return (user[msg.sender].AvailableGenerationPower);
Function GetTotalGenerationOfSomeone(address addr)
If(msg.sender==addr ll msg.sender == administrator)
return (user[addr].totalGeneratedPower);
```

Figure 14.15 Algorithm for retrieving available generation power. Pseudocode for querying users' current demand for power generation and the total amount of power generation.

Figure 14.16 Available power. Screenshot after implementing GetAvailableGeneratedPower.

```
Function GetInformation()
return(user[msg.sender].information);
```

Figure 14.17 Algorithm of GetInformation function. GetInformation function is used to retrieve dispatch information.

14.7.2 Implementation on Hyperledger Sawtooth

Hyperledger Sawtooth is a blockchain platform for businesses that allows them to build distributed ledger programs and networks. The ledgers are distributed, and smart contracts are secured. Sawtooth facilitates the creation of blockchain applications by decoupling the core technology from the software program. Sawtooth is a very flexible programming language that helps companies to make the best policy judgments possible. The core design of Sawtooth allows applications to tailor transaction rules, permissioning, and consensus algorithms to match the individual company requirements.

14.7.2.1 Architecture

The process involved in Hyperledger Sawtooth blockchain implementation is depicted in Fig. 14.18.

14.7.2.1.1 Global state

The primary objective of a distributed ledger like Sawtooth is to spread the ledger across participating nodes. The capacity of Byzantine consensus to maintain a consistent copy of data among nodes is a critical strength of blockchain technology. On each validator, Sawtooth stores the state for all transaction families in a single Merkle-Radix tree instance. The block validation procedure on each validator guarantees that identical transactions result in identical state changes and that the resultant data is identical for all network participants.

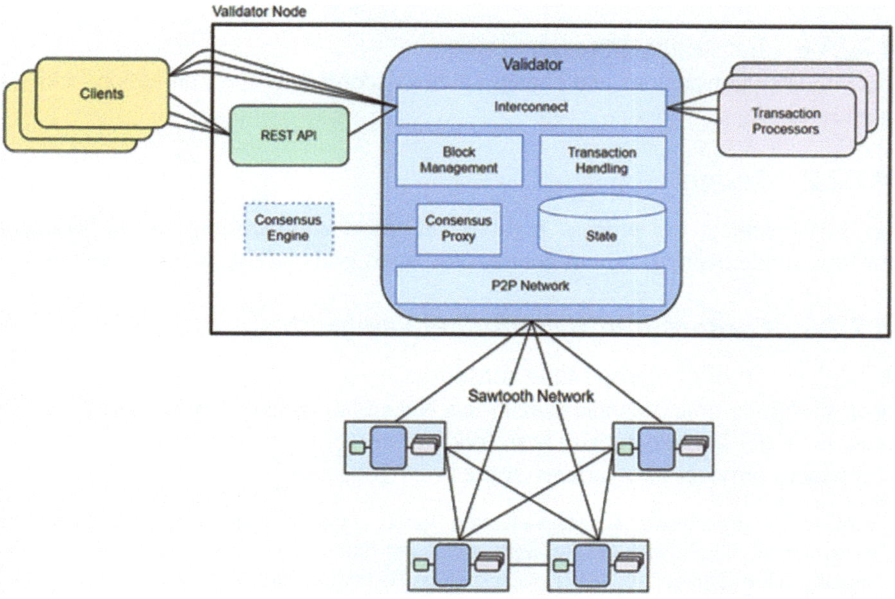

Figure 14.18 Hyperledger Sawtooth blockchain implementation. Various modules included in Hyperledger Sawtooth implementation are shown.

14.7.2.1.2 Clients

Hyperledger Sawtooth is based on the asynchronous client/server model. The following is the pattern:

- Clients establish connections to the server and submit queries.
- The server answers each request with 0 or more responses.
- Clients may submit numerous queries without waiting for a response from the server.
- Additionally, servers can provide numerous responses and are not required to wait for fresh requests.

14.7.2.1.3 Validator

This is the component responsible for verifying batches of transactions, grouping them into blocks, establishing consensus with the Sawtooth network, and managing communication between clients, transaction processors, and other network validators.

14.7.2.1.4 REST API

In Sawtooth, this is a fundamental component that conforms to HTTP/JSON standards when communicating with a validator. Sawtooth comes with a REST API that is utilized by clients, such as the Sawtooth command line interface. Developers may either utilize this REST API or create their own proprietary APIs to communicate with the client-validator (Surwase, 2016).

14.7.2.1.5 Consensus engine

This is the Sawtooth component that offers Sawtooth node-specific functionality. The consensus engine operates independently of the validator on the node and communicates with it via the consensus API.

14.7.2.1.6 Transaction processor

This validates transactions and changes states according to the transaction family's stated rules.

14.7.2.2 Design

Fig. 14.19 depicts the process involved in smart contracts put on Hyperledger Sawtooth implementing a smart contract for smart grids in Hyperledger Sawtooth.

14.7.2.3 Consensus in Hyperledger Sawtooth

14.7.2.3.1 Proof of elapsed time consensus

Proof of elapsed time, this mechanism is a Nakamoto style that is optimized for big networks. PoET is not definitive (can fork).

Sawtooth provides two variants of the PoET consensus:

- PoET-SGX implements a leader-election lottery system using a Trusted Execution Environment (TEE), such as Intel Software Guard Extensions (SGX). PoET-SGX is occasionally referred to as PoET/BFT because of its Byzantine Fault Tolerance.
- On systems lacking a TEE, the PoET simulator implements the same consensus process. PoET simulator is also referred to as PoET/CFT due to the fact that it is crash fault-tolerant.

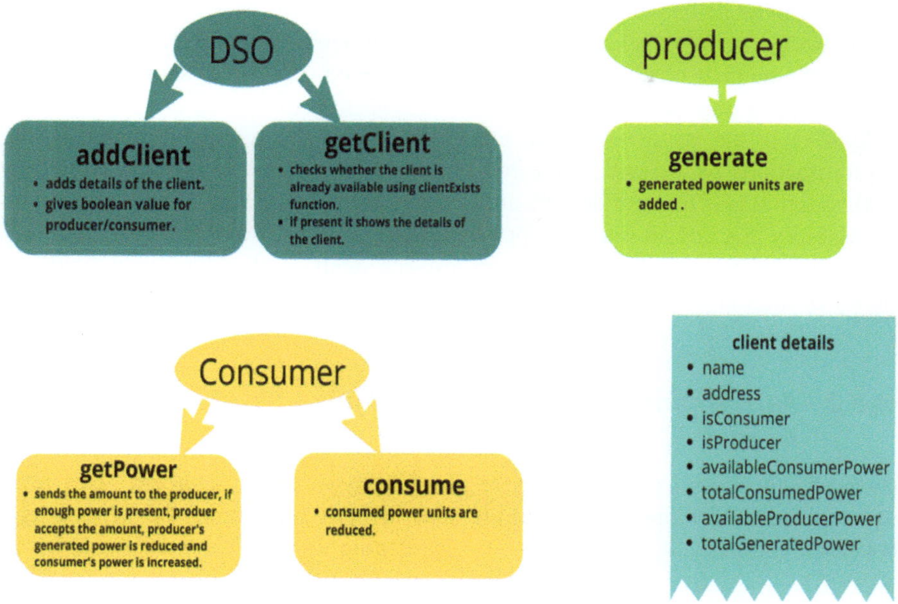

Figure 14.19 Smart contract on Hyperledger Sawtooth. Smart contracts for smart grids should have functionalities like adding a client, adding power generated, and consuming power.

14.7.2.3.2 Practical Byzantine Fault Tolerance consensus

Practical Byzantine Fault Tolerance, this is a voting-based consensus mechanism with finality that incorporates BFT.

14.7.2.3.3 Dev mode consensus

A simple random-leader consensus technique, this mechanism may be used to evaluate the performance of a transaction processor running on a single Sawtooth node. Dev mode consensus is not advised for a network with numerous nodes and should not be utilized in production.

14.7.2.3.4 Raft consensus

A consensus method that is optimized for small networks with restricted membership based on leaders, Raft is crash-tolerant.

14.8 Result analysis

14.8.1 Performance analysis of Ethereum using Caliper

The Hyperledger Project and the Linux Foundation both host Caliper (Ampel et al., 2019). It is used to assess performance and serves as a foundation for integrating several distributed ledgers. Inferring blockchain protocols, settings, and blockchain selection can be assisted by Caliper. This tool is helpful for developers and is useful for analyzing performance and impact. The following report in Table 14.1 summarizes the number of transactions per gas consumed. Fig. 14.20 shows the performance analysis done using Caliper.

As seen in the above table, the Ethereum blockchain is capable of processing two transactions per second. With this capacity, its implementation can serve only 6000 households if the transactions are updated every 8 minutes. This brings a limit to scalability that demands a new approach.

The graphical representation of performance analysis showing throughput and latency of the transaction is shown in Fig. 14.21.

14.8.2 Performance analysis of Hyperledger Sawtooth using Caliper

Sawtooth dev mode is a consensus engine for Hyperledger Sawtooth. This technique is meant for application testing and is built on top of the Hyperledger Sawtooth platform. It is useful when developing transaction processors, smart

Table 14.1 Gas consumed vs amount of transactions.

Amount of transactions	Gas consumed in normal transactions
1	129,520
2	200,758
4	286,908
6	343,630
8	475,758
16	932,653
24	1,364,612
32	1,942,418
40	2,342,097
50	2,726,303

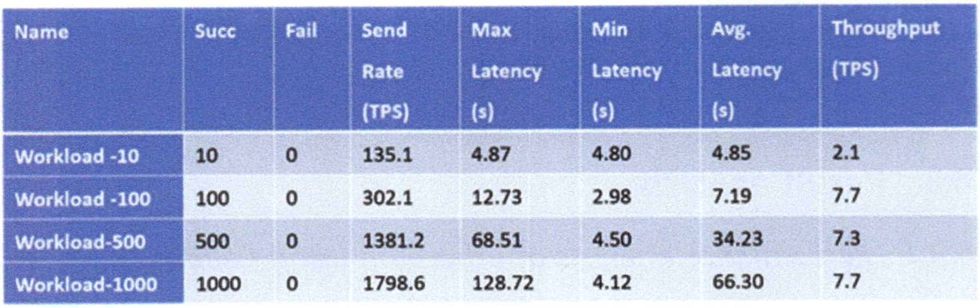

Name	Succ	Fail	Send Rate (TPS)	Max Latency (s)	Min Latency (s)	Avg. Latency (s)	Throughput (TPS)
Workload -10	10	0	135.1	4.87	4.80	4.85	2.1
Workload -100	100	0	302.1	12.73	2.98	7.19	7.7
Workload-500	500	0	1381.2	68.51	4.50	34.23	7.3
Workload-1000	1000	0	1798.6	128.72	4.12	66.30	7.7

Figure 14.20 Summary of performance metrics using Caliper. Performance analysis of Hyperledger Sawtooth using Caliper through various metrics is tabulated.

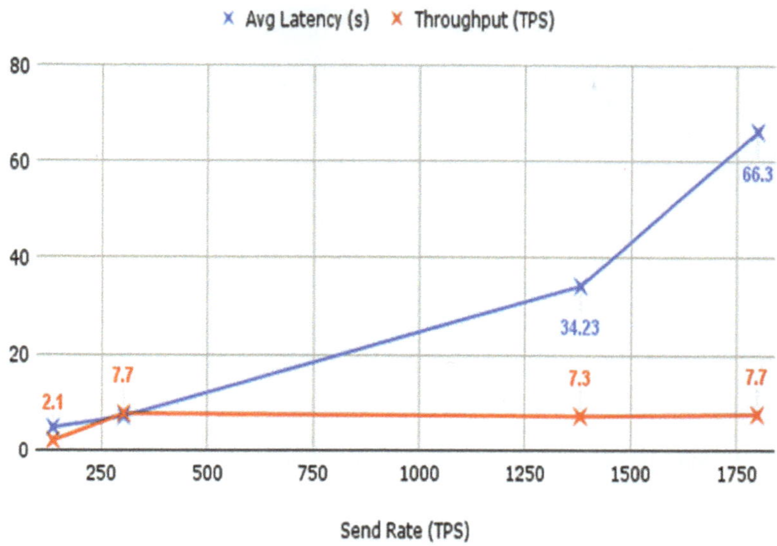

Figure 14.21 Graphical representation of the number of transactions per gas consumed. Graph plotted for average latency/throughput versus rate of transaction.

contracts, or other components because of its high commit rate. Dev mode also demonstrates how to use the Sawtooth consensus engine API for a lottery-style consensus implementation. Developers interested in implementing a consensus engine can use the dev mode source code as an example. Although the dev mode consensus engine can be used in a Sawtooth network, it has a very inefficient fork-resolution algorithm and makes no guarantees about crash fault tolerance. It should not be used in a production environment. Fig. 14.22 summarizes the performance of Hyperledger Sawtooth for our smart grid implementation.

Fig. 14.23 shows us the Throughput and Avg-Latency for invoking transactions in Hyperledger Sawtooth implementation obtained for corresponding workloads.

14.8.3 Comparison of Ethereum and Hyperledger Sawtooth implementations

EVM was utilized to create a virtual smart grid that maps smart contracts to Ethereum's blockchain in the Ethereum implementation. Each user is given a one-of-a-kind address that also serves as a wallet. The study involves implementing a smart contract function by any participant like a DSO, prosumer, or consumer. Every time a new contract is deployed, the number of possible transactions per second, the amount of gas consumption in each transaction, and the number of prospective transactions per second are all tracked. The second implementation was using Hyperledger Sawtooth, which provides a substantial speed and scalability advantage over Ethereum. Additionally, the following strategies were employed to bolster security and privacy.

Name	Succ	Fail	Send Rate (TPS)	Max Latency (s)	Min Latency (s)	Avg. Latency (s)	Throughput (TPS)
Workload - 10	10	0	10	0.44	0.44	0.44	22.9
Workload - 100	100	0	100	1.47	1.47	1.47	68.2
Workload- 500	500	0	500	9.14	9.14	9.14	51.3
Workload- 1000	1000	0	1000	27.66	27.66	27.66	36.2

Figure 14.22 Summary of performance metrics using Caliper. Performance analysis of Hyperledger Sawtooth using Caliper using various metrics.

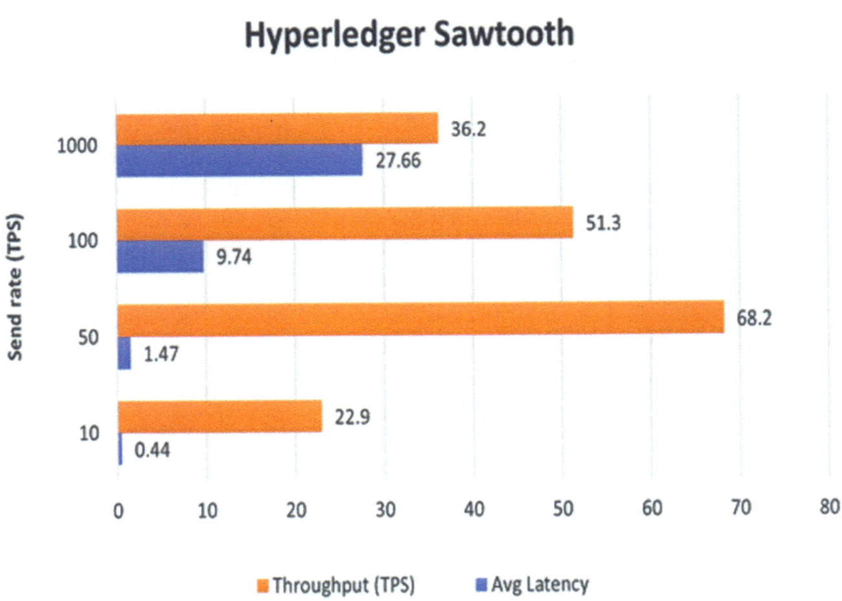

Figure 14.23 Performance analysis of Hyperledger Sawtooth. Comparison between Throughput and Avg-Latency vis-à-vis send rate (for invoking transactions) in Hyperledger Sawtooth implementation.

SHA-256 is used as the hashing algorithm in Hyperledger Sawtooth. It is a 256-bit digest-length cryptographic hash algorithm. It is a hash function that does not require a key. A message is processed in 64-round. When SHA-256 is used, there are 2 power 256 potential hash values, making it almost impossible for two distinct documents to have the exact same hash result. Hyperledger Sawtooth uses SHA-256 as its hashing algorithm because its 256-bit key is significantly more secure than other commonly used hashing algorithms.

14.9 Conclusion

Distributed ledger technology, despite its limitations, has the potential to transform the energy industry. We need to promote renewable energy sources to play a more active part in making our planet a cleaner and better place to live while also encouraging low-carbon energy production. An outline of blockchain and smart grid principles is provided in this chapter. Two methods with extensive system designs are then used to analyze a virtual smart grid capable of deploying smart contracts for market payment and related services. To execute the suggested design, Ethereum and Hyperledger Sawtooth platforms were used. This chapter compares the performance of the two methods. Ethereum is the most widely used blockchain-based system; however, it has limits in terms of scalability. Hyperledger Sawtooth, on the other hand, addresses scalability, but it compromises decentralization. Hyperledger Sawtooth is a young technology that has the potential to overcome system limitations in smart grids. Both approaches have distinct application scenarios based on their conclusions and practical consequences. Ethereum implementations provide methods for establishing a local P2P energy trading infrastructure among a society's households capable of sustaining its power generation and consumption. The residences will generate electricity mostly from renewable energy sources and will not rely on the national grid or any other party, except in the event of load shedding or a power outage on a cold, nonwindy winter day. In the case of Ethereum implementation, the throughput capacity may be simply regulated inside a microgrid. However, the capacity of a smart grid is the most critical aspect in determining its usage. The Hyperledger Sawtooth implementation paints a bigger picture of how future smart cities would integrate electric cars, IoT, and AI gadgets, necessitating a more regulated environment. In terms of security and privacy, Hyperledger Sawtooth serves as a better option as, by default, it has few security algorithms incorporated. Privacy in Ethereum implementation can be enhanced by third-party security algorithms. There is still much more research and development required to successfully integrate blockchain technology in the energy sector.

References

Abidi, A., Bouallegue, B., & Kahri, F. (2014). Implementation of elliptic curve digital signature algorithm (ECDSA). In 2014 Global Summit on Computer & Information Technology (GSCIT) (pp. 1–6). IEEE.

Agung, A. A. G., & Handayani, R. (2020). Blockchain for smart grid. *Journal of King Saud University-Computer and Information Sciences.*

Aitzhan, N. Z., & Svetinovic, D. (2016). Security and privacy in decentralized energy trading through multi-signatures, blockchain and anonymous messaging streams. *IEEE Transactions on Dependable and Secure Computing*, *15*(5), 840–852.

Ampel, B., Patton, M., & Chen, H. (2019). Performance modeling of hyperledger sawtooth blockchain. In 2019 IEEE International Conference on Intelligence and Security Informatics (ISI) (pp. 59–61). IEEE.

Bashir, I. (2017). *Mastering blockchain*. Packt Publishing Ltd.

Bekara, C. (2014). Security issues and challenges for the IoT-based smart grid. *Procedia Computer Science*, *34*, 532–537.

Butt, O. M., Zulqarnain, M., & Butt, T. M. (2021). Recent advancement in smart grid technology: Future prospects in the electrical power network. *Ain Shams Engineering Journal*, *12*(1), 687–695.

Gensollen, N., Gauthier, V., Becker, M., & Marot, M. (2016). Stability and performance of coalitions of prosumers through diversification in the smart grid. *IEEE Transactions on Smart Grid*, *9*(2), 963–970.

Gunduz, M. Z., & Das, R. (2020). Cyber-security on smart grid: Threats and potential solutions. *Computer Networks*, *169*, 107094.

Gungor, V. C., Sahin, D., Kocak, T., Ergut, S., Buccella, C., Cecati, C., & Hancke, G. P. (2012). A survey on smart grid potential applications and communication requirements. *IEEE Transactions on Industrial Informatics*, *9*(1), 28–42.

Khajeh, H., Laaksonen, H., Gazafroudi, A. S., & Shafie-khah, M. (2020). Towards flexibility trading at TSO-DSO-customer levels: A review. *Energies*, *13*(1), 165.

Mengelkamp, E., Notheisen, B., Beer, C., Dauer, D., & Weinhardt, C. (2018). A blockchain-based smart grid: Towards sustainable local energy markets. *Computer Science-Research and Development*, *33*(1), 207–214.

Miglani, A., Kumar, N., Chamola, V., & Zeadally, S. (2020). Blockchain for internet of energy management: Review, solutions, and challenges. *Computer Communications*, *151*, 395–418.

Mollah, M. B., Zhao, J., Niyato, D., Lam, K.-Y., Zhang, X., Ghias, A. M., Koh, L. H., & Yang, L. (2020). Blockchain for future smart grid: A comprehensive survey. *IEEE Internet of Things Journal*, *8*(1), 18–43.

Moubarak, J., Filiol, E., & Chamoun, M. (2018). On blockchain security and relevant attacks. In 2018 IEEE Middle East and North Africa Communications Conference (MENACOMM) (pp. 1–6). IEEE.

Mukhopadhyay, U., Skjellum, A., Hambolu, O., Oakley, J., Yu, L., & Brooks, R. (2016). A brief survey of cryptocurrency systems. In 2016 14th Annual Conference on Privacy, Security and Trust (PST) (pp. 745–752). IEEE.

Mylrea, M., & Gourisetti, S.N.G. (2017). Blockchain for smart grid resilience: Exchanging distributed energy at speed, scale and security. In 2017 Resilience Week (RWS) (pp. 18–23). IEEE.

Nakamoto, S. (2008). Bitcoin: A peer-to-peer electronic cash system. *Decentralized Business Review*, 21260.

Pee, S.J., Kang, E.S., Song, J.G., & Jang, J.W. (2019). Blockchain based smart energy trading platform using smart contract. In 2019 International Conference on Artificial Intelligence in Information and Communication (ICAIIC) (pp. 322–325). IEEE.

Pilkington, M. (2016). *Blockchain technology: Principles and applications. Research handbook on digital transformations*. Edward Elgar Publishing.

Shrivas, M.K., & Yeboah, D.T. (2018). The disruptive blockchain: Types, platforms and applications. In Fifth Texila World Conference for Scholars (TWCS) on Transformation: The Creative Potential of Interdisciplinary.

Siddiqui, F., Zeadally, S., Alcaraz, C., & Galvao, S. (2012). Smart grid privacy: Issues and solutions. In 2012 21st International Conference on Computer Communications and Networks (ICCCN) (pp. 1–5). IEEE.

Strasser, T., Andrén, F., Kathan, J., Cecati, C., Buccella, C., Siano, P., Leitao, P., Zhabelova, G., Vyatkin, V., & Vrba, P. (2014). A review of architectures and concepts for intelligence in future electric energy systems. *IEEE Transactions on Industrial Electronics, 62* (4), 2424–2438.

Surwase, V. (2016). REST API modeling languages-a developer's perspective. *International Journal of Engineering, Science and Technology*, *2*(10), 634–637.

Wang, H., & Wu, T. (2017). Cryptography in Blockchain. *Journal of Nanjing University of Posts and Telecommunications (Natural Science Edition)*, *6*, 61–67.

Yoldaş, Y., Önen, A., Muyeen, S., Vasilakos, A. V., & Alan, İ. (2017). Enhancing smart grid with microgrids: Challenges and opportunities. *Renewable and Sustainable Energy Reviews*, *72*, 205–214.

Zhang, H., Liu, B., & Wu, H. (2021). Smart grid cyber-physical attack and defense: A review. *IEEE Access*, *9*, 29641–29659.

Blockchain in monitoring and automation of distribution systems 15

César Cazal (Andrés) and Ferdinanda Ponci
Institute for Automation of Complex Power Systems, RWTH Aachen University, Aachen, Germany

15.1 Introduction

Blockchain constitutes a real game-changer technology, building trust between parties that would not otherwise easily cooperate.

This technology makes itself heard by reducing the costs and intervention of third-party authorities involved in the validation of agreements and transactions (Henten & Windekilde, 2020). Therefore, as a transactional technology, it gained increased acceptance in different environments, from banks, legal agreements, and even by replacing money for its token market value. The rising use of this technology has affected directly as well as indirectly the technology and applications for power systems, starting from its physical impact from the mining activity on the global energy consumption, given that the energy requirements for the mining activity for just one of the cryptocurrencies are comparable to the energy consumption of a country the size of Argentina (Martin & Nauman, 2021). On the other hand, even when the effects of the technology have been considerable, its adoption in power systems as an automation technology has just been hyped in recent years, emphasizing its applications in transactional as well as financial aspects. Nevertheless, lately, new applications are finding a niche in the power automation sector, addressing problems from the novel perspectives of the technology. Exploring these characteristics is necessary to break the common perception of technology as a purely IT one, to introduce the reader to the trust world of the blockchain and let him dive into the development of applications for power systems.

In this chapter, we will introduce some fundamental power system concepts to explain some distribution network monitoring and automation problems to be solved with the use of blockchain technology. In addition, some blockchain fundamentals will be presented for a better understanding of the philosophy behind the thriving development of blockchain applications for the modern energy grid.

15.2 Power system monitoring on distribution grids and Local Energy Communities

Power system monitoring is the measurement of the different electrical variables and component status present in the grid, where measurements are taking place at very

Blockchain-Based Systems for the Modern Energy Grid. DOI: https://doi.org/10.1016/B978-0-323-91850-3.00014-7

regular intervals of time. From a pragmatic viewpoint, this information is used for billing metering, load prediction, Supervisory Control and Data Acquisition systems display (SCADA), etc., and is often used for the control and automation of the network to maintain the correct, efficient, and secure operation of the electrical infrastructure for energy supply. According to Kundur (Kundur & Balu, 1994), these operational states of transmission and distribution systems can be classified as:

- **Normal**: All the variables are within the expected limits and operational range.
- **Alert**: Some variables can be on the operational limit, and the system is at the risk of falling outside the limits, that is into the emergency operational state in the presence of small disturbances or even into an "In Extremis" state as a consequence of Big Disturbances.
- **Emergency**: Some variables escape the expected operational limits, and the system experiences extreme risks of affecting the energy supply, the stability of the system, or endangering the electrical devices present in the system.
- **In Extremis**: Cascade outages and major shutdowns take place due to the extremely unstable condition of the system. The countermeasures are intended to save as much of the system as possible from reaching a total blackout.
- **Restorative**: Under the restorative state, countermeasures as preventive as well as corrective actions are taken to place the system back into a normal operational state.

Making correct and fast decisions to respond to the different changes in the grid depends on the monitoring infrastructure to support the decision-making process. Without proper knowledge of what is going on in the grid, we cannot know the current grid operational status, neither the variables states. Considering that operational voltages in distribution grids are relatively insignificant compared to transmission grids as well as the high number of nodes that require monitoring functions and the reduced material consequences of failures in the distribution grids, it is possible to understand the poor investment in Monitoring, Control, and Automation equipment. This situation used to be a small problem considering the passive character of the grid. Nevertheless, the growing popularity of Distributed Energy Resources (DER) and the need for active as well as efficient control of the grid behavior has increased the interest in monitoring and automation at these voltage ranges with cost-effective solutions. Due to the scattered nature of the grid, distributed technologies have gained special attention over the years, opening an important niche for technologies exploiting this particular characteristic. One of these is the blockchain technology, given that this is a promising technology to exchange information between heterogeneous actors, to be addressed in Section 15.3.

15.2.1 Characteristics of the distribution grid and Local Energy Communities

Until recently, medium and low voltages were considered passive, that is constituting the last piece of the energy supply chain before reaching the client end with an electromechanical measurement device, which periodically required an in-person read to submit the electric bill. Due to the strong separation between generation, transmission, and distribution networks, often separated in terms of Power and

Voltage priorities, the last-mentioned has always felt ignored as occupying the last position in the list of priorities for allocation of resources. Hence, most of the technologies on the power system used to be designed for the first two.

The evolution of DER and Smart Grids is dramatically changing the priority associated with distribution grids, where proper coordination of DER now enables a bidirectional Power Flow on Medium and Low Voltage networks, that is allowing to locally supply the energy demand and even feeding power upstream, from Distribution toward the Transmission System, changing the classification of distribution networks to active. This radical game change is also considered the future of energy systems: the energy no longer requires to be generated and transported for hundreds (or even thousands) of kilometers to reach the end-user. Now, this technology makes it possible to locally generate energy and supply the needs of local populations, reducing in this way the strong dependence on costly and complex transmission lines as well as big centralized power plants, encouraging the local investment in sustainable energy generation according to the natural resources present in the region. DER technology makes the system more interdependent between regions, but, on the other hand, a reduction in the power exchanged by transmission lines weakens the overall system stability, demanding a higher effort on monitoring, control, and automation to sustain the energy transition. Because the focus of the chapter is placed on distribution grids and Local Energy Communities, the examples associated with distribution grids are going to be set forth around the single network diagram Fig. 15.1. As we can see in the model, integration of Photovoltaic Park (PVP1) and Wind Parks (WP1) are considered in the example model, as well as Battery Storage Power Stations (BSPS1), to show how DERs might be present and integrated into current distribution grids. Also, measurements of voltage and current are placed on more than 30% of the buses: see buses 2, 6, 9, 11, and 14.

On the other hand, according to Eurelectric (2021), an association for the European electricity industry that represents more than 3500 companies in power generation, distribution, and supply in over 32 European countries, a Local Energy Community can be defined as: "an SME (small and medium enterprise) or a non-profit organization, the shareholders or members of which cooperate in the generation, distribution, storage or supply of energy at the local level, including across borders whose simple majority is constituted by legal or physical members of the local community". These represent local community efforts to satisfy the local energy demand, store energy for its use during production shortage hours, and/or sell any remaining surplus to the rest of the network. These communities represent a special case of the low voltage and medium voltage (LV/MV) distribution network, whose operation is self-governed by the local management, and decisions are taken based on the community's interests.

As these communities participate in the energy market, running the local network also demands extra effort to connect with the rest of the energy world, even when the main objective of these communities is to fulfill the local energy demand, which does not always depend on synchronous combustion generators based on fossil fuel technologies. Considering the presence of days without wind or sun (or even both at the same time), fulfilling the demand on islanded networks represents

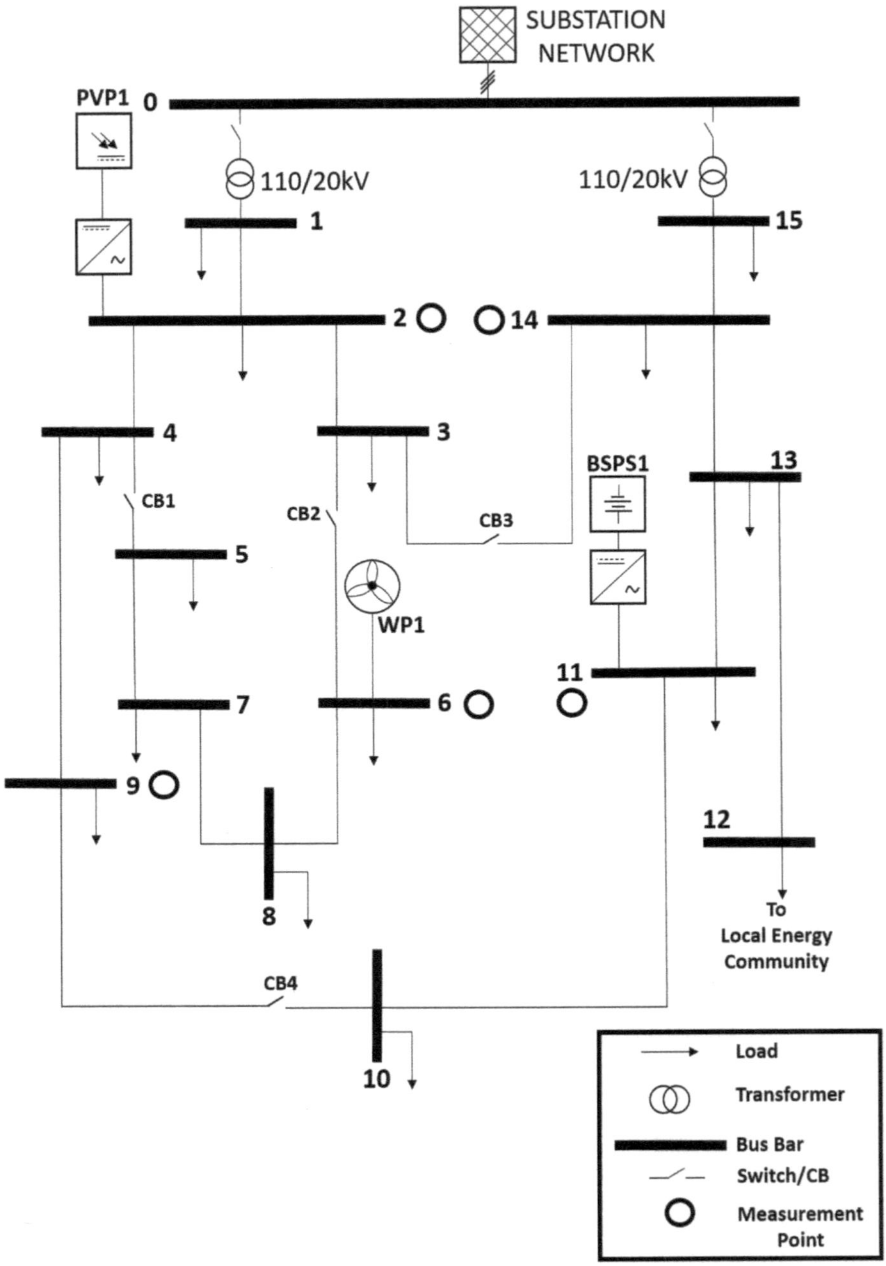

Figure 15.1 MV distribution network example.

a complex tax; that's why there still exists the need of remaining part of the bigger grid. Hence, the demand for energy from external sources is still present, and, in consequence, a need to monitor and control the supplied energy and the overall demand with proper coordination of the local resources, as well as adapting these according to the global energy requirements, continues to perplex the distribution networks. As an example of Local Energy Communities, as depicted in Fig. 15.2, we have presented an LV distribution grid with Photovoltaic, Wind, and Battery Storage Power Units intended to feed the local community. Also, measurements of

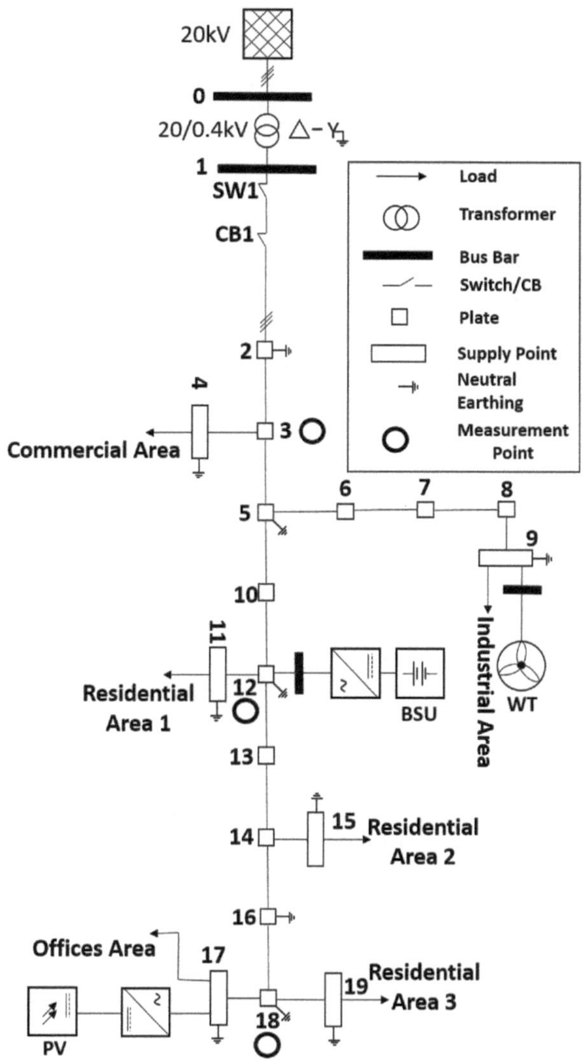

Figure 15.2 LV Local Energy Community example.

voltage and current are placed on more than 30% of the main points for estimation errors: under plates 3, 12, and 18.

15.2.2 Decentralized state estimation on distribution grids

Different approaches can be taken to collect the information from the different nodes of the distribution grid: from the deployment of measurement devices at every node to the calculation of the values by having just one measurement point. Nevertheless, the deployment of measurement devices on every distribution point will not be cost-efficient enough to justify its implementation; neither would the calculation of values be accurate enough to fulfill contemporary standards. A good solution to comply with contemporary accuracy requirements (yet its deployment cost is going to be indirectly discussed within the chapter) is the State Estimation technique: it consists of applying an algorithm to "estimate" the state of the system, having previous knowledge of the system's model and real-time measurements or pseudo-measurements from the field. In our case, we will put emphasis on decentralized solutions to easily exploit recent technological concepts like blockchain and Substation Automation Units (SAU) for distribution.

For medium and large distribution networks, the information to run the state estimation might be insufficient due to the investment required on every node, as well as the processing infrastructure required to deal with the important information volume, making the state estimation often financially unfeasible as the already present smart meter infrastructure cannot be exploited due to data privacy policies on consumer ends. Current solutions to deal with these problems are workarounds due to the lack of information (e.g., pseudo-measurements and/or centralized machine-learning state estimation techniques), where new possible blockchain-based solutions that can change the way we collect the information from smart meters in the next couple of years are to be presented in the following subsections.

The method to be presented is a solution to the processing infrastructure that suggests a horizontal scaling on every "concentration point" to split the estimated area into smaller ones (i.e., adding multiple machines with the same characteristics, rather than just one "bigger" machine whose cost involved might not be directly proportional to the processing power because of the complexity associated on centralized solutions) (Sitaram & Manjunath, 2012). This method is the Multi-area State Estimation and is easily applicable to small distribution networks and Local Energy Communities as well. The reason for implementing the method in Local Energy Communities is that unlike public electric grids with their produce-and-distribute model, and for reasons inherent in its collaborative character, Smart Meter values used for billing purposes might be exploitable under the agreement implicit in the advantage ensuing from the trust that binds the community (Moroni et al., 2019), as well as the DER's measurement data to gain access to enough information for the deployment of the State Estimation, a situation that might not be the case for public distribution networks due to data privacy as well as diversified ownership of the resources in the network (e.g., generators can be owned by third parties as Local Energy Communities while the distribution network runs under the Distribution System Operator).

15.2.2.1 Network architecture for decentralized state estimation

Before diving into the associated algorithm, we would like to introduce the network architecture for the decentralized State Estimation on distribution networks based on the grid introduced in Figs. 15.1 and 15.2.

It is important to remark that it is necessary to get access to enough information on variables along the network to perform a correct state estimation. To optimally place the measurement devices to gather enough information to carry out the state estimation, it is necessary to perform an optimization study, where the expected values along the network are to be simulated under different conditions and, based on that, place an optimal quantity of measurement devices in the right nodes to estimate with the least error the whole set of variables of the network. As this study is not the main point of the chapter, we can apply a rule of thumb for this: apply measurement devices in around 30% of the network nodes, to be placed on the feeders and branches within which more power is expected to flow, as there might be expected most of the estimation errors, for example, important load congestion point, important load behavior variation, DER units modifying a commonly expected power flow on passive networks (see Fig. 15.1). In Fig. 15.2, we have only one measurement point because, as the generation resources are from the community, we expect to get some information from the DERs.

Now, the division of areas and good communication between the areas are required for a Multi-area State Estimation. The blockchain concepts for monitoring and automation applied in the chapter are suitable for any fully distributed peer-to-peer system (e.g., IoT of devices or actors in the energy community); so, to apply a generic concept, but at the same time specific to the energy sector, a decentralized architecture is suggested, based on the concept of "SAU" (Angioni et al., 2017), which is a good option to reduce the data concentration to a single point. As presented in Fig. 15.3, this concept is carried out by processing the unit dedicated to running the monitoring, control, and automation algorithms, which has access to the cyber-physical resources via a communication layer. This layer is strongly suggested to be TCP-based due to its extended media compatibility (e.g., over Ethernet, Fiber-Optics, 3G, 4G, 5G, radiofrequency, Power Line Communication, etc.), virtualization capabilities, and long-time testing, implementation, and development. The SAU then processes the input data from sensors and the current actuator states—to then apply monitoring, control, protection, and automation algorithms—and communicates the results either to the Load Control Center to represent the required data on a SCADA system or to the actuator devices in the network to modify its current state.

By implementing a fully distributed architecture, we can have remote variable measurement as well as remote actuation, let alone the possibility to split the processing areas and reassign devices within them in case of necessity. In our example,

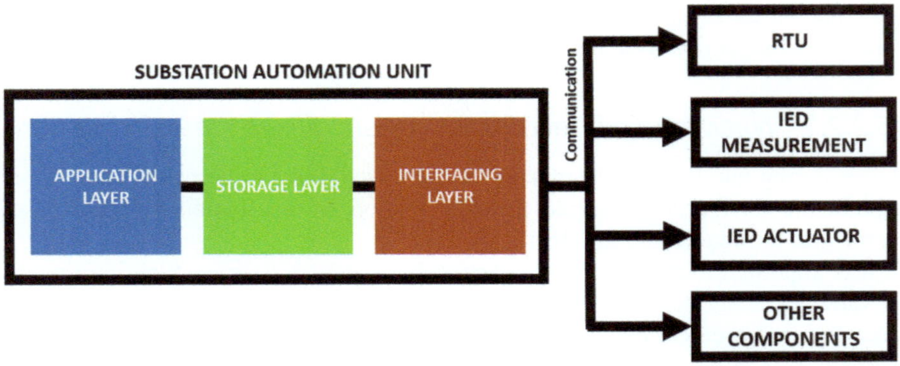

Figure 15.3 Substation automation unit.

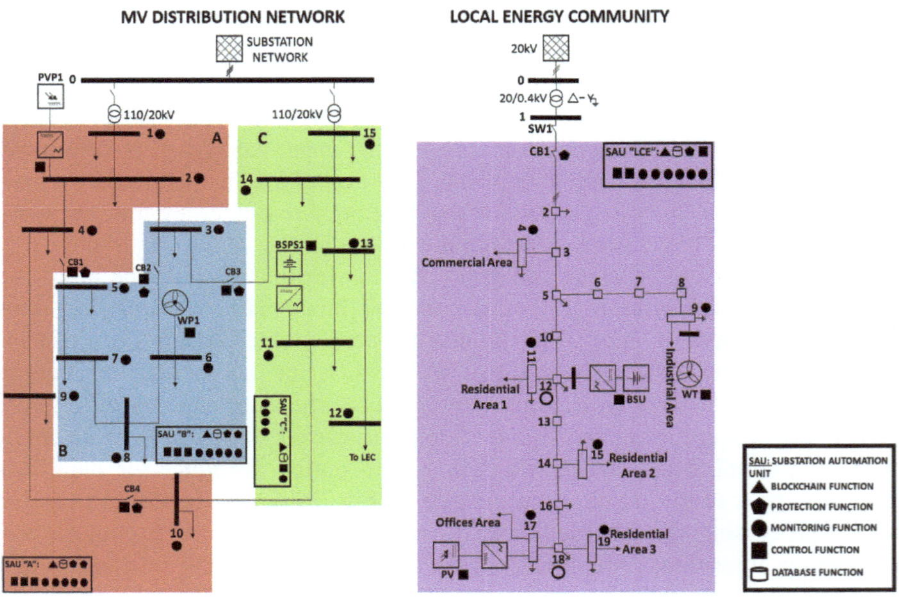

Figure 15.4 Example of area divisions and substation automation unit implementation on distribution grids and Local Energy Communities.

as we can see in Fig. 15.4, we have implemented the SAU concept to distribute the processing load into different areas, as well as the different functions associated with the different components present in the respective areas.

This concept allows information collection, exchange, and precise monitoring algorithms along different voltage and power levels, starting from low voltage grids, going through the medium voltage grid to even transmission systems at high voltage levels, as shown in Fig. 15.5.

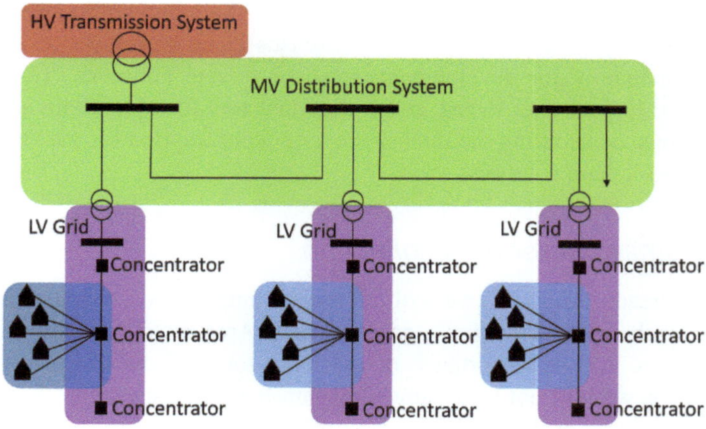

Figure 15.5 Information exchange along with different voltage levels.

15.3 Blockchain-based services for distribution grid monitoring and automation

In this section, our main study focuses on "Blockchain as a structural solution," suggesting in this way blockchain's advantages as solutions to problems we think are not the same advantages if other technologies were in place. We will first see the associated problems and then present the possible solution using the blockchain, which, in most cases, is in the process of future adoption by the industry, gaining enough maturity to be transferred to the public.

15.3.1 Blockchain-based monitoring services

Along with the whole State Estimation process, different blockchain applications represent feasible solutions to different challenges to enable the full implementation of monitoring services on distribution networks. The main reason this might represent a feasible solution compared to other technologies is the trust component associated and the cryptographic background on top of which blockchain has been built. It is important to add that the blockchain examples we are considering are based on Hyperledger Fabric (Hyperledger Fabric: A Blockchain Platform for the Enterprise, 2020), as most of the applications we want to address do not necessarily require tokenization, while, at the same time, their permissioned characteristic brings a higher degree of security needed for them to operate within the critical infrastructure needs as the energy supply of a whole city (Gai et al., 2019). As this is a permissioned blockchain solution, it also incorporates a strong advantage in the consensus mechanism for transaction validation: it incorporates, as a first layer, credentials of participants to be checked and, then, making use of "pluggable" consensus mechanisms, it applies signature policies to accept or deny the request.

The consensus mechanism can be selected once the blockchain architecture is defined but comes by default with Raft consensus or no consensus at all, the last one an option mainly for development, tests, or private networks. This makes a strong case for Hyperledger Fabric as an important tool for blockchain adoption on energy solutions, considering the intensive energy consumption for the validation of PoW-based blockchain solutions.

15.3.1.1 Smart meter data collection

As we have already discussed, we will put special emphasis on blockchain methods to feed and exchange information regarding Multi-Area State Estimation for distribution networks. As the importance of the main topic of the chapter is put into smart contracts, we will not study any algorithm for state estimation. Nevertheless, if the reader wants a deeper understanding of the different algorithms, he can find them in the literature: we have, for example, M. Pau et al. (2017), which propose an efficient method for processing load-splitting between nodes on Multi-area State Estimation. In case we take our time to dive into the literature associated with the topic, it would be a foregone conclusion we would have reached that most state estimation algorithms make use of an iterative process to minimize the estimation error. If the fundamentals of State Estimation Theory are required, the reader can gain a deep understanding from books regarding power system state estimation (Abur & Gómez Expósito, 2004).

As the previous section already makes clear, an important problem for the state estimation of distribution networks is the lack of data, even when the adoption of smart meters on these networks can represent more than 80% of the population in some countries (European Commission DG ENER; Joint Research Centre, 2014), enough to provide accurate and reliable monitoring of LV and MV networks. The real problem, as already mentioned, lies in data privacy: the consumer is reluctant to share his private data (Jakobi et al., 2019).

If we go back to the first successful blockchain network, Bitcoin, we might find out more about one of its most renowned properties: pseudo anonymity (Fabian et al., 2016). Bitcoin, as well as most of the other blockchain technologies (if not all of them), makes strong use of encryption to ensure a safe, trusted, and distributed operation, which, with good use of VPN, can even take care of hiding the IP address.

This means that a physical person or entity can have a public address, which is linked to a "wallet," an intermediary identity management software that manages credentials, private keys, and metadata associated with the owner (see Fig. 15.6). As long as the person does not reveal that he or she is the actual owner of a specific public address, their identity and transactions are safe. An analog result can also be replicated on a distribution network by making use of a private/public key combine associated with a blockchain identity or wallet for every smart meter to gain access to information through secure and anonymous transactions, preserving in this way the privacy of the user with a secure, trusted and distributed solution.

As mentioned before, the data privacy problem is solvable with blockchain technology but, once implemented, presents an important scalability issue:

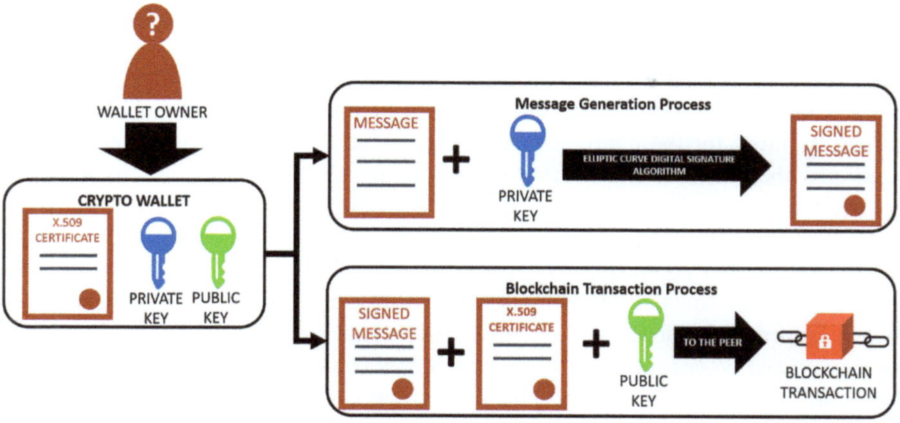

Figure 15.6 Wallet operations.

If we consider a 250000-inhabitants city, where 20% of the apartments are commercial, or offices and the average house has around 2.5 inhabitants/apartment, we have that the number of smart meters in a city like this would be around 125000. If all the smart meters in the city reported every 15 minutes with a single-precision floating-point format, we would have that the resource occupancy would be:

- **Voltage (3-phases)**: 32 bits × 3 = 96 bits
- **Active Power (3-phases)**: 32 bits × 3 = 96 bits
- **Reactive Power (3-phases)**: 32 bits × 3 = 96 bits
- **Timestamp**: 64 bits.

We would have that every smart meter would report 352 bits of read/write sets apart from the variable name for every variable changed (around 100 bytes), the endorsing peer's signatures (min 64 bytes each, at least one needed), and the Channel ID (recommended around 16 bytes). We have that every transaction would take around 160 bytes. With these calculations for storage space, we would have the required storage space in bytes per hour ($S_{\mathrm{B/h}}$) every 15 minutes as:

$$S_{B/h} = \text{Quantity of meters} \times \text{Bytes} \times \frac{\text{Writes}}{\text{h}} \tag{15.1}$$

$$S_{B/h} = 125,000 \times 180B \times \frac{1}{0.25h} = 90,000,000\ B/h \quad S_{GiB/y} = 751.87\ GiB/year \tag{15.2}$$

If we consider that every node on the blockchain might have a hard drive of 512 GiB, without considering the fact that handling 125000 transactions every 15 minutes might be, alone, unfeasible for some public blockchain technologies like Ethereum, where the usual bandwidth is around 20 transactions per second (Mechkaroska et al., 2018) (requiring around 104 minutes just to store 15 minutes of information for a single city of 250000 inh.), and if we are to consider the replication of the ledger along with that of the whole network, where all the nodes would be totally filled with one-time-use information over the duration of less than a year, and that without considering the presence of other transactions and the signatures required for the confirmation of every block, and this considering a small-sized city, then the fact is that these values might look small compared to what will be required for significantly larger cities.

Now, if we consider the concept of SAU and LV data concentrators, previously introduced, we can coordinate the read of samples to make them happen all at the same time on the smart meter end and, with permissioned blockchains (like Ethereum PoA or Hyperledger Fabric, the last one to be considered for our example), we can split the storage and processing load by creating a new channel for a small sector of users. A channel is just a group of blockchain peers that can speak together, and they store their own ledger and manage their own smart contracts as if they were on a different blockchain service but with the difference that they can share the same consensus service (or Ordering Service, for Hyperledger Fabric) with the rest of the blockchain. This might call for strong bandwidth requirements, so opting for a dedicated consensus service might considerably reduce the number of transactions as well as the bandwidth required by every node to operate. In the same SAU, we can have different blockchain services and applications to interact with them, assigning a particular blockchain public address for every smart meter and bus bar, even when an SAU might manage more than one bus bar. Also, Hyperledger Fabric implements a service called Private Data Collection (Thummavet, 2019), which makes the data visible only to the sender and the receiver of the transactions, where publishing only a hash SHA256 of the transaction (which can have some "salt" in the middle, that is some random value to avoid brute force attacks to reveal the data content) means we will have, again, around 105 bytes registered in the ledger (according to some tests, for one signature), independent from the size of the data as what we store is only the transaction notarization, while the private data is deleted after a certain number of blocks. If we apply the SAU concept, then after dividing the network into groups of around 1000 smart meters for each concentrator service, we would have, based on that:

$$S_{\mathrm{B/h}} = 1,000 \times 180B \times \frac{1}{0.25\mathrm{h}} = 720,000\ \mathrm{B/h} \quad S_{GiB/y} = 5.63\ \mathrm{GiB/year} \tag{15.3}$$

At this point, we can also consider information replication: we can participate on different blockchains or channels to replicate the information of a zone on different concentrators in case we lose one concentrator. If we consider a 5 times redundant service (that's a significant amount of redundancy if we don't expect too many failures in our system when 3 is already enough), we might have around 5 times this amount, totalizing 29.37 GiB/year of data, then for 512 GiB of storage space, if we consider some other transactions as registration of smart meters as requiring additional storage space, then the smart meters would easily give us more than 17 years of service, possibly, far beyond the expected lifecycle of the hardware. The idea here is that once we have calculated the required nodes' values and/or properties, we can freely feed smart meter data without breaking the rights for data privacy. If we consider around 25 transactions per second (with up to 20k transactions per second achievable) (Gorenflo et al., 2019), considering low communication bandwidth by, for example, GSM technologies (already discontinued in some places of the world), then we would feed all the information needed from the 5000 smart meters in less than 3.5 minutes. From this setup, we are not only feeding the State Estimation Algorithm with reliable, fresh data, but we can also apply algorithms for topology change tracking, control, protection algorithms (i.e., overvoltage, distribution transformer overload, etc.), which would trigger alarms in case of electrical faults or loss of power in the network, etc. The monitoring data of every LV distribution bus bar can then be stored in an MV SAU service to feed it with data to provide MV monitoring, control, protection, and automation services. In Figs. 15.7–15.9, we present the described algorithms:

ALGORITHM 1: Smart Meter Registration (sm_register)

Input: sm_wallet_id,public_key (pk), bus_id, branch_id (optional)
Output: successful_registration ← bool

```
if ( ( check (sm_wallet_id,pk) & (NOT ( is_registered (sm_id) ) ) ) ) then
      sm_register ( sm_wallet_id , pk , bus_id, (branch_id) )
      successful_registration ← true
else
      successful_registration ← false
end if
return (successful_registration)
```

Figure 15.7 Algorithm 1: Smart meter wallet registration for anonymous data exchange.

ALGORITHM 2: Smart Meter Private Data Write (sm_write_measure)

Input: sm_wallet_id,V3ϕ, P3ϕ, Q3ϕ,time
Output: write_success ← bool

```
if ( check_privileges (sm_wallet_id) & (new_time_for_collection -
   get_private_data(sm_wallet_id,last_time)>10min) ) then
      put_private_data(sm_wallet_id,V3ϕ, P3ϕ, Q3ϕ,time)
      write_success ← true
else
      write_success ← false
end if
return (write_success)
```

OBS: as discussed, a simple "write" function is enough as long as the wallet ID is not associated to a real clue about the physical person or smart meter, put and get private data functions are only to not maintain sensitive data in the Blockchain

Figure 15.8 Algorithm 2: Smart meter data write.

ALGORITHM 3: Smart Meter Private Data Read (sm_read_measure)
Input: concentrator_id, sm_wallet_id, collection_start_time
Output: V3ϕ, P3ϕ, Q3ϕ ← floating point matrix

```
1: if (check_privileges (concentrator_id)&
   (get_private_data(sm_wallet_id,last_time -collection_start_time>0ms ))
2:       data ← get_private_data(sm_wallet_id,last_time)
3: else
4:       data ← Not a Number (nan)
5: end if
6: return (data)
```

OBS: a simple "read" function is enough in case put private data function was not used to store the data in the Blockchain

Figure 15.9 Algorithm 3: Smart meter data read.

It is important to add that, as previously mentioned, there is no need to collect all the measures on all the points of the network to have an accurate state estimation with more than the minimum necessary nodes for reliable execution. In addition, in the case of a reduced percentage of smart meters reports in any node, better and more accurate techniques for pseudo-measurement injection can be achieved. Finally, blockchain layer 2 solutions are good ideas to dramatically reduce the amount of one-time-use information within the blockchain. The concepts and algorithms proposed here have been tested in a laboratory environment, where the best results obtained were under information notarization through the Private Data Collection.

15.3.2 Cryptographic signature authentication for the exchange of monitoring measurement data and critical automation data

The main objective of this service can be said to be the exact opposite in reference to Private Data Collection: what we want here is to be able to explicitly associate a message with a physical actor, where this actor can be a person, a measurement device, an SAU, etc., to associate the message to a physical element of the network.

Under some circumstances, we might want to guarantee that the message that came from the point of the network had not been tampered with on its way to the receiver. This can be done without the need of the blockchain by implementing a Cryptographic Signature algorithm like RSA or Elliptic Curve Digital Signature Algorithm (Ali, 2015), but this process commonly associates a virtual identity to a specific address without revealing the true identity of the owner. The whole idea behind using blockchains in the process seeks to add a further traceability level to the message, so we can know if the real message comes from an intermediary device or the measurement device itself.

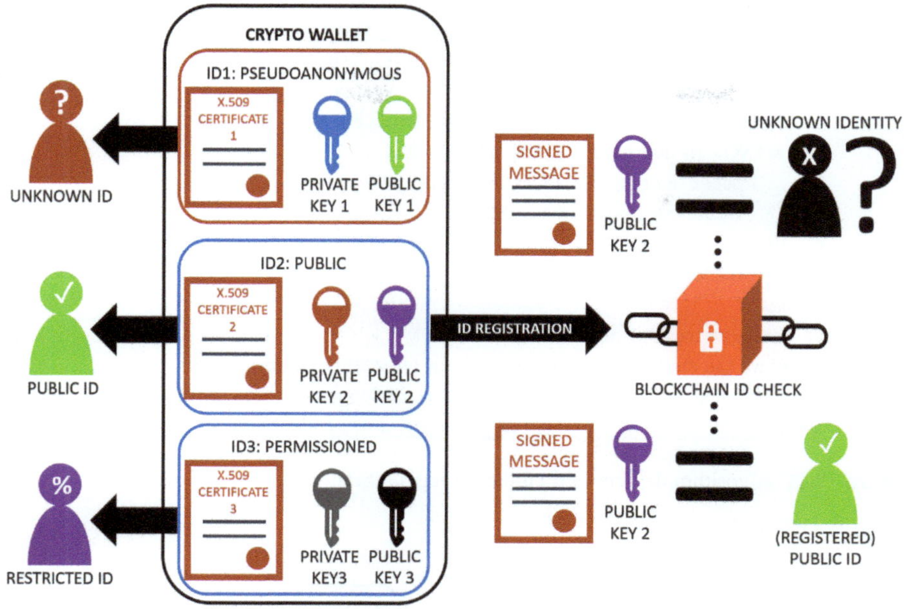

Figure 15.10 ID Check Blockchain service for public key and certificate registration to track message signatures.

Together with the data collection feature previously mentioned, different ID elements should be created for the public identity, independent from the first "anonymous ID" associated with the collection of private data, that is a totally new group of the crypto certificate, private key, and public key, so, then, the public key and/or the certificate issued to the ID by the certificate authority should be registered into the blockchain as shown in Fig. 15.10.

This procedure can be used to associate some data whose origin we would like to guarantee, such as may be, one-time annual access to energy measures from smart meters, the current value of critical or public measurement units (e.g., medium voltage connections to industry customers), or any data signed and exchanged by critical infrastructure as SAUs State Estimation results, network reconfigurations, triggering of protection devices, or parameters to DERs. The discussed proposal can be found in Fig. 15.11.

15.3.2.1 *Cryptographic signature authentication for measurement equipment and smart meters*

The idea behind this service is to discourage data tampering as much as to boost the degree of automation in the system at the same time, and thereby it is suggested to adopt smart meters with integrated crypto technologies, which nowadays seems to be a topic balanced on the pivot of uncertainty in this sector of the market, where a debate regarding the future of smart meters is part of the last part of this section

ALGORITHM 4: Public ID Registration (public_id_register)

Input: real_id, wallet_id, certificate, public key(pk) (optional)
Output: successful_registration ← bool

```
if ( ( check (certificate,real_id) & check (certificate,wallet_id)
   & (NOT (is_registered (wallet_id) ) ) ) ) then
       public_id_register( wallet_id, real id , pk )
       successful_registration ← true
else
       successful_registration ← false
end if
return (successful_registration)
```

OBS: we can check the identity just after the registration, for every time some degree of authentication or validation is needed with a check_id(pk) command by providing the wallet id, public key or the real_id as input.

Figure 15.11 Algorithm 4: Required public ID registration.

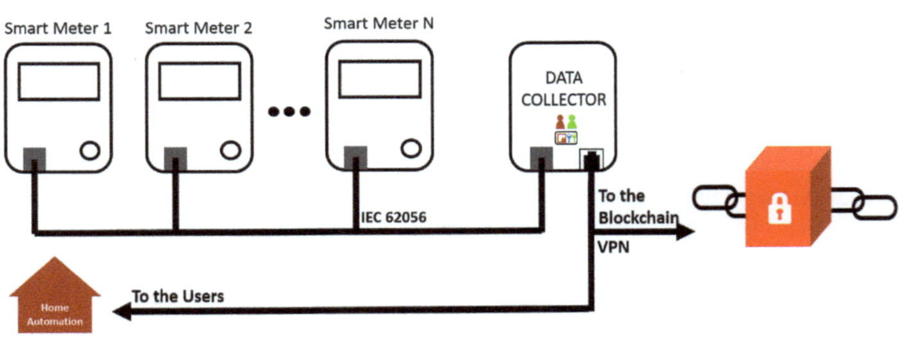

Figure 15.12 Automated data collection.

discussed. At the moment, we should conform with local data collectors placed inside buildings, in charge of polling the data from smart meters and publishing it on the blockchain. This idea presents some data privacy threats for non-cryptographic data, but, with the arrival of the IEC62056 protocol, crypto-signed data can already be implemented, and with the help of data concentrators, where an application layer can be provided to communicate with the blockchain and provide energy-related information (see Fig. 15.12).

On the other hand, energy prices and energy sources (e.g., renewable energy) can be fed by the DSO into the blockchain to inform the customer and encourage some consumption behaviors. To create trust, we validate the ID of the data source, creating in this way a bidirectional channel and getting social benefits, as well as economic ones, from the information exchange.

An important point to add is that the energy report should be limited to a certain number of readings per year as the idea is to preserve the consumer's energy profile and habits. So, if we are using the energy data for billing purposes, we should perform billing through a smart contract in charge of reading the value once a month in the worst case, and, so, if we combine the Private Data Collection service with the ID registration for energy billing to associate a cumulated energy consumption to the discrete integration of power over time, then for reasons attributed to the instantaneous integration errors due to the time window for measurement report, we might be facing enough errors to discourage the association of a particular user from a pool of users to an individual energy consumption bill; hence, the lesser the number of yearly energy readings, the more private remains the data.

15.3.2.1.1 Distributed state estimation with blockchain information exchange

In this case, we are considering that every area shown in Fig. 15.13 is operated by different regional load centers or even different DSOs. To create trust, we propose the SAU ID blockchain parameters registration to automatically adopt constant communication methods between different SAUs (Asefi et al., 2021) and perform the state estimation and topological state of the network. The idea of using the blockchain for state estimation information exchange and notarization might delay the state estimation and overflow the SAUs' storage units, as for every iteration, we require a new set of values for every area when, sometimes, hundreds of iterations might be required to obtain the estimation result for a specific moment (depending on how finely tuned the estimation variables are). By storing some authentication parameters (i.e., public key and certificate), sender and receiver IDs, we can perform the data exchange, making the cyber-security of the communication network the main goal of this approach. By using a blockchain approach, we are not only requiring the registration and validation of SAUs but also the information provisioning is only performed between registered peers and not disclosed to foreign actors, even when they might be physically connected to the network. In this case, we suggest not only registering the authentication information onto the blockchain but also registering their real ID and physical capabilities to obtain better coordination between areas, as presented in Fig. 15.14.

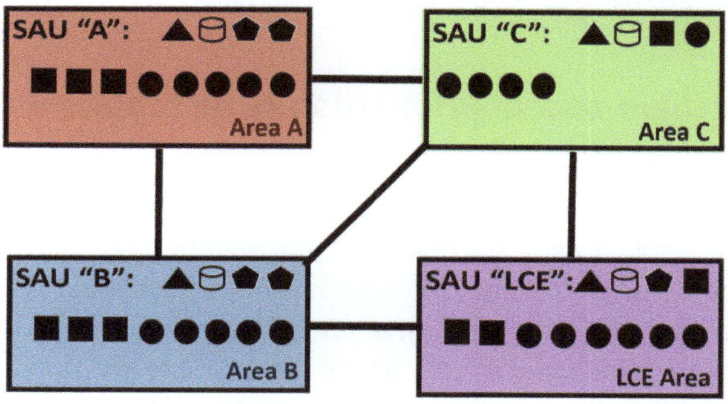

Figure 15.13 Multiarea state estimation information exchange.

```
ALGORITHM 5: Data exchange connection (create_data_connection)
Input: id_receiver, id_sender , variables list (optional)
Output: successful_connection ← bool
if ( ( is_registered (id_receiver) & is_registered (id_sender) &
   (id_sender ≠ id_receiver )&(requester = id_receiver) ) then
      create_data_connection( id_receiver, id_sender, (variables_list) )
      successful_connection ← true
else
      successful_connection ← false
end if
return (successful_connection)
```

Figure 15.14 Algorithm 5: Data exchange connection creation.

```
ALGORITHM 6: State Estimation Data Send (estimation_data_send)
Input: id_receivers, id_sender , variable_values_matrix
Output: data_sent ← bool
if (check_connection (id_receivers,id_sender) & (requester = id_sender))
      estimation_data_send (id_receiver, id_sender, variable_values_matrix) then
      data_sent ← true
else
      data_sent ← false
end if
return (data_sent)
OBS: variable_values_matrix size can easily be above 32 bytes,
hence it might be a good idea to send the exchanged information as
put_private_data, i.e., only a hash of the information will remain in the
blockchain while the information will be deleted after some number of blocks.
```

Figure 15.15 Algorithm 6: State estimation secure data exchange.

In case the number of iterations is reduced to the tens or twenties, exchanging the data directly through the blockchain might be a good idea always considering the amount of data to be exchanged, as we can see in Fig. 15.15.

15.3.3 Blockchain-based automation in distribution systems

Up to this moment, we have discussed the necessary hardware architecture for the successful deployment of distributed systems in electrical grids and some blockchain-based applications, starting from the sample collection process, device identification for security reasons, the creation of connections for the Multi-Area State Estimation, and the secure information exchange along with the network, all necessary to run monitoring services, where the process given to the information from this point is left to the imagination of the reader, who can feed their own algorithms and applications with all the resources presented up to this point.

From this point, we will discuss more about applications using the blockchain as a fundamental artifice for its operation, whose main purpose is not the generation nor the secure exchange of information within the network but to ensure the secure, correct, and efficient operation of the Local Energy Communities and distribution grid, in general.

15.3.3.1 Consensus groups for automation

It is important to make a differentiation here as most of the blockchain platforms make use of consensus as an underlying layer for the blocks' approval; so, when we refer to it, we will speak about "Blockchain's consensus" and our consensus application for Automation as "Consensus Group."

What we on this application want is to make use of the blockchain itself to create a smart contract for consensus on top of the blockchain, whose rules can be changed over time, and different operational policies can be defined with previous validation. In the process, we can define a small group of participants, and the decision of incorporation can be approved by the different actors participating already in the seed group of at least three actors.

To understand a particularity here, we need to quickly review the CAP theorem's Trilemma (Gilbert & Lynch, 2002), and the tradeoff of always picking two out of three properties: Consistency, Availability, and Partition Tolerance (see Fig. 15.16). As a simple analogy, "Consistency" implies that "all active participants will receive the most recent values", "Availability", that "every request received will be attended" and "Partition Tolerance", also known as "Fault Tolerant", means "our system can tolerate the presence of faulty nodes". In the opposite corners, we have the tradeoffs if we choose the other two properties (e.g., Consistent and Available, hence, the tradeoff would be "Intolerant to Faults", i.e. in case there is a fault in a node, no consensus can be reached). It is important to consider that, actually, this is not a black and white situation and these properties can be chosen as fractions. Another particular characteristic to consider while designing a consensus groups is that common Blockchain systems make use of the Partition Tolerant property as per default.

The point to consider here is that the new Consensus Group will "implicitly" share the same properties as the blockchain's consensus if we do not modify them. This is because the Consensus Groups' algorithm working as a smart contract will depend on how the blockchain itself works, that is we can define our smart contract will be safe and always

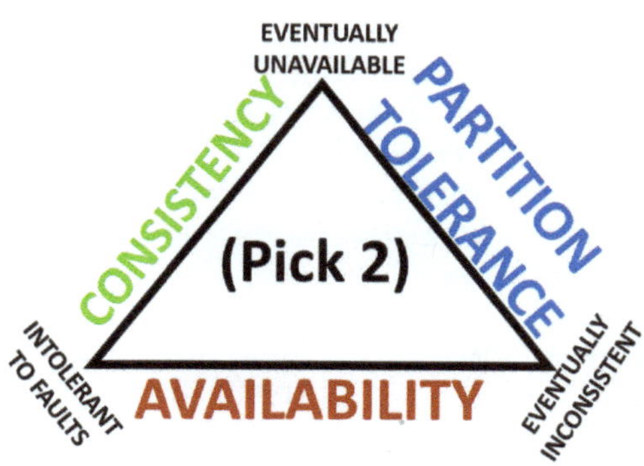

Figure 15.16 CAP theorem's Trilemma.

available, but the underlying blockchain's consensus won't actually take all the requests always in case it does not contemplate "availability" as a property. On the other hand, some degree of independence exists in our Consensus Group working on top of the blockchain. With the help of an external application feeding the smart contract, we can bend these rules as, finally, our application is a virtual one. We can create our own rules different from the consensus used by the blockchain, but it would be good practice to implement a Consensus Group's algorithm at least similar to the consensus algorithm implemented by the blockchain, where the two properties exploited can be the same for both algorithms, considering that some blockchain platforms like Hyperledger Fabric make use of "pluggable" consensus algorithms (Hyperledger Fabric: Pluggable Consensus, 2020), with Raft as a default.

On top of our Consensus Groups, different policies and rules might be necessary, for example, the registration of credentials and keys, definitions in case of lack of consensus, or how the requests in terms of input and output are going to be addressed.

A good application for Consensus Groups is the creation of small redundant measurement networks and the possibility of deciding the state of some physical resources or topological changes in case there is a need. Specifically, addressing a distributed automation based on SAUs, we are led by the possibility of controlling some critical resources redundantly to represent a shield against cyber-physical attacks against the network; so, no changes might be possible to the state of these critical resources in case there is no consensus between different actors.

Finally, for the creation of Consensus Groups, depending on the number of participants and how often the information is exchanged, there might be no need to implement the Consensus Group smart contract on a different channel unless an intensive data exchange is expected between peers. Here, a very basic Consensus Group creation is presented, where we can take total advantage of the underlying consensus algorithm, as can be seen in Figs. 15.17–15.20.

An interesting proposal for special policies could be the application of sensor fusion techniques or security applications where the decision of multiple peers is required in case the decision might affect several cyber-physical variables beyond the influence area of the SAU resource owner.

ALGORITHM 7: Consensus Group Creation (consensus_group_create)

Input: group_id, id_list, special_policies (e.g. timeout, new_proposal_wait)
Output: group_created ← bool

```
1: if ((requester = inside_the_list (id_list))&
   (NOT ( is_registered(group_id)))&list_length(id_list)>2) then
2:        consensus_group_create(id_list)
3:        define special consensus policies or particular parameters
          in case they exist using (group_id,special_policies)
4:        group_created ← true
5: else
6:        group_created ← false
7: end if
8: return (group_created)
```

Figure 15.17 Algorithm 7: Consensus group creation.

```
ALGORITHM 8: Proposal request for consensus (consensus_proposal)
Input: group_id, proposal
Output: proposal_id ← long long integer
if ((requester = inside_the_list (group_id)) then
        proposal_id ← consensus_proposal (group_id,proposal,current_time)
else
        proposal_id ← Not a Number (nan)
end if
return (proposal_id)
OBS: the participating peers should poll for any pending proposal
```

Figure 15.18 Algorithm 8: Consensus Vote Proposal Request.

```
ALGORITHM 9: Consensus vote (consensus_vote)
Input: vote_value, group_id, proposal_id
Output: vote_submited ← bool
if ((requester = inside_the_list (group_id)) then
        consensus_vote (vote_value, requester, group_id, proposal_id)
        vote_submitted ← true
else
        vote_submitted ← false
end if
return (vote_submited)
```

Figure 15.19 Algorithm 9: Consensus Vote Submission.

```
ALGORITHM 10: Consensus result (consensus_result)
Input: group_id, proposal_id
Output: vote_result ← vote variable (e.g. integer, real, bool, etc.)
if ((requester = inside_the_list (group_id)) then
        vote_result ← consensus_result (group_id,proposal_id)
else
        vote_result ← Not a Number (nan)
end if
return (vote_result)
```

Figure 15.20 Algorithm 10: Consensus Proposal Result Request.

15.3.4 Remote operation of automation devices

A small portion of this chapter considers the Remote Variable modification request. For instance, in case we would like to proceed to the maintenance of some area outside the area of influence of our SAU, for example, if we want to perform the maintenance of the circuit breaker (CB) on the feeder of the Circuit Breaker of the Local

ALGORITHM 11: Remote Variable Modification (remote_var_change)

```
Input: id_requester, id_remote_device , new_remote_variable
Output: request_submitted ← bool
1: if ((requester = id_requester) &
   check_privileges(id_requester,id_remote_service)) then
2:       remote_srvc_request (id_requester, id_sender , new_remote_variable)
3:       request_submitted ← true
4: else
5:       request_submitted ← false
6: end if
7: return (request_submitted)
```

Figure 15.21 Algorithm 11: Remote variable modification request.

Energy Community, we would have to request the de-energization of upstream MV feeders of the DSO providing service to that specific part of the network (but please, always take care of the five golden rules for these procedures); if we want to request the change of some physical variable or topology in the network whose active element is not present within the local area of influence of the SAU or, if we take it the other way around, we would like to modify a local physical variable whose change might affect other areas. In any case, we can perform the request remotely. An example of this remote request, where the privileges of the requester are previously checked, can be found in Fig. 15.21. Here we can add a line to wait for the request's answer, or we can poll for the answer from the remote device.

15.3.4.1 Blockchain-based virtual redundancy

This application is quite fresh and particular between blockchain applications and is still under further improvement in the institute for Automation of Complex Power Systems in the RWTH Aachen University (Sadu et al., 2021), under the project BRILLIANT.

An important advantage of blockchains is data replication along with the whole network and the creation of trust among different actors. This not only creates trust but maintains cohesion beyond the independence of every network actor. If we exploit these possibilities for automation, we can not only control a big system by controlling its small pieces but also carry out the aggregation of resources joined to the efforts to overcome complex challenges and problems beyond the influence of every SAU while, at the same time, an increased complexity brings with it a higher failure probability on any of the components present in the network, and this by considering the individual probability of failure. This problem might discourage the investment in new distribution network technologies as the investment required might eventually demand a considerable maintenance cost with the loss of the very functionalities it was deployed to fulfill. The use of redundant elements should be considered to overcome the failure of processing units in critical resources. As presented in Fig. 15.22, this concept consists of the replication of the critical

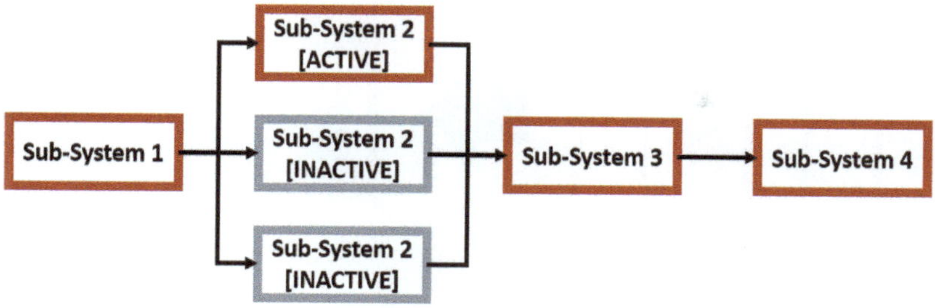

Figure 15.22 Voting-logic-based redundant system.

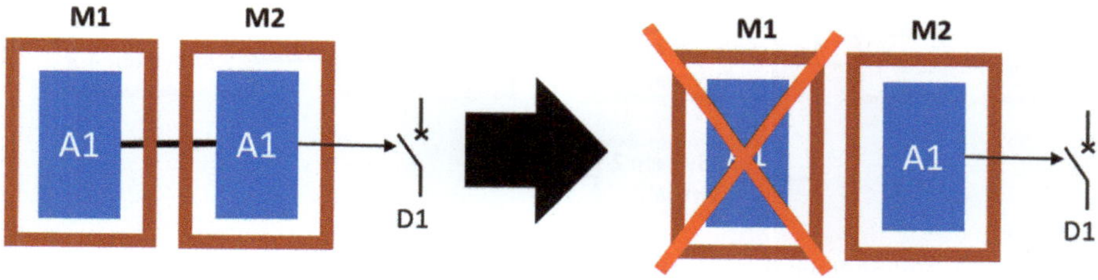

Figure 15.23 Common physical redundant scheme.

components of the system, where logic is to be implemented for the correct function. This logic is commonly known as "voting logic." This consists of having a majority of valid votes to validate an output result. An odd number of voters are required to implement voting-logic-based redundant systems, where the simplest voting logic system consists of a single voter deciding the output on the system with at least one other voter that remains inactive until the main voter goes inactive. In that circumstance, one of the inactive elements then becomes active and starts deciding the output of the subsystem.

This is commonly done by physical replication of the subsystem, as can be seen in Fig. 15.23. In this case, both systems are dedicated to the same task, being identical, having the same characteristics, resources, and functions running.

On the other hand, while analyzing the resource requirements of all the functionalities deployed on every control unit, based on experience, it is possible to consider that, due to the technological advances in the last years and the fact that control units do not process computationally intensive information like audio or video data, the aggregation of the different algorithm resource requirements does not finally exploit all the resources present in a computer (be it RAM, processor, storage capacity, communication bandwidth and, even, fewer GPUs in case they are present), hardly reaching in the worst case a 33% of resource occupancy. In consequence, we can assume there will be the presence of a different amount of idle resources on every SAU in the network. Due to the distributed characteristics of our network and the possibility of having information replication, we can exploit this possibility by virtualizing the resources on the network and reassigning them to

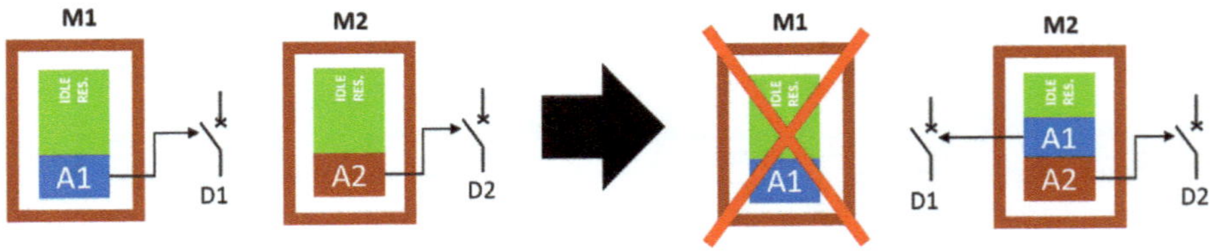

Figure 15.24 Virtual redundant scheme proposed.

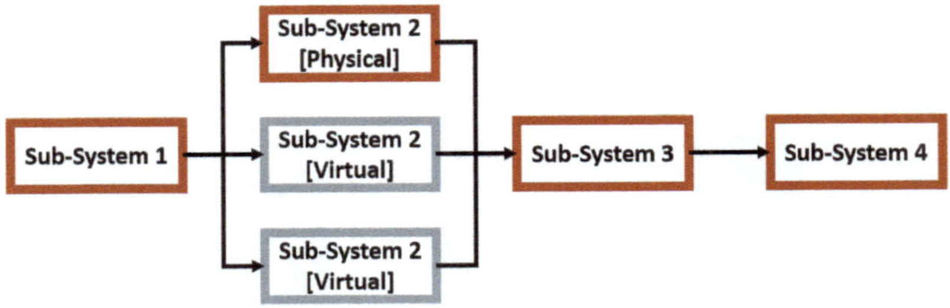

Figure 15.25 Voting logic for virtual redundant system.

different elements, so, in case any SAU fails, we can virtualize the machine and redeploy it into another SAU of the system with the closest characteristics to the optimal ones, as presented in Fig. 15.24.

In this case, the voting logic for Virtual Redundancy would, in consequence, look like the system presented in Fig. 15.25.

15.3.4.1.1 Application example

Our practical example of Virtual Redundancy will be based on the control, automation, and monitoring service running on SAU technologies with area distribution for Multi-Area State Estimation and distributed area of influence for control and automation strategies as the one presented in Fig. 15.4. This system is divided into three areas with different requirements and algorithms, so, in case we lose connection to one of the SAUs (e.g., SAU "A," as shown in Fig. 15.26), we would not only lose the possibility of performing accurate monitoring on the whole network but also lose the coordination of DERs to run Virtual Power Plants algorithms. The wide variety of failure causes is not considered here, which can go from partial errors to cyber-physical attacks against a node in the system. In any case, a human operator can be quickly deployed for the restoration of the service, which, in case of severe damage to the SAU, might lead to days or even weeks of service outage in case the total SAU replacement is needed. This kind of problem might discourage the technology adoption on distribution networks, where costs must always justify the benefits of the adoption.

In case the system makes use of Virtual Redundancy, it means we have virtualized all the functions. Hence, an optimal allocation to one of the other two SAUs

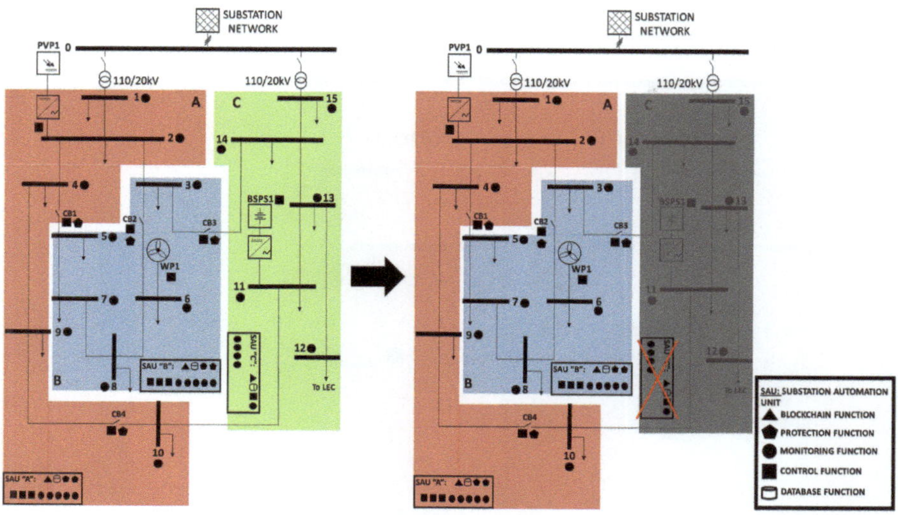

Figure 15.26 Services lost on MV distribution network's area C due to SAU "C" failure.

present in the network can be done. In this case, if we assume the optimal SAU for the migration is the SAU "C" and, so, this actor has the algorithm of SAU "A" already waiting for deployment, then given the right moment, we can perform the migration of SAU "A" into SAU "C," and restore the operation and functions of the elements present in the area "A," as shown in Fig. 15.27.

15.3.4.1.2 Application layer interacting with the blockchain

A point not directly related to the blockchain smart contract itself is the application interacting with it, its tools and components, all of the variables for different solutions or approaches. That's why we will provide a short description of the solution proposed as the first proof of concept presented about Virtual Redundancy based on a blockchain service.

As it may already be supposed, there is a need for a virtualization layer to bring up some degree of hardware independence to our system and facilitate the algorithm migration. In this case, the architecture of the application interacting with the blockchain can be found in Fig. 15.28.

The application is mounted on top of Calvin, an open framework based on Python for the development of IoT applications (Persson & Angelsmark, 2015). This makes it a perfect tool to deal with agent-based automation, which can manage the application, bringing a virtual machine to the automation, thereby granting the application some degree of independence from hardware platforms and operative systems. Also, by the use of commands through TCP ports, we can control the runtime deployment and its migration to other agents.

For the communication transactions layer, we have Node-Red (Node-RED: Low-Code Programming for Event-Driven Applications, n.d.), which is a flow-based

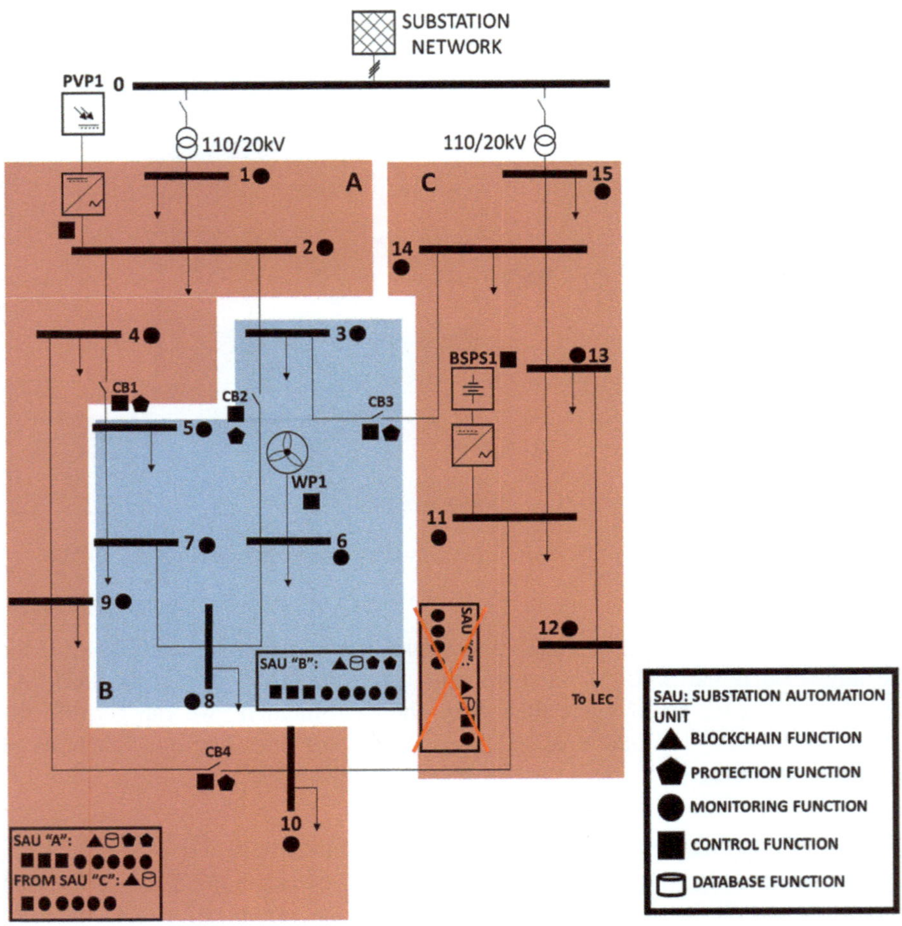

Figure 15.27 Area "C" service restoration after SAU "C" functions migration to SAU "A."

programming tool to manage the transactions between Calvin IoT and Hyperledger Fabric, our blockchain service provider framework.

15.3.4.1.3 Virtual redundancy algorithm

A Multiple Attribute Decision-Making algorithm is applied to allocate the algorithms. First, an Analytical Hierarchical Process breaks the problem into a hierarchical structure based on the requirements of every machine (Saaty, 1987):

Create a pairwise comparison matrix A, as shown below, whose values are taken from Table 15.1. This table is used together with the pairwise comparison matrix A, where the importance interpretation value depends on "how important one attribute is in relation to another attribute," for example, the importance value of RAM over Internet Bandwidth can be 7, which will mean that Internet Bandwidth over

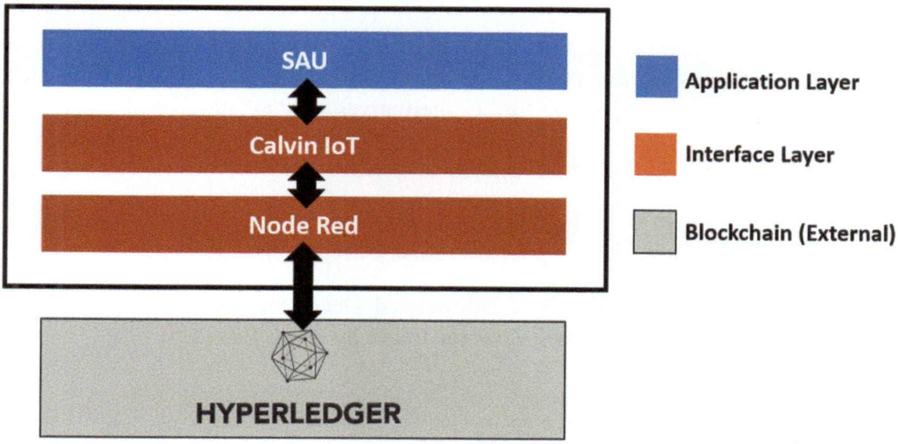

Figure 15.28 Application interacting with the Blockchain for the Virtual Redundancy Service.

Table 15.1 Importance interpretation values to fill the pairwise comparison matrix.

Importance interpretation	Equal importance	Moderate importance	Strong importance	Very strong importance	Extreme importance	Intermediate values
Importance Intensity	1	3	5	7	9	2,4,6,8

RAM will be 1/7 in this case. The matrix A is an $m \times m$ matrix where m is the number of application criteria (requirements). In matrix A, each item represents the importance of the jth attribute in relation to the ith attribute.

$$A = \begin{bmatrix} 1 & a_{12} & \cdots & a_{1m} \\ \dfrac{1}{a_{12}} & 1 & \cdots & a_{2m} \\ \vdots & \vdots & \ddots & \vdots \\ \dfrac{1}{a_{1m}} & \dfrac{1}{a_{2m}} & \cdots & 1 \end{bmatrix} \tag{15.4}$$

Calculate the Priority Vector (PV_j), as shown in the equation below for each application, depending on the requirements of the application. It can be done based on the Pairwise Comparison Matrix of Table 15.1.

$$PV_j = \frac{\sum_{l=1}^{m} \overline{p_{jk}}}{m} \quad \text{where} \quad \overline{p_{jk}} = \frac{a_{jk}}{\sum_{l=1}^{m} a_{lk}},$$

$$\text{and } "m" \text{ is represents the number of application criteria.} \tag{15.5}$$

Now, based on the obtained Priority Vector (which should be provided by the Virtual Service Requester machine), a TOPSIS algorithm (Yue, 2011) is applied to rank all the machines capable of handling our algorithm's requirements. At this step, we calculate the decision matrix D for "t" alternatives and "m" number of criteria:

$$D = \begin{bmatrix} d_{11} & \dots & d_{1m} \\ \vdots & \ddots & \vdots \\ d_{t1} & \dots & d_{tm} \end{bmatrix} \tag{15.6}$$

Now, we calculate the Weighted Normalized Decision Matrix by applying the equation and using the particular PV vector found by using Eq. 15.5.

$$v_{ij} = PV_J r_{ij} \text{ where } r_{ij} = \frac{d_{ij}}{\sqrt{\sum_{i=1}^{m} d_{ij}^2}} \tag{15.7}$$

Calculate the positive ideal solution (A^+) and the negative ideal solution (A^-), where B is the benefit attribute and C is the cost attribute. We have then found the distance to the optimal solution given every particular solution. The best alternative

ALGORITHM 12: Virtual Redundancy Subscription(virt_redundancy_subsc)

Input: id_requester,algorithm_requirements_matrix,machine_resources_matrix

Output: algorithm_receiver_id←variable used for ID registration(e.g. HASH)

```
1:  if ((requester = id_requester)&
    check_correct_data(algorithm_requirements_matrix)
    &check_correct_data(machine_resources_matrix)) then
2:      virt_redundancy_subsc (id_requester, algorithm_requirements_matrix,
        machine_resources_matrix)
3:      Request a Round Trip Time measure from all the machines
        registered in the idle_resources list
4:      Perform Optimization Algorithm using the list of other machines
        registered machine_resources_matrix and this machine's algorithm
        requirements
5:      Request a Resources Update to the Optimal Machine
6:      If the resources are still similar, then notify the receiver machine
7:      algorithm_receiver_id ← Result of Optimal Machine
8:      Update the list of machine_resources_matrix on the network and
        list of id_requester machines (virtual redundancy service requesters)
9:  else
10:     algorithm_receiver_id ← Not a Number (nan)
11: end if
12: return (algorithm_receiver_id)
```

OBS: To send the algorithm we can make use of put and get private data or by requesting the IP.A good practice is to not directly make use of the IP

Figure 15.29 Algorithm 12: Smart contracts for Virtual Redundancy Subscription.

ALGORITHM 13: Virtual Redundancy Heartbeat (virt_redundancy_heartbeat)

Input: id_requester, public key(pk), polling_period
Output: migration_triggered ← bool

```
if ((requester = id_requester) &
is_registered(id_requester,id_requesters_list)) then
    Report service start to the requester machine and start heartbeat_timer
    do:
        Query Heartbeat and the current variables state of the requester
        or read them from the Blockchain using the variables and time of
        submission digitally signed, together with the provided pk
        if ( heartbeat_timer.timeout = true ) then
            heartbeat_present ← false
        else
            heartbeat_present ← true
            heartbeat_timer ← 0
        end if
    while (heartbeat_present = true)
    Report the need of algorithm deployment to the receiver machine
    through direct communication or through the blockchain
    Update the list of requester machines, and, in case the requester
    machine was hosting other machine's virtual redundant algorithms,
    redeploy them into another optimal machine
    Update the list of machine_resources_matrix and list of id_requester
    machines
    migration_triggered ← true
else
    Report polling service failure to the requester machine
end if
return (algorithm_receiver_id)
```

OBS: The hearbeat and variables states can be reported via VPN to another peer in case the requester machine also work as a blockchain peer or it can be published in theblockchain as private data. The service can be automatically done as an endless routine, checking all the machines heartbeats and variables update present in the id_requesters_list instead of using a dedicated service

Figure 15.30 Algorithm 13: Smart contract for Virtual Redundancy Heartbeat Polling Service.

would be the closest to the positive ideal solution (A^+) and farthest to the negative ideal solution (A^-).

$$A^+ = \{v_1^+, \ldots, v_m^+\} \text{ where } v_j^+ = \{\max(v_{ij}) \text{ if } j \in B;\ \min(v_{ij}) \text{ if } j \in C\} \quad (15.8)$$

$$A^- = \{v_1^-, \ldots, v_m^-\} \text{ where } v_j^- = \{\min(v_{ij}) \text{ if } j \in B;\ \max(v_{ij}) \text{ if } j \in C\} \quad (15.9)$$

Now we calculate the relative closeness to the ideal solution:

$$C_t^+ = \frac{S_t^+}{S_t^+ + S_t^-} \quad (15.10)$$

ALGORITHM 14: Virtual Redundancy Recovery Report (virt_redundancy_recovery)

Input: id_requester, public key(pk)

Output: migration_reverted ← bool

```
if ((requester = id_requester) &
is_registered(id_requester,id_requesters_list)) then
    id_current_machine = virt_redundancy_recovery(id_requester,pk)
    Request the current state of the migrated algorithms to the machine
    currently running them with "id_current_machine"
    Wait until the current variable states are received over
    direct TCP communication or put/get private data
    Wait until the current variables are loaded
    Report of successful actor redeployment to the current algorithm owner
    (Optionally, perform again the subscription and the optimization algorithm
    in order to request a new receiver machine)
    Update the list of machine_resources_matrix and list of id_requester
    machines (to include again the redeployed machine)
    Request the lists update to the previous host
    if (no error reported) then
        migration_reverted ←true
    else
        migration_reverted ←false
    end if
else
    migration_reverted ←false
end if
return (migration_reverted)
```

OBS: In this case, we supose the recovered machine and the host machines will both remain as service subscriber and optimal host. If this is not the case, the optional step can be performed

Figure 15.31 Algorithm 14: Smart Contract to report the service recovery of an actor after a critical failure.

where

$$S_t^+ = \sqrt{\sum_{i=1}^{m}\left(v_{ij} - v_j^+\right)^2} \text{ and } S_t^- = \sqrt{\sum_{i=1}^{m}\left(v_{ij} - v_j^-\right)^2} \qquad (15.11)$$

Now, we take the higher C_t^+.

This procedure is to be performed every time we require to choose the optimal machine to run our local algorithm. Based on this, the smart contracts associated with the Virtual Redundancy Service are presented in Figs. 15.29 and 15.30.

In case the service is reestablished on a previously lost SAU, we can then migrate back the algorithm by applying the algorithm presented in Fig. 15.31.

15.3.5 A broader picture of monitoring and automation from the blockchain technology perspective

After discussing smart meter Private Data Collection as well as ID registration, we might be asking if there is any advantage in the adoption of any of these

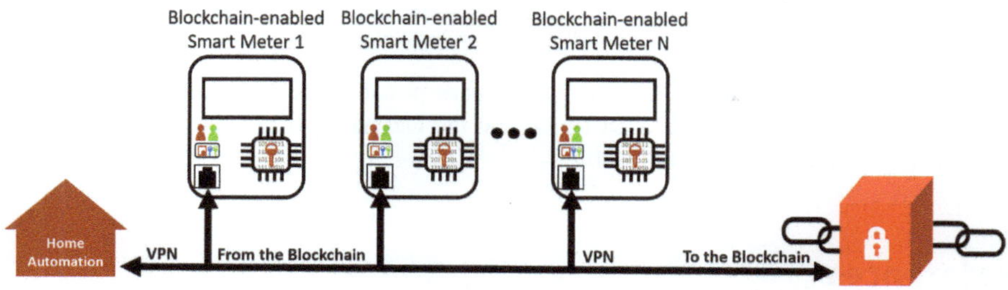

Figure 15.32 Blockchain-enabled smart meters with a blockchain application layer.

technologies, as most smart meters do not have the possibility of adding an application layer to manage the blockchain interactions. That is exactly why these applications pose a strong possibility to adopt crypto-friendly technologies in smart meters, as shown in Fig. 15.32, to truly exploit the "smart" capabilities proper to their names, as these technologies not only enable the collection of consumer data but also enable the communication with the DSOs to encourage the consumption during renewables production peak and bring a more efficient climate-friendly behavior to their operation. Important questions come up regarding the topic: why call for a smart meter-driven world when these might not cohere either with the privacy law (by adopting higher-level programming languages) or with the secure and widely used communications standards that might allow us to change the future of energy systems? Current standards in the general industry are presently thriving by bringing optimal and efficient processes into use that make use of important improvements from the last years—like new communication technologies and "smart" machines—which can interact with internal as well as external factors to optimize resource consumption for the maximum production and quality. We are positioned at the limit of a precipice where the time is to finally decide the future of the current climatic crisis, where the only positive path toward a less carbon-driven civilized world relies either on resource-efficient processes or on energy rationalization (the last one considered a threat to economic development). As a consequence, when we are currently investing big efforts in green, renewable technologies for energy production, why not guide the consumers' efforts toward key "smart" efficient technologies, protecting users' privacy at the same time as providing a positive environmental impact?

Now, with the arrival of the blockchain, we expect to see a new scenario to reach the holy grail of a totally decarbonized society driven by the era of the information, permeating every level of the electrical grid, especially with the easiness provided by blockchain interoperability, which can not only create trust between competing companies, but also unite different types of network operators, Local Energy Communities and even nations, all of them collaborating in a way to reach the goal of a fully cooperative grid among nations, and why not, the entire world.

Acknowledgment

This book chapter and whole content have been made possible through the BRILLIANT project, which is funded by the European Union and the German Federal State of North Rhine-Westphalia under grant no. EFRE-0400353.

References

Abur, A., & Gómez Expósito, A. (2004). *Power system state estimation: Theory and implementation* (1st ed.). Marcel Decker, Inc.

Ali, A. I. (2015). Comparison and evaluation of digital signature schemes employed in NDN network. *International Journal of Embedded Systems and Applications*, *5*(2), 15–29. Available from https://doi.org/10.5121/ijesa.2015.5202.

Angioni, A., Kulmala, A., Della Giustina, D., Mirz, M., Mutanen, A., Ded'e, A., Ponci, F., Shengye, L., Massa, G., Repo, S., & Monti, A. (2017). Design and implementation of a substation automation unit. *IEEE Transactions on Power Delivery*, *32*(1), 1133–1141. Available from https://doi.org/10.1109/TPWRD.2016.2614493.

Asefi, S., Madhwal, Y., Yanovich, Y., & Gryazina, E. (2021). Application of blockchain for secure data transmission in distributed state estimation. *IEEE Transactions on Control of Network Systems*, *XX*(XX), 1–10. Available from https://arxiv.org/abs/2104.04232.

Eurelectric. (2021). *Local energy communities*. Eurelectric. <https://www.eurelectric.org/media/2458/local-energy-communities-final.pdf>.

European Commission DG ENER; Joint Research Centre. (2014). *Country fiches for electricity smart metering*. European Commission. Available from https://ses.jrc.ec.europa.eu/smart-metering-deployment-european-union. (2014) Accessed 20.05.21.

Fabian, B., Ermakova, T., & Sander, U. (2016). Anonymity in Bitcoin? – The users' perspective. In: *Conference: International conference on information systems 2016 (ICIS 2016)*.

Gai, K., Wu, Y., Xu, L., & Zhang, Y. (2019). Permissioned blockchain and edge computing empowered privacy-preserving smart grid networks. *IEEE Internet of Things Journal*, *6* (5). Available from https://doi.org/10.1109/JIOT.2019.2904303.

Gilbert, S., & Lynch, N. (2002). *Brewer's conjencture and the feasibility of consistent, available and partition-tolerant and web services*. Massachusetts Institute of Technology (MIT). Available from https://users.ece.cmu.edu/~adrian/731-sp04/readings/GL-cap.pdf.

Gorenflo, C., Lee, S., Golab, L., & Keshav, S. (2019). FastFabric: Scaling Hyperledger Fabric to 20,000 transactions per second. In: *2019 IEEE international conference on blockchain and cryptocurrency (ICBC)*. Available from https://doi.org/10.1109/BLOC.2019.8751452.

Henten, A., & Windekilde, I. (2020). Blockchains and transaction costs. *Journal of NBICT*, *01*(1), 33–52. Available from https://doi.org/10.13052/nbjict1902-097X.2020.002.

Hyperledger Fabric. *A blockchain platform for the enterprise*. (2020). <https://hyperledger-fabric.readthedocs.io/en/release-2.2/>.

Hyperledger Fabric. *Pluggable consensus*. (2020). <https://hyperledger-fabric.readthedocs.io/en/release-2.2/whatis.html#pluggable-consensus>.

Jakobi, T., Patil, S., Randall, D., Stevens, G., & Wulf, V. (2019). It is about what they could do with the data: A user perspective on privacy in smart metering. *ACM Transactions on Computer-Human Interaction*, *26*(1). Available from https://doi.org/10.1145/3281444.

Kundur, P., & Balu, N. (1994). (1st ed.). *Power system stability and control*, (Vol. 1). McGraw-Hill Education.

Mechkaroska, D., Dimitrova, V., & Popovska-Mitrovikj, A. (2018). Analysis of the possibilities for improvement of BlockChain technology. In: *26th Telecommunications forum TELFOR 2018*. Available from https://doi.org/10.1109/TELFOR.2018.8612034.

Martin, K., & Nauman, B. (2021). *Bitcoin's growing energy problem: 'It's a dirty currency. Financial Times*. Available from https://www.ft.com/content/1aecb2db-8f61-427c-a413-3b929291c8ac.

Moroni, S., Antoniucci, V., & Bisello, A. (2019). Local energy communities and distributed generation: Contrasting perspectives, and inevitable policy trade-offs, beyond the apparent global consensus. *Sustainability, 2019*(11), 16. Available from https://doi.org/10.3390/su11123493.

Node-RED. *Low-code programming for event-driven applications*. (n.d.). <https://nodered.org/> Accessed 23.07.21.

Pau, M., Ponci, F., Monti, A., Sulis, S., Muscas, C., & Pegoraro, P. A. (2017). An efficient and accurate solution for distribution system state estimation with multi-area architecture. *IEEE Transactions on Instrumentation and Measurement*, *66*(5), 910–919. Available from https://doi.org/10.1109/TIM.2016.2642598.

Persson, P., & Angelsmark, O. (2015). CALVIN - Merging Cloud and IoT. *Procedia Computer Science*, *52*, 210–217. Available from https://doi.org/10.1016/j.procs.2015.05.059.

Saaty, R. W. (1987). The analytic hierarchy process—What it is and how it is used. *Mathematical Modelling*, *9*(3), 161–176. Available from https://doi.org/10.1016/0270-0255(87)90473-8.

Sadu, A., Jindal, A., Lipari, G., Ponci, F., & Monti, A. (2021). Resilient design of distribution grid automation system against cyber-physical attacks using blockchain and smart contract. *Blockchain: Research and Applications*, 100010. Available from https://doi.org/10.1016/j.bcra.2021.100010.

Sitaram, D., & Manjunath, G. (2012). *Chapter 9—Related technologies. Moving to the cloud: Developing apps in the new world of cloud computing*. Elsevier. Available from https://doi.org/10.1016/B978-1-59749-725-1.00009-3.

Thummavet, P. (2019). Demystifying Hyperledger Fabric (2/3): Private Data Collection. 2019. (Accessed 25 July 2021).

Yue, Z. (2011). A method for group decision-making based on determining weights of decision makers using TOPSIS. *Applied Mathematical Modelling*, *35*(4), 1926–1936. Available from https://doi.org/10.1016/j.apm.2010.11.001.

Blockchain-enabled energy sector management

16

Desh Deepak Sharma[1] *and Ravinder Kumar*[2]
[1]MJP Rohilkhand University, Bareilly, India, [2]Ajay Kumar Garg Engineering College, Ghaziabad, Uttar Pradesh, India

16.1 Introduction

With the application of information and communication technology, digitization is transforming the whole power system. In the energy sector, new business models are being developed with the use of blockchain technology, focusing on providing a secure connection for energy trading among various entities in a peer-to-peer (P2P) network. A comprehensive overview comprises blockchain core architectures, blockchain, and the internet of things (BIOT), blockchain in healthcare, blockchain in business, and blockchain in the vehicular industry (Andoni et al., 2019; Casino et al., 2019). The blockchain architecture topic covers the requirements of a decentralized system, public blockchain, permissioned blockchain architecture, and platforms based on execute-order-validate. Next, there is a discussion on the need, integration, model, various application areas of BIOT. The blockchain in healthcare comprises discussion on secure health information exchange between stakeholders in order to ensure privacy, improve the integrity of health data, decentralized model of insurance, smart contracts between doctor and patient for medical records, and interoperability issues in healthcare applications. In the business sector, the requirement is the application of blockchain in cloud sourcing, the application of property registration systems, and the adoption of blockchain in the banking sector. Blockchain technology can also be applied in tracking vehicles transporting goods (2019b).

The multimicrogrid system comprises various microgrids interconnected via a distribution network. The multimicrogrid system can cater to the increasing electricity demand. The P2P energy trading on the decentralized platform among microgrids requires new approaches to be developed. The owners of microgrids may have different objectives for management of control. The centralized authority cannot work properly in this scattered multimicrogrid framework (Li et al., 2019). A new mechanism and algorithms have to be designed for decentralized energy trading among various microgrids. In a microgrid, most of the electricity demand is catered by local generation using distributed generation (DGs). Obviously, the decentralized structure of the microgrid based on local generation should be utilized adequately for P2P energy trading. This structure minimizes the dissipation of power transmitted, avoids cascade failure, and reduces dependency on the power

Blockchain-Based Systems for the Modern Energy Grid. DOI: https://doi.org/10.1016/B978-0-323-91850-3.00006-8

grid. Blockchain technology should be used for designing an adaptable, reliable, and secure decentralized system of microgrids involving various entities. Smart contracts which are reliable, secure, and agreeable to all stakeholders have to be developed among various local sellers and numerous buyers (Sabounchi & Wei, 2017).

A simple microgrid comprises distributed power sources, energy storage devices, loads, and monitoring and protection devices. In interconnected microgrids, a single microgrid can share extra power generation with the other microgrids. In the interconnected microgrid, the major challenges are (1) the mechanism to share energy must be secure, (2) the data must be secure, and (3) social and economic preferences must be given (Yin et al., 2021). Within the microgrid, a secure and reliable decentralized renewable energy trading platform has been suggested. The P2P energy trading platform in the microgrid may utilize a private blockchain network that comprises authorized nodes only . The nodes in the microgrid can be classified as consumers who consume energy and prosumers should be capable of producing renewable energy. An energy trading framework based on Ethereum's smart contract has been developed between consumers and prosumers (Kang et al., 2018). The blockchain-based energy trading algorithm consisting of two layers is proposed for interconnected microgrids. A two-phase consensus algorithm comprising of practical byzantine fault tolerance (pBFT) and a modified proof-of stack (PoS) has been incorporated into blockchain-based energy trading algorithms (Masaud et al., 2020). Blockchain technology is preferably suitable for microgrids compared to the power grid, given the microgrids are located within a limited geographical area and, hence, the transmission losses are reduced (Onyeka Okoye et al., 2020). The multimicrogrid system is now an important part of the future power grid. Networking and communication are the major part of the whole multimicrogrid system and, hence, the system becomes vulnerable to cyberattacks (Hu et al., 2019). Blockchain technology and a multisignature approach are implemented to facilitate nodes to negotiate energy prices, and transactions are executed in a secure way. However, implementing flexible prices is a cumbersome task. In the microgrid, the pricing and balance of the transactions are achieved securely by implementing the blockchain-based continuous double auction mechanism (Zhang et al., 2019). In the microgrids, the P2P power flows must consider the power losses. The proportional sharing rule index method has been suggested for the appropriate estimation of power losses during the energy transaction (DI Silvestre et al., 2020). The schemes regarding the design, development, and implementation of event-based local energy management with smart contracts for microgrids incorporate blockchain technology and IEC 61850 communication protocols (Gayo et al., 2020). Blockchain technology is applied to detect cyber attacks like false data injection attacks in the cyber-physical system of a microgrid. The cyber-physical system of the microgrid facilitates the data exchange among multiple DGs, energy storage devices, plug-in electric vehicles, smart meters, control units, and various loads. The event messages in vehicular networks are made secure and confidential by using blockchain technology (Ghiasi et al., n.d.). Blockchain technology is considered for the shared voltage regulation as an ancillary service in the microgrid.

The blockchain technology-based platform is considered for the technological and financial aspects, along with the ancillary services in the microgrid having a distribution network for energy transactions (Di Silvestre et al., 2019). This chapter proposes the trust value evaluation of the nodes using beta distribution. The real-time solar photovoltaic (PV) power generation data through inverters are used for the simulation using python programming. The results are obtained and discussed in the simulation section.

16.2 Blockchain technology

Blockchain technology is basically a distributed time-stamped data structure that can be appended with the new blocks in a chain. This technology facilitates developing a distributed P2P network in which nontrusting members can interact without any central authority. The blockchain architecture topic covers the requirements of a decentralized system, public blockchain, permissioned blockchain architecture, and platforms based on the execute-order-validate method of operation. All the transactions are secured with cryptography and are immutable, and these accumulated transactions are stored in different blocks. Each block comprises the stored data, the hash value of the previous block, and the hash value of the current block. Bitcoin, Ethereum, and Hyperledger are examples of blockchain technology. In Bitcoin and Hyperledger blocks, the transactions are structured in Merkle Tree and Ethereum blocks considering Trie, respectively. Blockchain technology considers various consensus algorithms such as proof-of-work (PoW), Proof-of-Byzantine-Fault Tolerance (PBFT), Proof-of-Stake (PoS), Proof of Elapsed Time (PoET), etc.

In blockchain technology, the cryptography algorithms such as hash function and Merkle Tree are being used for the encryption of data. A transaction is a process under which the ledger is updated between two nodes, and the block comprises Ids, data, hash keys, timestamps, and requisite information.

Blockchain technology is based on a decentralized approach and shows various applications in numerous domains such as cryptocurrency, internet of things (IoT), financial services, risk management, social services, real estate, energy sector, health sector, etc. A long chain of blocks is formed that is free from tampering and, consequently, an integrated and immutable database is constructed Yang et al. (2019). The P2P networks, however, show a wide range of applications while avoiding the central authority with the power to intervene. But, in P2P networks, stability and node reliability are some challenges and, consequently, performance may degrade. The network bandwidth adjustment is another challenge as information is shared in a sequential way. A method or technology has to be designed to measure and analyze suspicious or malicious nodes (Hu et al., 2019). Basically, the blockchain is a distributed ledger in which all the executed transactions are tracked. The ledger is scattered over all the nodes of the network. The main objectives of the ledger are to show an immutable log of all transactions and in which all the transactions must be transparent. The blockchain technology is categorized into public blockchain, private blockchain, and consortium blockchain as shown in Table 16.1 and Fig. 16.1.

Table 16.1 Main blockchain categorization.

	Public blockchain	Private blockchain	Consortium blockchain
Participants	All	Control by central authority	All are grouped in a consortium
Type	The nonrestrictive, permission-less distributed ledger system	Operative only in a closed network. Small and restrictive network.	A semidecentralized type where more than one organization manages a blockchain network, more than one organization can act as a node
Examples	Bitcoin and Litecoin	Hyperledger projects	Energy Web Foundation, R3
Consensus mechanism	PoW	Deployed for voting, supply chain management, digital identity, asset ownership	PoW, PoS
Advantages	Trustable Secure Open and transparent	Speed: transactions per second is higher; Scalability Open and transparent	Enhancement of functioning with the incorporation of organization of different disciplines.
Disadvantages	Lower Transactions per Second Needs trust building; Lower Security Centralization	Needs trust building, Lower Security Scalability issues, Lightning network, High energy consumption	There may be problems in the constructing of consortiums.

PoW, proof-of-work; *PoS*, proof-of-stake.

16.2.1 Consensus process

The data should be verified by a fast and efficient consensus algorithm. A new block carries the transaction, and information about the transaction is broadcast to other consensus nodes. Each transaction has to be verified with the following criteria:

1. The data of the transaction should be appropriate.
2. Input conditions and output deductions must have some value and comply with the defined size and should obey predefined rules of the smart contract.

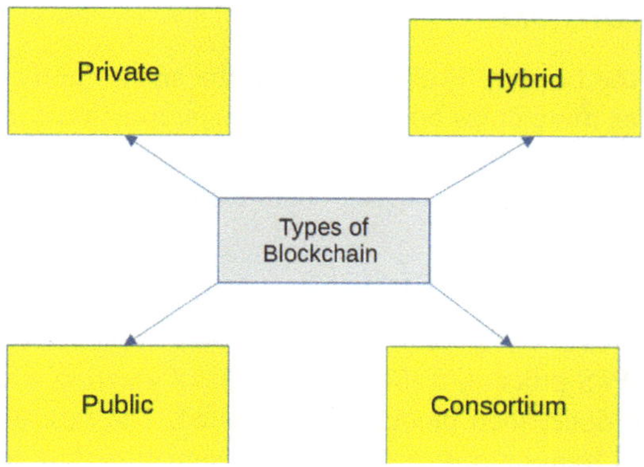

Figure 16.1 Types of blockchain.

3. The hash value must be positive and should not be zero.
4. If the new transaction is found already available, then it must be avoided.
5. The signature in the transaction must be made legal.

The blocks of data comprise the details of the smart contract finalized among buyers and sellers in energy trading. In the blocks, the details of the executed contract contain the amount and price of energy to be traded, the hash key from the previous block, and a new hash key obtained by using an appropriate algorithm such as SHA -256 hashing algorithm. The distributed ledger comprises various blocks, and each appended block has to be validated. The popular validation algorithms are different consensus methods such as PoW, PoS, practical BFT (pBFT), etc. The pBFT algorithm is used in the validation process of the blocks. In the distributed platform, this pBFT is able to do validation even while the faulty and malicious nodes are present (Castro & Liskov, 1999).

16.2.1.1 Proof-of-work

It is a sort of consensus mechanism in which the work of the participating node is to identify if that work is quite worthy, and if it so, then that node is awarded more transactions for adding to the blockchain. In this mechanism, the main objective is to form a single state of the ledger by a group of participating nodes and, hence, a leader node is chosen who can create a new block to make a chain of blocks. For developing distributed timestamp transaction data on P2P networks, the PoW consensus approach has been implemented. PoW with hash value is employed in the blockchain for finding nonce value while the hashed additional block is included such that the hash has to be smaller than the existing target value. The miner finds this nonce value and creates a block on the distribution layer network. A PoW is a consensus approach in which participating nodes of a network try to solve a mathematical puzzle or play a game to win. In blockchain technology, this mechanism validates transactions and new tokens. However, a PoW mechanism needs a huge amount of energy during processing as more miners are added to the network (Block chain explained, 2020)

16.2.1.2 Proof-of-stake

In cryptocurrency, the basis of PoS depends on the stack the miner possesses, that is the miner, who has more coins compared to others, may have the right to validate or mine a block. In the PoS, the mining power is assigned according to the percentage of coins possessed by a miner (Block chain explained, 2020).

16.2.1.3 Delegated proof-of-stake

The DPoS algorithm was designed by Larimer, who implemented it in the project BitShares. In this DPoS process, if the first process of electing witnesses (i.e., block producers) has been done, then blocks are generated. The task of the block producers is to witness the transactions, and verify the signature, but not to participate in the trading. A block is generated by them within the span of every 3 seconds. A witness tries to perform the aforementioned processes; if this witness is not capable of carrying out the processes, then these processes are done by another witness. The more are the blockchain stacks possessed by a witness, the more are the chances for them to become a witness who is trusted by the nodes in the network. In this mechanism, the voting rights are possessed by a small number of voters who have more influence in voting as compared to the other nodes (Yang et al., 2021).

16.2.1.4 Delegated proof-of-stake with downgrade

The public blockchain technology may face the problem where it may tend to be centralized and insecure. In the delegated proof-of-stake with downgrade algorithm, first, some consensus nodes are selected and, after the verification of the blocks consensus is achieved, the malicious nodes are relegated. In the network, the trading nodes are those that create transactions, and the consensus nodes are those that create transactions and verify blocks. Two environments are developed, such as the trading environment, which comprises trading nodes, and another environment, such as the consensus environment, which comprises consensus nodes. The consensus nodes that do not generate transactions else participate in the consensus process. The trading nodes are not part of the consensus process; else, they generate, encrypt, sign transactions, and store blockchain data (Yang et al., 2021).

16.2.1.5 Proof-of-concept

The PoC develops some method, approach, or mechanism to identify the feasibility and realization of the proposal. A PoW finds feedback from all the stakeholders involved in the project. In blockchain technology, some software codes (chain code, etc.) are used for the design of smart contracts while invoking specific functions through applications.

16.2.1.6 Practical Byzantine fault tolerance

In the distributed network, a consensus has to be built among peers or nodes complying with some rules or protocols to achieve an objective during regular intervals.

Most of the nodes should obey these rules. There is the possibility that some of the nodes may misbehave and/or may disagree or fail to follow the rules. So, these nodes cannot build consensus among other nodes. A system is said to be Byzantine Fault Tolerant if two-thirds of nodes continue to build consensus. So, BFT is the property of the network such that the system can function properly while one-third of the nodes may fail or act maliciously. The consensus protocols which are used in blockchain technology, such as POW, PoS, or proof-of-authority have characteristics of some sort of BFT. Assume $|F| = 3f + 1$, where f is the maximum number of nodes that may be faulty. The performance of the system may degrade if the faulty nodes are more than f (Castro & Liskov, 1999).

16.2.1.7 Raft algorithm

This is one of the consensus algorithms in which each node may find either of three: state leader, follower, or candidate. In this approach, time is divided into fixed durations named as terms that are numbered by integer values. As a term starts, candidates try to become leaders in an election. Obviously, a winning candidate is selected as a leader in an election for all the trailing terms. All other nodes are considered follower nodes who will become candidates if they do not follow the leader. The candidates concentrate on sending requests to other nodes to become leaders. The candidate gets elected leader on getting votes from most of the nodes (Huang, Ma et al., 2020; Paris & Long, 2015).

16.3 Forking and double-spending in blockchain technology

16.3.1 Forking

Forking in blockchain technology is when two nodes (or miners) find a block at the same time; consequently, there are possible two separate networks that may be created. This condition of forking must be avoided in the blockchain (Fig. 16.2).

16.3.2 Double spending

The double-spending problem is a condition where the seller is shown to spend the same transaction amount twice in two different transactions. The situation of double-spending happens when the blockchain network is disrupted by some malicious node or hackers who either send a copy of the transaction which looks legitimate or they erase transactions. The situation of double-spending must be avoided in blockchain technology.

16.4 Smart contracts in blockchain technology

Smart contracts have two main features: (1) value and (2) state. Smart contracts are designed such that all nodes in the blockchain system accept and sign these smart

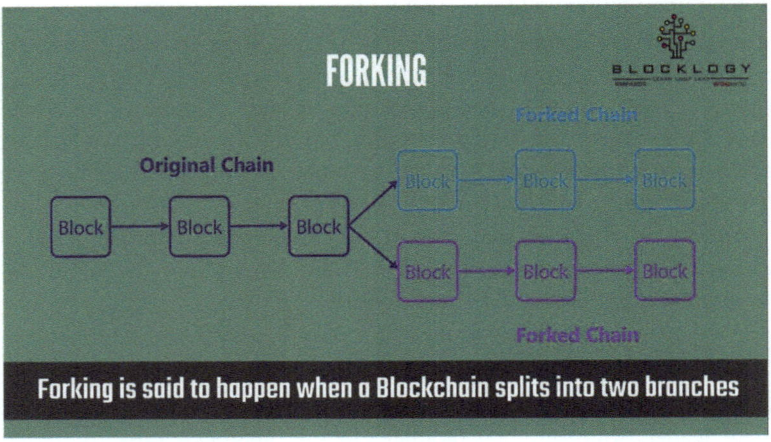

Figure 16.2 Occurrence of forking in blockchain.

contracts, which are broadcast on P2P networks. The miners obtain rewards by the incentive mechanism for verifying the transactions contained in the smart contracts. Then, this executed contract, after verification, is converted into a block chained into distributed ledger via a consensus mechanism in the network. The Ethereum and Hyperledger Fabric are two operational platforms for the execution of smart contracts. Ethereum, a public blockchain platform, provides two sorts of contracts: externally owned contracts and contract accounts. The externally owned contracts are driven by the public keys without code; on the other hand, contract accounts comprise coded public keys. Hyperledger Fabric is a permissioned blockchain that provides an open-source blockchain platform which is scalable, modular, general-purpose, and with limited availability (Wang et al., 2019).

Smart Contracts implemented in the blockchain-based system are described by the functions as given in Table 16.2 below:

The smart contract has the following structure: basically, some executable instructions that are carried out while getting a message, the simple smart contracts are considered in first-generation blockchains like bitcoin and in second-generation blockchains like Ethereum, where data associated with the smart contracts can be manipulated or stored on the blockchain. Thus smart contracts are defined to be based on some rules.

A general data structure can be written in the following way,

Struct block {
Information ID
Data
Hash keys
Time stamp
Sender address
Receiver address
State of information (other requisite information)}

Table 16.2 The functions of the blockchain.

1.	Accept deduction	A user request is received and input data is represented by requisite semantic or numerical form
2.	Confirm deduction	Confirm signal is generated if the input data obeys the predefined rules
3.	Drop deduction	A submitted transaction can be denied to be considered if confirm deduction fails

16.5 Layers and nodes in blockchain technology

Blockchain technology mechanism has been divided into six layers: network, data, consensus, incentive, contract, and application layers (Bao et al., 2019).

1. *Network layer*: Obviously, most blockchain networks are based on P2P networks in which nodes have identical features. The blocks are forwarded or broadcast by the nodes on the P2P network, and the data associated with blocks are processed by a consensus mechanism. The optimization in the network should be included in the blockchain network. First, the main part of the network layer is a node, where a simple node initiates and forwards a transaction which is executed and comprised in a ledger, and some specific node termed as a miner has the right to keep the ledger and allows participants to append the blocks in the ledger. The transaction pool is described as the awaited transactions to be executed and obtained from other nodes. The node sends the information about the new transaction to the peers, and the node which receives the information may reply back only if it does not possess that information.
2. *Data Layer*: The data comprise the transactions, nonce, timestamp, Merkle Tree/hash values, and other requisite data. The data are enclosed in the blocks and stored in the various nodes. Two blocks are interconnected by hash values as a block contains a hash value of the previous block. In this process, a chain of blocks is formed. The new data structure methods have to be incorporated and implemented.
3. *Consensus layer*: This layer comprises the algorithms and rules so that a consensus can be built up among different nodes. An appropriate node is chosen that has the capability to generate and append the next block to the distributed ledger. The situation must be avoided such that nodes may face multiple different copies of ledgers. In the PoW consensus algorithm, the miners strive to create the next block. In PoS, nodes are given priority based on stake or money. In this layer, block data contained by scattered nodes are verified by an appropriate consensus algorithm. The popular consensus algorithms are PoW, PoS, delegated proof-of-work (DPoW), practical Byzantine Fault Tolerance (pBFT), Ethash(similar to PoW), and Casper (similar to PoS). PoS uses less energy as compared to PoW and is based on the amount of money owned by the miner. Ethereum is now using Casper in place of Ethash. PBFT is used by Hyperledger Fabric.
4. *Economic layer*: This layer concentrates on the economy of the system, such as digital currency, trading issues, and incentives and rewards given to nodes, etc.
5. *Contract layer*: The transaction scripts and smart contracts are the two technical developments covered at this level. The transaction scripts cover the programming attributes and avoid the infinite loop developments. Smart contracts are basically some rules or protocols to execute simple or complex trading. The transaction scripts and smart contracts are openly available to all the nodes.

6. *Application layer*: This application layer is divided into two categories: extended techniques and industrial applications. The extended techniques are related to improvements in performance and security issues in the blockchain.
7. *Incentive layer*: This layer corresponds to the fixing of the rewards and incentives given to the participating nodes. The miners are awarded if miners append the new blocks to the distributed ledger. An appropriate mechanism and process of rewarding must be designed. Hence, the balance of the miners is updated after rewarding.

16.5.1 Types of network nodes

There are the following types of network nodes in the network (Table 16.3).

16.6 Merkle tree and pool mining in blockchain technology

16.6.1 Merkle tree

Merkle tree (also known as a Hash tree) is a kind of data structure that is being used in the verification of data and synchronization. The structure is in the form of a tree in which all the leaf nodes are considered with the same label. A large data set enveloped in a tree can be validated by comparing the Merkle root/root node. A hash function is basically an algorithm to generate a hash value. A hash key uses mathematical algorithms to compress the longer-length data into fixed-length data. However, if data changes, obviously, the hash value will also be changed. The change in the data can be determined by computing a hash value. The data integrity in the blocks is represented by Hash in the blockchain technology.

Properties of Hash function

1. It is deterministic as similar or identical inputs generate the same key every time.
2. The hash function shows uniqueness as the same hash value cannot be generated by two or more than two differently charactered data. Every change in the data always generates different Hash values, that is two differently charactered data cannot be represented by the same hash value.

Table 16.3 Types of network nodes.

	Nodes	Task
1.	Leaf nodes	To send information to the core node and start transaction
2.	Core node	To manage leaf nodes and to forward transactions to the other nodes of a cluster
3.	Routing node	To choose these node out of core node, to store routing information, and forward transaction to other nodes

Table 16.4 Hash algorithms.

Sr. no.	Hash algorithms	Bit size
1.	Message digest version 5 (MDP5)	128
2.	Secure Hash Algorithm version 1 (SHA1)	160
3.	Secure Hash Algorithm version 2(SHA2)	256(SHA256) and 512(SHA512)
	Secure Hash Algorithm version 3(SHA3)	256, 384 512

3. A hash function has irreversible property as from one available hash value, it is very cumbersome to obtain the original string.
4. Hash value possesses a constant length, whatever the size of input data.
5. Different hash algorithms are freely available to create hash values (Table 16.4).

16.6.2 Pool mining

There is the possibility that the incentive mechanism for rewarding the miners may be unstable; hence, Pool Mining has been suggested as a solution. In this pool mining process, miners form the group and, hence, the pool is formed. In Pool Mining, there are many pools for mining in the network. Miners have the choice to migrate from one pool to another pool. However, the communication delay may pose problems in the Blockchain Network. The delay in the network creates nonlinearity in the system. The pay per last N shares is a rewarding process in which submitted shares are awarded for being distributed to the miners (Majeed et al., 2020).

16.7 Trust enhanced blockchain technology

The local trust value is evaluated in two parts: trust value Tv_{pq} and modified trust value mTv_{pq} between two nodes. The direct trust value is defined:

$$Tv_{pq} = \begin{cases} \frac{A_{pq} - B_{pq}}{A_{pq} + B_{pq}}, A_{pq} + B_{pq} \neq 0 \\ 0, A_{pq} + B_{pq} = 0 \end{cases} \tag{16.1}$$

where A_{pq} is the number of data transmissions between nodes p and q which are not failed, and B_{pq} is the number of data transmissions that are failed. The trust value will be zero, which indicates that previously there has been no interaction between the two nodes. The indirect trust value shows the long-term historical behavior of the nodes and reflects the trust value more effectively.

$$IT\ v_{pq} = \sum\nolimits_{m \in M} T\ v_{pm}\ T\ v_{mq} \tag{16.2}$$

where Tv_{pm} is the local trust value between node p and main node m in one network cluster. Tv_{mq} is the direct trust value between main node m and node q. M is the set of main nodes which carry major information within the cluster.

16.7.1 Global trust value evaluation

The local trust value is evaluated within the cluster under consideration. The Global Trust Value Evaluation has been suggested for all clusters. Let Mi be the set of main nodes of ith cluster; then, Global Trust Value has been suggested as shown below:

$$GT\ v_{pq} = \sum_{i}^{n} \sum_{m \in M_i} T\ v_{pm}\ T\ v_{mq} \tag{16.3}$$

where n is the total number of clusters.

The most untrusted node is considered a malicious node as this node broadcasts untrusted transactions.

16.8 Blockchain-based federated learning

Cognitive computing was developed by designing computerized models by mimicking the human thought process in a complex situation. Cognitive computing has many applications in Industry 4.0. In cognitive computing, the data are gathered from various resources; however, the privacy and security issues of end devices are matters of concern. The main drawbacks in data-driven cognitive computing are three types: (1) low efficiency, (2) privacy leakage, and (3) insufficient incentive. To overcome these drawbacks decentralized blockchain-based cognitive computing model (D2C) has been suggested. The performance of cognitive computing that can be low efficiency can be enhanced by using Blockchain Technology. With the usage of this technology, (1) for accumulation of different decentralized models, the central server is not required, (2) an incentive or award mechanism is included for nodes that carry high performance or high-quality data, and (3) cognitive models can be averted from poisoning attacks. Federated learning, a machine learning approach, facilitates decentralized digital devices to learn robust learning models while local data samples are not shared, that is only the model is shared, not the raw data. However, during the learning process, this learning approach needs repeated communication among various nodes (Dong Li & Cao, 2021).

16.9 Blockchain modeling

BlockSim is a software tool to provide a platform for the simulation of the discrete-time dynamic system models for the blockchain systems. It is an appropriate tool for the analysis of the various kinds of blockchains and blockchain modules as Network Modules, Consensus Modules, and Incentive Modules. The Base model comprises mainly, model three abstraction layers such as network, consensus, and incentive layers. The Base Model is the main part of BlockSim software, and it analyzes and models the different building blocks such as blocks, transactions, nodes, incentives, etc. BlockChain Simulator *BlockClass* defines the structure of the blocks with seven attributes such as depthID, previousID, timeStamp, size, miner ID, Nonce and Transactions. *BlockClass* also contains the process needed to generate and receive the blocks. BlockSim has the objectives of Generality, Extensibility, and Simplicity. Generality pertains to the design, configurations, and system of blockchain. Extensibility covers different kinds of features of the blockchain system, while the simplicity of BlockSim means it should be simple to understand and apply (Carlos Sérgio Figueira Faria, 2018). BlockSim Net is a convenient, uncomplicated, and effective network-based BlockSim Simulator that is used in blockchain security research, beta testing, and optimization benchmarking. BlockSim Net can simulate a BlockChain system in which numerous distributed miners participate at the network level. Also, this simulator is able to detect Network-based attacks and is very effective at simulating both Bitcoin and Ethereum-based test systems (Agarwal et al., 2020).

16.10 Blockchain-enabled energy trading

Each block of the blockchain comprises (1) identification ids of buyers and sellers, (2) the amount of power to be traded, (3) transaction price per kW, (4) the time stamp when the execution of the transaction takes place, and (5) a hash value linking to the previous block's hash. Furthermore, to be particular, the transaction has to be validated with some adequate algorithm. The validation time is defined as the difference between transaction submission time and confirmation time. Obviously, the average validation time (seconds) increases as the number of microgrids increases as more blocks representing executed contracts are added to the chain (Masaud et al., 2020).

The distributed renewable energies such as solar power and energy storage devices are penetrating into the conventional power system. In the blockchain-based energy trading system, where the decentralized structure of blockchain energy system is suggested for energy trading, each participant/node shares the information among different participants. Each energy user decides to exchange the amount of energy such that energy can be purchased from a conventional power system; prosumers can sell energy to other users and store energy using energy storage devices. The prosumers can (1) buy energy from the conventional power system, (2) buy renewable energy from other users, (3) store their harvested energy, and (4) sell energy to other users (Li et al., 2019).

The digital tools and big data management using blockchain would prepare a P2P trading platform and will enhance a prototype for flexibility in the energy trading market with the blockchain-based smart contract and smart billing. The blockchain platform architecture mixes both state-of-the-art decentralized databases for their capacity to deal with the very high amount of data and a blockchain protocol for trust-less technical consensus ensuring data reliability and consistency. Blockchain-based projects must be scalable and check data switch compatibility. Bidding and switching flexibility services among providers and procurers must be compatible with blockchain technology. A TSO-DSO flexibility market platform has been initiated to comprise blockchain-based smart contracts and smart billing. Both P2P market demos will be compatible with the market design analysis. Blockchain technology should be adopted as a means for secure, reliable, and transparent cooperating agreements and information sharing, exploiting its decentralized approach. Some innovations are required to be done for the design of the TSO-DSO flexibility market with blockchain-based smart contracts and smart billing. Blockchain-based local market platform, along with the use of smart contracting and billing systems, will engage consumers in cost-efficient electricity trading that unleashes local flexibility potential while exploiting the existing familiarity of end-users in using P2P technologies.

A hybrid data-sharing model should be designed as both a traditional distributed database system for speed and cost-effectiveness anchored to a blockchain-based platform for timestamps, audit trails, reliability, and critical availability are needed. All processing is done via middleware, and settlement will be handled autonomously via a smart contract. All data can be stored and shared via an Ethereum or Hyperledger network. With this technique of blockchain, data handling and decision sharing become easier and, hence, flexible resources can be controlled in a transparent manner.

16.10.1 Blockchain-based CrowdSourcing for energy trading

CrowdSourcing is any sourcing approach that has the faceless crowd for its main provisioner of a need for anything such as funds, energy, etc., to be collected within a period of time and is used in various domains such as medicine, cyber-physical system, and other engineering systems. The objective of crowdsourcing is based on the crowd as a potentate of unused resources that is open to use provided the crowd wills it and is applied at the level of the system. Energy Crowdsourcing is defined as an operational model for collective production and consumption in energy markets. In the energy sector, the crowdsourcer is one type of participant who may be a utility company or any other system operator. The objectives of CrowdSource Energy System (CES) are plug-in electric vehicles, charging and discharging a battery, deferring loads, and supplying the power network with renewable energy via solar panels. The objectives of the smart inverters, plugs, and meters are connected with power grids and a distributed blockchain network. The blockchain platform should find the interested CES users and issues related to solar panels as capable of resolution. The computational capability evaluates the performance of the consensus algorithm, and the security issues of the users are the prime importance (Sharma et al., 2019).

16.10.2 Tokenization in energy trading

The power ledger platform, which is scalable and adaptable, provides interoperability among different market management and units and electricity (kWh), and seamless energy is traded by pre-purchased tokens. The Power Ledger Tokens have been introduced by the Power Ledger company. The physical energy produced by renewable energy sources and measured by smart meters can be represented by equivalent energy tokens. Under the blockchain platform, in energy trading, buying and selling energy can be represented by energy tokens which are becoming popular day by day in the marketplace. The objective of tokenization is to create new business models and new markets.4New, a UK startup, initiated an energy token termed KWATT, which means 1kbusinessW of electricity per year of waste by a power plant. The Impact PPA offered two energy tokens one as MPAQ used to attract potential investors to invest in energy for the communities who are not accessing energy, and the second NRG used to procure green energy as required. The SolarDAQ employs the DAQ tokens for purchasing and trading the energy on the blockchain platform. Energi Mine considers Ethereum-based tokens named as EnergiTokens or EKTs.

Power tokens are the compatible blockchain tokens used by entities involved in energy trading. The two tokens such as Sparkz and POWR are introduced by Power Ledger company for market flexibility on the blockchain platform. The Sparkz and POWR, low-level tokens, represent units of electricity in the volatile electricity market. The two-layer blockchain platform keeps a constant exchange rate between local electricity market prices and the tokens. Within the incentive programs, prosumers/nodes are awarded tokens that can be converted into local currencies.

The Greeneum token was launched by Greeneum, and it is being used for awarding carbon credits and certificates for using green energy. RecycleTocoin was introduced by Blockchain Development Company for awarding the recycling of plastic, steel, and aluminum cans. The Ecocoin is a cryptocurrency used to reward for doing sustainable work such as buying vegetarian lunches and cycling. Pylon, introduced by Pylon Network, is used for rewarding green energy generation by Pylon Network. The Energy Chain is a blockchain platform for the distributed energy resources developed by EnLedger. Energy coin is used for possession of distributed energy resources, credits, certificates, and smart metering. Energy Efficiency Coin (EECoin) offers the operation of the EnergyChain Platform. Enervails offers NRGcoin to develop effective incentive and reward mechanisms for systematic and structured usage of renewable energy on the local side. SolarCoin was launched by SolarChange and recognized by the International Renewable Energy Agency, and used in solar energy production (Andoni et al., 2019).

16.10.3 Microgrid network management

For the scattered and interconnected microgrids, a two-layer energy trading algorithm is suggested using blockchain technology. A distributed multiagent system is assumed for the isolated networked microgrids while the node of the network represents the agent. Agents of the system comprise renewable energy sources and power demand. In the first layer, microgrid energy surplus or deficit (i.e., seller or buyer) has been identified. In advance, each microgrid (MG) settles a selling or buying contract for execution.

Next, this executed contract is verified by the requisite blockchain-based appropriate algorithm, and the contract is stored in a block. The practical Byzantine Fault Tolerance (pBFT) and a modified PoS algorithm are used for achieving the higher security and reliability in the smart contract for P2P energy trading (Masaud et al., 2020).

A distributed system of various networked microgrids is considered. Each microgrid may comprise distributed renewable generation, energy storage devices, and power demand. Some of the microgrids may not possess storage distributed devices and controllable loads. The microgrids participate in the P2P energy trading to cater to the load demand. For the load balancing, a mechanism has to be developed such that a microgrid does not have to depend on its flexible devices such as energy storage devices, fuel cells, etc., and controllable loads. Instead, they should trade energy with other interconnected microgrids through the local controller. These local controllers record and manage the renewable energy generation and power demand information. The local controller sends the information to the trading agent to buy and sell the energy. The objective of P2P energy trading is to manage energy surplus and deficit while the high and low load demand is balanced.

The local controller identifies the energy surplus or deficit for trading by

$$P_{\text{bal}}(t) = P_{\text{ren}}(t) + \alpha\, P_{\text{bt}}(t) - P_{\text{load}}(t) \tag{16.4}$$

where P_{load} is load demand, P_{bt} is the power of the battery and $P_{\text{ren}}$is the renewable energy generation in the microgrid, $\alpha = 1$if the battery is discharging, $\alpha = -1$ if the battery is charging. If $P_{\text{bal}} > 0$ then there is a condition of surplus energy and if $P_{\text{bal}} < 0$ then energy is a deficit. According to these conditions, the microgrids can sell or buy energy in the networked system.

The ith microgrid decides its selling price for the smart contract as given:

$$P\,R_s^i = P\,R_{\text{bs}}^i + \left(C_{\text{ren}}^i\, P_{\text{ren}}^i + C_{\text{bt}}^i\, P_{\text{bt}}^i\right) + \max \sum\nolimits_j^n A_{i,j} C_{i,j} P_{\text{bal}}^j \tag{16.5}$$

where $P \quad R_s^i$ is the selling price, $P \quad R_{\text{bs}}^i$ is the base selling price determined by the operator, C_{ren}^i, C_{bt}^i are the operation and maintenance costs of the renewable generation and battery energy storage system, respectively, P_{bt}^i is the power obtained from the battery. $A_{i,j}$ is the interaction matrix for P2P trading between microgrid i and j. $C_{i,j}$ is the trading cost between microgrid i and j.

The ith microgrid decides its buying price by

$$P\,R_{\text{by}}^i = \min\left(\sum\nolimits_j^n A_{i,j}\, C_{i,j}\, P_{\text{bal}}^j + C_{\text{sd}}^i \Delta P_D^i\right) \tag{16.6}$$

The ith microgrid interacts with other microgrids for buying energy in P2P energy trading and it tries to minimize buying price. ΔP_D^i is the load shedding and C_{sd}^i is the cost for load shedding. ΔP_D^i is also to be minimized.

The selling price of the microgrid is estimated at different time intervals based on the operation and maintenance cost, and trading cost with other different microgrids. The interaction matrix is being developed based on the information of the distances between microgrids. The lesser the distance the higher the weightage assigned to the interaction matrix. Each microgrid generates a timestamp and hash key associated with the selling price. For developing a smart contract, the data comprises selling price, elements for interaction matrix, sender address, etc.

The multimicrogrid system can function properly with faulty nodes as per the following steps:

1. A primary microgrid sends a request for energy trading to the other MGs.
2. The other MGs hold the request for analysis.
3. Other MGs execute the request and reply with an appropriate answer for the energy trading request.
4. The primary MG waits for answers from other MGs with the same result.

The $3f+1$ is the minimum number of nodes that allow a multiagent/multimicrogrid heterogeneous system to work safely and reliably while f nodes are faulty. If a total number of nodes are n then the possible communication has been done among (n-f) nodes, and f nodes are not responding. Furthermore, nonfaulty nodes must be more than faulty nodes such that $n-2f > f$ and, hence, $n>3f$ (Castro & Liskov, 1999). Among different microgrids, a microgrid having more participants is assigned a large weighting factor. The MG, which can trade energy highest, is chosen as the validator MG. This validator (primary) MG designs a block for the execution of a smart contract for energy trading. The MG verifies the block to be added to a distributed ledger by using pBFT consensus approach. Malicious action is identified if two blocks do not have a similar property (Masaud et al., 2020).

16.10.4 Demand response scheme

Under this scheme, the customers adjust their load demand as required by the power grid. The customers adjust their demand by modifying their load profile. Smart meters play a big role in Dr events by monitoring and recording various parameters. For the execution of the Dr event, the parameters which are considered are day, time window, and total load reduction or increase. Based on this information, using blockchain technology, smart contracts are designed for the request to customers for the execution of Dr. The network operators initiate the smart contract and validate whether the customers comply with the request (DI Silvestre et al., 2020). The Customer Baseline Load(CBL) is the typical consumption profile of a customer and is used to analyze the customer's participation in the Dr program (Mohajeryami et al., 2017). For a customer, the vector of 24 hours power base consumption is considered as,

$$P_b = [\overline{P}_{b,1}, \overline{P}_{b,2}, \ldots, \overline{P}_{b,t}, \ldots\ldots, \overline{P}_{b,24}] \tag{16.7}$$

where $\overline{P}_{b,t}$ be the dynamically averaged load over various days at time t of the day d, obtained by

$$P_{b,t} = \frac{1}{X} \sum_{j \in \text{high}(X,Y,d)} P_{b,t,j} \; \forall\, t \in [1, 2, \ldots\ldots, 24] \tag{16.8}$$

This dynamically averaging number is known as the HighXofY method. The days are chosen such that on these days, no Dr programs and curtailment programs are conducted, excluding holidays as well. The X are the days on which there is the highest consumption, and these days are chosen from days on which the customer is informed of the execution of Dr events.

16.10.5 Proposed trust value evaluation of the node

The Bayesian framework represents a suitable method to reduce error in the estimation of probabilities. By using more relevant information and data, the estimated probability is updated. In the machine learning approach, the beta distribution is used for learning unknown mean and variance. The Bayesian framework is used for finding the probability distribution θ_i^l, which represents the opinion of l about i. The beta distribution beta(α, β) has been used for interaction among the networked nodes. The probability function of the beta distribution is as shown below:

$$P(x) = \frac{(x)^{(\alpha-1)}(1-x)^{(\beta-1)}}{\text{beta}(\alpha, \beta)} \tag{16.9}$$

$$P(x) = \frac{\Gamma(\alpha)\Gamma(\beta)}{\Gamma(\alpha+\beta)}(x)^{(\alpha-1)}(1-x)^{(\beta-1)} \tag{16.10}$$

At different conditions of interactions (α, β) are updated in the beta probability distribution. The opinion representing the reputation of the node based on trading capacity is defined as:

$$R\,E_j^i = \{\text{opinion of node } i \text{ about } j\}$$

Let k represent a node possessing a trust value $Tv_k \geq Tv^{\circ}$ (a threshold) based on energy trading capability. RE_j^i is the first-hand opinion of node i about j based on the feasible mutual interaction for trading. RE_j^k is the reputation value of reputed and trusted node k about j based on the trading capacity. The node i collects and updates opinion of node k based on trading capacity. The opinion of the node is expressed in terms of parameters of beta probability distribution as given below:

$$\theta_j^k \sim \text{beta}\left(\alpha_j^k,\ \beta_j^k\right),\ R\,E_j^k = \{\alpha_j^k,\ \beta_j^k\} \tag{16.11}$$

At time t, the feasible interaction for energy trading between node i and node j for energy trading can be obtained with (α_j^i, β_j^i) that are given in the following equations:

$$\alpha_j^i(t) = \frac{\alpha_j^i + \sum_{k \in N_k} E[\theta_j^k]\alpha_j^k}{1 + \sum_{k \in N_k} E[\theta_j^k]} \tag{16.12}$$

$$\beta_j^i(t) = \frac{\beta_j^i + \sum_{k \in N_k} E[\theta_j^k]\beta_j^k}{1 + \sum_{k \in N_k} E[\theta_j^k]} \tag{16.13}$$

where N_k comprises all the nodes whose trust value is more than a predefined threshold value, that is Tv°. These nodes are able to provide the requisite amount of energy. This suggested approach facilitates to focus and combine the estimated opinion of a node about another node. The feasible interaction between node i and node j for energy trading can be obtained with parameters (α_j^i, β_j^i) that are updated with the following equations:

$$\alpha_j^i(t+1) = \alpha_j^i(t) + \Delta\, \alpha_j^i(t) \tag{16.14}$$

$$\Delta\, \alpha_j^i(t) = \frac{\alpha_j^i + \sum_{k \in N_i} E[\theta_j^k]\Delta\alpha_j^k}{1 + \sum_{k \in N_i} E[\theta_j^k]} \tag{16.15}$$

$$\beta_j^i(t+1) = \beta_j^i(t) + \Delta\, \beta_j^i\,(t) \tag{16.16}$$

$$\Delta\, \beta_j^i(t) = \frac{\beta_j^i + \sum_{k \in N_i} E[\theta_j^k]\Delta\beta_j^k}{1 + \sum_{k \in N_i} E[\theta_j^k]} g \tag{16.17}$$

At different time intervals, if the parameter α increases, the trust value of the corresponding node increases, while if β increases, the trust value of the node reduces. The proposed methodology is applied to PV Solar systems. The local controller identifies the surplus energy available from PV solar panels through inverters. For the development of a smart contract, the selling or buying price of the energy has to be decided. Possibly, the tokenization of the price named as a solar coin can be done.

16.11 Simulation

The MJP Rohilkand, Bareilly (India), has a grid-connected 1 MW of rooftop solar power plant, which comprises 27 networked inverters installed on six buildings which are assumed as a microgrid. So, six microgrids are considered for feasible energy transactions management. These 27 inverters are interconnected by a cyber-physical system, and communication among the devices is proposed to be encrypted with blockchain technology. A remote monitoring system has been installed to monitor and record the timely global horizontal irradiance, which is basically the total amount of light obtained by the ground a per square meter, and Plane of Array, the irradiation received from the sun directly multiplied by the cosine of the angle of the incidence, and the power outputs (kW) of the inverters (Table 16.5).

The power (kW) obtained from these inverters is being monitored and recorded in 30 minutes. The data is being collected at different times t, and the most trusted inverter has to be identified. The real-time recorded data for the day, July 25, 2021,

Table 16.5 Rated capacities of solar power inverters.

Invertor no.	**Inverter no. 1**	**Inverter no. 2**	**Inverter no. 3**	**Inverter no. 4**	**Inverter no. 5**	**Inverter no. 6**	**Inverter no. 7**	**Inverter no. 8**	**Inverter no. 9**	**Inverter no. 10**
Maximum capacity (kW)	50	30	50	50	30	50	20	50	30	30
inverter no.	Inverter no. 11	Inverter no. 12	Inverter no. 13	Inverter no. 14	Inverter no. 15	Inverter no. 15	Inverter no. 17	Inverter no. 18	Inverter no. 19	Inverter no. 20
Maximum capacity (kW)	30	30	30	50	50	20	20	20	50	50
inverter no.	Inverter no. 21	Inverter no. 22	Inverter no. 23	Inverter no. 24	Inverter no. 25	Inverter no. 26	Inverter no. 27			
Maximum capacity (kW)	50	50	50	30	30	30	50			

Table 16.6 Mean and standard deviation of generated power by inverter.

Inverter no.	Inverter 1	Inverter 2	Inverter 3	Inverter 4	Inverter 5	Inverter 6	Inverter 7	Inverter 8	Inverter 9
Mean	156.92	98.30	163.90	155.89	61.44	157.35	49.05	50.05	101.90
SD	70.10	44.08	66.60	64.97	50.81	66.85	25.84	26.28	47.67
Inverter No.	Inverter 10	Inverter 11	inverter 12	inverter 13	Inverter 14	Inverter 15	Inverter 16	Inverter 17	Inverter 18
Mean	97.09	97.24	95.38	95.38	114.18	160.60	160.57	60.76	55.43
SD	43.03	43.21	40.82	54.82	66.12	66.16	25.62	22.99	23.56
Inverter No.	Inverter 19	Inverter 20	Inverter 21	Inverter 22	Inverter 23	Inverter 24	Inverter 25	Inverter 26	Inverter 27
Mean	56.10	118.17	151.14	95.38	152.03	155.21	101.76	103.00	94.50
SD	51.45	77.74	49.01	67.85	68.82	43.673	44.25	41.83	51.45

has been considered and analyzed. This data comprises the power (in kW) available via 27 inverters. In MATLAB, the mean and standard deviation of the kW power available from the inverter is obtained. The mean and standard deviation (SD) are obtained and shown below in Table 16.6:

Anaconda open-source ecosystem is used for python programming for generating the time stamp and hash key. The program for obtaining the hash key and timestamp for the blockchain-based smart contract to be developed for the solar panel inverter is shown below.

Time Stamp and Hash keys of four solar power inverters are generated as shown below:

```
Solar panel Inverter: hash key for Inverter 1
372a408a0f6d81e2cae139bb04b97b6d0dd2cc1b909b2a667659948aa53e445d
timestamp 2021-07-27 23:31:19.071570 Solar panel Inverter: hash key for
Inverter
2f724f1f9f22e43a4838af72e64f55ee653a6e0cf6c7d19bed84d4b3c6941dbf9
timestamp2021-07-27 23:31:19.071570 Solar panel Inverter: hash key for
Inverter 3
ad57b67d7fad02cb560380ca829a99dea104c9f06ce99f73dcbbed6df998295c
timestamp 2021-07-27 23:31:19.071570 Solar panel Inverter: hash key for
Inverter 4
0fe908adae7ecb10000d6e309e80605ca72c4ae567c8ab994ee669d31a19ef76
timestamp 2021-07-27 23:31:19.071570
```

The maximum (163.90 kW) and minimum (49.0 kW) mean are shown by inverter 3 and inverter 7, respectively. While the maximum and minimum standard deviations are 77.74 and 22.99 shown by inverter 21 and inverter 18, respectively. The beta distribution of the four inverters is shown in Fig. 16.3.

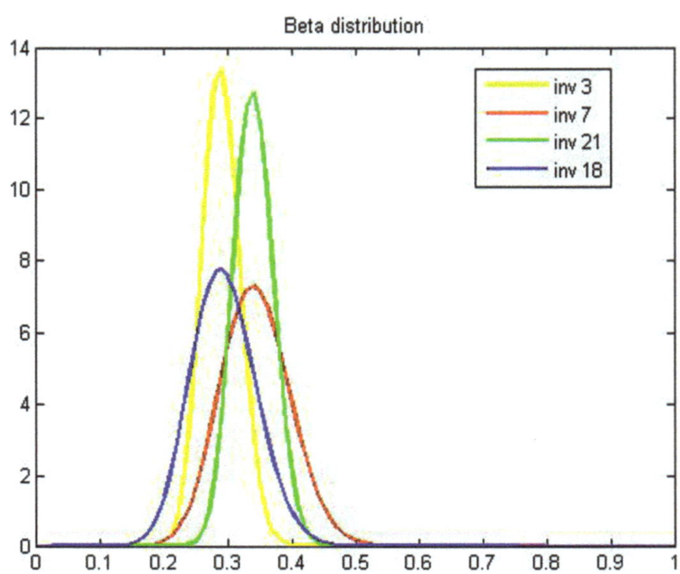

Figure 16.3 Beta distributions.

16.12 Conclusion

This chapter covers the basic concept of blockchain technology and its components. The different consensus algorithms such as PoW, PoS, byzantine, and raft algorithms are discussed. The different sorts of layers and nodes have been identified. The forking and double-spending phenomena have been discussed as phenomena to be avoided. The reputation of the nodes is analyzed by assigning the trusted node, and, hence, malicious nodes can be identified. The implementation of blockchain technology in real-time in the energy sector has been identified and analyzed. Energy crowdsourcing in real-time energy management has been discussed. The different sorts of tokenization used in energy trading have been covered. The recorded data of solar PV inverters have been considered for simulation, and results are obtained with the beta distribution. The hash keys for the four inverters with SHA 256 algorithm have been obtained using python programming. (Andoni et al., 2019; Bandara et al., 2018; Castro & Liskov, 1999; Charapko et al., n.d.; De Villiers & Cuffe, 2020; Di Silvestre et al., 2018, 2019, 2020; Gayo et al., 2020; Ghiasi et al., n.d.; Hu et al., 2019; Huang, Ma et al., 2020; Huang, Zhang et al., 2020; Kang et al., 2018; Li et al., 2019; Liu et al., 2019; Long et al., 2018; Mangia et al., 2020; Masaud et al., 2020; Mohajeryami et al., 2017; Mollah et al., 2021; Musleh et al., 2019; Natoli & Gramoli, 2016; Onyeka Okoye et al., 2020; Paris & Long, 2015; Paudel et al., 2020; Pyoung & Baek, 2020; Sabounchi & Wei, 2017; Sharma et al., 2019; Siano et al., 2019; Wang et al., 2018, 2019; Yang et al., 2021; Yin et al., 2021; Zhang et al., 2016, 2019; Zhou et al., 2019).

References

Agarwal N., R. Prashanthi, O. Biçer, A. Küpçü. (2020). BlockSim-Net: A network based blockchain simulator. *Semantic Scholar*.

Andoni, M., Robu, V., Flynn, D., Abram, S., Geach, D., Jenkins, D., McCallum, P., & Peacock, A. (2019). Blockchain technology in the energy sector: A systematic review of challenges and opportunities. *Renewable and Sustainable Energy Reviews, 100*, 143–174. Available from https://doi.org/10.1016/j.rser.2018.10.014.

Bandara, E., NG, K., Zoysa, K.D., Fernando, N., Thakara, S., Maurakirinathan, P., & Jayasuriya, N. (2018). Mystiko-blockchain meets big data. In *Proceedings of the IEEE international conference on big data*.

Bao, Z., Pu, M., Wang, B., & Dong., B. (2019). A privacy protection scheme of microgrid direct electricity transaction based on consortium blockchain and continuous double auction. *IEEE Access, 7*, 151746–151753.

Block chain explained. (2020). <http://nvestopedia.com/terms/b/blockchain.asp>.

Carlos Sérgio Figueira Faria. (2018). *BlockSim: Blockchain simulator*. Tecnico Carlos.

Casino, F., Dasalkis, T. K., & Patsakis Os, C. (2019). A systematic literature review of blockchain-based applications: Current status, classification and open issues. *Telematics and Informatics, 36*, 55–81.

Castro, M., & Liskov, B. (1999). Practical Byzantine fault tolerance. In *Proceedings on operating systems design and implementation*.

Charapko, A., Ailijiang, A., & Demirbas, M. (n.d.). *Bridging paxos and blockchain consensus*. Available from https://doi.org/10.1109/Cybermatics_2018.2018.00260.

De Villiers, A., & Cuffe, P. (2020). A three-tier framework for understanding disruption trajectories for blockchain in the electricity industry. *IEEE Access*, *8*, 65670–65682. Available from https://doi.org/10.1109/ACCESS.2020.2983558.

Di Silvestre, M. L., Gallo, P., Ippolito, M. G., Musca, R., Riva Sanseverino, E., Tran, Q. T. T., & Zizzo, G. (2019). Ancillary services in the energy blockchain for microgrids. *IEEE Transactions on Industry Applications*, *55*(6), 7310–7319. Available from https://doi.org/10.1109/TIA.2019.2909496.

Di Silvestre, M. L., Gallo, P., Ippolito, M. G., Sanseverino, E. R., & Zizzo, G. (2018). A technical approach to the energy blockchain in microgrids. *IEEE Transactions on Industrial Informatics*, *14*(11), 4792–4803. Available from https://doi.org/10.1109/TII.2018.2806357.

DI Silvestre, M. L., Gallo, P., Sanseverino, E. R., Sciume, G., & Zizzo, G. (2020). Aggregation and remuneration in demand response with a blockchain-based framework. *IEEE Transactions on Industry Applications*, *56*(4), 4248–4257. Available from https://doi.org/10.1109/TIA.2020.2992958.

Dong Li, Z. L., & Cao, B. (2021). *Blockchain-based federated learning methodologies in smart environments. Cluster Computing*. Spinger.

Gayo, M., Santos, C., Sanchez, F. J. R., Martin, P., Jimenez, J. A., & Tradacete, M. (2020). Addressing challenges in prosumer-based microgrids with blockchain and an IEC 61850-based communication scheme. *IEEE Access*, *8*, 201806–201822. Available from https://doi.org/10.1109/ACCESS.2020.3036340.

Ghiasi, M., Dehghani, M., Niknam, T., Fard, A.K., Siano, P., & Alhelou, H.H. (n.d.). *Cyber-attack detection and cyber-security enhancement in smart DC-microgrid based on blockchain technology and hilbert huang transform.*

Hu, B., Zhou, C., Tian, Y. C., Qin, Y., & Junping, X. (2019). A collaborative intrusion detection approach using blockchain for multimicrogrid systems. *IEEE Transactions on Systems, Man, and Cybernetics: Systems*, *49*(8), 1720–1730. Available from https://doi.org/10.1109/TSMC.2019.2911548.

Huang, D., Ma, X., & Zhang, S. (2020). Performance analysis of the raft consensus algorithm for private blockchains. *IEEE Transactions on Systems, Man, and Cybernetics: Systems*, *50*(1), 172–181. Available from https://doi.org/10.1109/TSMC.2019.2895471.

Huang, K., Zhang, X., Mu, Y., Rezaeibagha, F., Du, X., & Guizani, N. (2020). Achieving intelligent trust-layer for internet-of-things via self-redactable blockchain. *IEEE Transactions on Industrial Informatics*, 2677–2686. Available from https://doi.org/10.1109/tii.2019.2943331.

Kang, E.S., Pee, J., Song, J., & Jang,. W. (2018). A blockchain-based energy trading platform for smart homes in a microgrid. In *Proceedings of the third international conference on computer and communication systems*.

Li, T., Zhang, W., Chen, N., Qian, M., & Xu, Y. (2019). Blockchain technology based decentralized energy trading for multiple-microgrid systems. In Proceedings of the third IEEE conference on energy internet and energy system integration: ubiquitous energy network connecting everything, EI2 2019 (pp. 631–636). Institute of Electrical and Electronics Engineers Inc. Available from https://doi.org/10.1109/EI247390.2019.9061928.

Liu, C., Chai, K.K., Zhang, X., & Chen, Y. (2019). Enhanced proof-of-benefit: A secure blockchain enabled EV charging system. *IEEE VTC*.

Long, C., Wu, J., Zhou, Y., & Jenkins, N. (2018). Peer-to-peer energy sharing through a two-stage aggregated battery control in a community microgrid. *Applied Energy*, *226*, 261–276. Available from https://doi.org/10.1016/j.apenergy.2018.05.097.

Majeed, U., Kim, K., & Hong., C. S. (2020). Mining pool selection strategy in blockchain networks: A probabilistic approach. *KIISE Transactions on Computing Practyice*, *26*(6), 280–285.

Mangia, M., Marchioni, A., Pareschi, F., Rovatti, R., & Setti, G. (2020). Chained compressed sensing: A block-chain-inspired approach for low-cost security in IoT sensing (Vol. 7, pp. 2102–2116).

Masaud, T. M., Warner, J., & El-Saadany, E. F. (2020). A blockchain-enabled decentralized energy trading mechanism for islanded networked microgrids. *IEEE Access*, *8*, 211291–211302. Available from https://doi.org/10.1109/ACCESS.2020.3038824.

Mohajeryami, S., Doostan, M., Asadinejad, A., & Schwarz, P. (2017). Error analysis of customer baseline load (CBL) calculation methods for residential customers. *IEEE Transactions on Industry Applications*, *53*(1), 5–14. Available from https://doi.org/10.1109/TIA.2016.2613985.

Mollah, M. B., Zhao, J., Niyato, D., Lam, K.-Y., Zhang, X., Ghias, A. M. Y. M., Koh, L. H., & Yang, L. (2021). Blockchain for future smart grid: A comprehensive survey. *IEEE Internet of Things Journal*, *8*(1), 18–43. Available from https://doi.org/10.1109/jiot.2020.2993601.

Musleh, A. S., Yao, G., & Muyeen, S. M. (2019). Blockchain applications in smart grid-review and frameworks. *IEEE Access*, *7*, 86746–86757. Available from https://doi.org/10.1109/ACCESS.2019.2920682.

Natoli, C., & Gramoli, V. (2016). The blockchain anomaly. In Proceedings of the IEEE fifteenth international symposium on network computing and applications, NCA 2016 (pp. 310–317). Institute of Electrical and Electronics Engineers Inc. Available from https://doi.org/10.1109/NCA.2016.7778635.

Onyeka Okoye, M., Yang, J., Cui, J., Lei, Z., Yuan, J., Wang, H., Ji, H., Feng, J., & Ezeh, C. (2020). A blockchain-enhanced transaction model for microgrid energy trading. *IEEE Access*, *8*, 143777–143786. Available from https://doi.org/10.1109/ACCESS.2020.3012389.

Paris, J.-F., & Long, D.D.E. (2015). *Pirogue, a lighter dynamic version of the Raft distributed consensus algorithm*.

Paudel, A., Sampath, L. P. M. I., Yang, J., & Gooi, H. B. (2020). Peer-to-peer energy trading in smart grid considering power losses and network fees. *IEEE Transactions on Smart Grid*, *11*(6), 4727–4737. Available from https://doi.org/10.1109/TSG.2020.2997956.

Pyoung, C. K., & Baek, S. J. (2020). Blockchain of finite-lifetime blocks with applications to edge-based IoT. *IEEE Internet of Things Journal*, *7*(3), 2102–2116. Available from https://doi.org/10.1109/JIOT.2019.2959599.

Sabounchi, M., & Wei, J. (2017). Towards resilient networked microgrids: Blockchain-enabled peer-to-peer electricity trading mechanism. In Proceedings of the IEEE conference on energy internet and energy system integration, EI2 2017—Proceedings (Vols. 2018, pp. 1–5). Institute of Electrical and Electronics Engineers Inc. Available from https://doi.org/10.1109/EI2.2017.8245449.

Sharma, P. K., Kumar, N., & Park, J. H. (2019). Blockchain-based distributed framework for automotive industry in a smart city. *IEEE Transactions on Industrial Informatics*, *15*(7), 4197–4205. Available from https://doi.org/10.1109/TII.2018.2887101.

Siano, P., De Marco, G., Rolan, A., & Loia, V. (2019). A survey and evaluation of the potentials of distributed ledger technology for peer-to-peer transactive energy exchanges in local energy markets. *IEEE Systems Journal, 13*(3), 3454–3466. Available from https://doi.org/10.1109/JSYST.2019.2903172.

Wang, S., Ouyang, L., Yuan, Y., Ni, X., Han, X., & Wang, F. Y. (2019). Blockchain-enabled smart contracts: Architecture, applications, and future trends. *IEEE Transactions on Systems, Man, and Cybernetics: Systems, 49*(11), 2266–2277. Available from https://doi.org/10.1109/TSMC.2019.2895123.

Wang, W., Yang, S., Noor, C., Chen, M., Guo, K.H., & D. van. (2018). *Blockchain based smart contract for energy demand management* (Vol. 2018).

Yang, J., Paudel, A., & Gooi, H. B. (2021). Compensation for power loss by a proof-of-stake consortium blockchain microgrid. *IEEE Transactions on Industrial Informatics, 17*(5), 3253–3262. Available from https://doi.org/10.1109/TII.2020.3007657.

Yang, W., Aghasian, E., Garg, S., Herbert, D., Disiuta, L., & Kang, B. (2019). A survey on blockchain-based internet service architecture: Requirements, challenges, trends, and future. *IEEE Access, 7*(1), 75845–75872.

Yin, F., Hajjiah, A., Jermsittiparsert, K., Al-Sumaiti, A. S., Elsayed, S. K., Ghoneim, S. S. M., & Mohamed, M. A. (2021). A secured social-economic framework based on PEM-blockchain for optimal scheduling of reconfigurable interconnected microgrids. *IEEE Access, 9*, 40797–40810. Available from https://doi.org/10.1109/ACCESS.2021.3065400.

Zhang, C., Wu, J., Cheng, M., Zhou, Y., & Long, C. (2016). *A bidding system for peer-to-peer energy trading in a grid-connected microgrid, . Energy procedia* (Vol. 103, pp. 147–152). Elsevier Ltd.

Zhang, S., Pu, M., Wang, B., & Dong, B. (2019). A privacy protection scheme of microgrid direct electricity transaction based on consortium blockchain and continuous double auction. *IEEE Access, 7*, 151746–151753. Available from https://doi.org/10.1109/ACCESS.2019.2946794.

Zhou, J., Ye, B., Qu, L., Wang, Y., Orgun, M. A., & Lie, L. (2019). A proof of trust consensus protocol for enhancing accountability in crowdsourcing services. *IEEE Transactions on Services Computing, 12*(3), 429.

Index

Note: Page numbers followed by "*f*" and "*t*" refer to figures and tables, respectively.

E

FSC
www.fsc.org
MIX
Papier | Fördert gute Waldnutzung
FSC® C083411

Kolibri 360 GmbH
Ferdinand-Jühlke-Straße 7
99095 Erfurt, Deutschland
produktsicherheit@kolibri360.de